THE HIMALAYAN DILEMMA

THE HIMALAYAN DILEMMA

Reconciling development and conservation

Jack D. Ives and Bruno Messerli

The United Nations University

ROUTLEDGE
London and New York

First published in 1989 by
Routledge
11 New Fetter Lane, London EC4P 4EE
29 West 35th Street, New York
NY 10001

Printed in Great Britain

British Library Cataloguing in Publication Data

Ives, Jack D.
The Himalayan Dilemma:
Reconciling development and conservation
1. Asia, Himalaya. Economic development.
Environmental aspects
I. Title II. Messerli, Bruno
330.954′052

ISBN 0–415–01157–4

Library of Congress Cataloging-in-Publication Data

Ives, Jack D.
The Himalayan Dilemma:
Reconciling development and conservation
Bibliography: p.
Includes index.
1. Human ecology—Himalaya Mountain Region. 2. Environmental policy—Himalaya Mountains Region.
I. Messerli, Bruno. II. Title.
GF696.H55194 1989 333.73′0954
88–17382
ISBN 0–415–01157–4

'As the dew is dried up by the morning sun,
So are mankind's sins at the sight of Himalaya'

The *Puranas*, scriptures of ancient India

Dedicated to the mountain subsistence farmers, men and women of the Himalaya, the best hope for resolution of the dilemma.

Contents

Figures

Tables

Foreword

No region of the world excites the imagination and calls up visions of the exotic more than the Himalaya. Their soaring peaks and fertile valleys have nourished some of the world's most ancient cultures and religions. But recently, there have been disturbing signs of trouble in Shangri La – alarming reports of widespread environmental degradation which is said to be producing dire and imminent threats to the future of the region and the contiguous lowland areas.

This book challenges the assumptions arising from these alarmist reports that the Himalayan region is in an advanced stage of irreversible environmental destruction. The authors, Jack Ives and Bruno Messerli, bring to this subject the insights and experience of two of the world's leading experts on mountain environments and have an intimate knowledge of the Himalayan region. They have analysed its problems with the disciplined objectivity of science and demonstrated that their roots are basically socio-economic and political rather than narrowly environmental.

Their analysis and the conclusions to which it points, have profoundly important implications for the future of this region and, indeed, for our understanding of the stresses and risks to which other major ecological systems are being subjected as a result of the rapid growth of population and the intensity of human activity. Ives and Messerli make clear the systemic nature of the cause and effect relationships which determine the nature and direction of the changes taking place in the region and the intrinsic complexity of these relationships. They point to no quick or easy solutions for the Himalayan dilemma and argue persuasively for a long-term approach, based on defining multiple solutions in which uncertainty is likely to be a continuing reality. This, they argue, will require fundamentally new and different thinking about the development of the region and the critical role of the indigenous subsistence farmer. They argue cogently for the kinds of policies and programmes that are sensitively attuned to, and supportive of, these people who are the prime actors at the interface of the man-nature relationship on which the region's future depends.

This book represents one of the finest examples I know of science with a human face. The authors clearly have a deep affinity for the region and its people. Their excellence as scientists lends to this book an authority and credibility that is unique in its field. Its value is multiplied immeasurably by the thoughtful and incisive manner in which the authors have translated their findings into a set of guidelines for the politicians, officials and practitioners on which the future of the region so largely depends.

All those who love the Himalayan region and are concerned about its future owe a great debt of gratitude to Jack Ives and Bruno Messerli. Not only have they produced a book that is thoroughly engaging, brilliantly written and enjoyable to read, but in it they present the most up to date, comprehensive and thoroughly researched analysis available today of the complex forces that are shaping the future of the Himalaya. The conclusions and recommendations that result from this provide the basis for the new attitudes, policies and practices which can produce a new era of hope for that unique, beautiful and imperiled region, and its wonderful people.

Maurice F. Strong
President, World Federation of United Nations Associations
May, 1988

PREFACE

This book seeks to examine the basis of the widely supported prediction that the Himalayan region is inevitably drifting into a situation of environmental *supercrisis* and collapse, a process of thought to which we refer as the Theory of Himalayan Environmental Degradation. For comparison, drought-ridden Africa today is defined as being in a state of supercrisis. The costs of the on-going relief action in Africa, and the attempts to avert future recurrences, are enormous. And these are merely the *material* costs; those relating to the large-scale human misery are unquantifiable and unthinkable. Terrible as this African tragedy undoubtedly is, if the many supporters of the Theory of Himalayan Environmental Degradation are proven correct over the next several decades, then the new world-scale supercrisis, in which the Himalayan region will have become enmeshed, will be much more severe. This evaluation is based simply upon consideration of the much larger numbers of people who will be involved, since the densely populated plains of the Ganges, Brahmaputra, Indus, upper Jinsha Jiang (Yangtze), and other major rivers are tied to the mountains by a welter of physical and human processes, both actual and perceived.

It is all too easy to urge that prevention is more effective, and less costly, than cure, and infinitely more humane. But simply to call for a world-level effort to avert a portending catastrophe in the Himalayan region ignores the sea of uncertainty that surrounds it. By this we mean that our understanding of the processes operating in the region, whether geophysical, environmental, social, economic, or political, is tenuous at best. Differentiation between cause and effect is thwarted by lack of data in all fields, by unreliable, even manufactured, data. This situation, in our estimation, is further confounded by a pervasion of assumptions, conflicting convictions, and latter-day myths, most of which lead into the perceived, and preconceived, downward spiral of environmental disaster.

At the risk of being accused of arrogance, or over-ambition, we will introduce and discuss in some detail the Theory of Himalayan Environmental Degradation. We will try to separate out the several linkages between its component parts and subject them to critical examination. We will come to no firm or unassailable conclusions, nor propose any clear-cut panaceas. We hope to demonstrate, however, that many of the linkages of the Theory are untenable, while others can be neither supported nor disproved in terms of the available body of information. We will argue from this, nevertheless, that the overall theoretical construct is unrealistic and that, since much development and foreign aid policy is based upon it, an urgent re-think is in order. This is required if only because current policies may well be contributing to problems that the people of the region are undoubtedly facing, based as they often are upon unacceptable simplification and extrapolation across one of the world's most varied and complex regions. We will then move on to make some recommendations that we hope will lead to a reduction in the degree of uncertainty, and we will provide an outline research strategy that, if carried forward, may further clarify the issues that affect the lives of a significant proportion of humankind.

Two recent examples from the mass media give a strong impression of the potential dangers of allowing the uncertainty and misunderstanding to proceed unchecked – and these dangers have political overtones, as well as physical and socio-economic ones. In an October 1987 article in *India Today* under the title: 'Bihar Floods: Looking Northward,' the author states:

> Each time north Bihar is devastated by floods, the state Government performs two rituals. It holds neighbouring Nepal responsible and promises to implement a master plan for flood control. ... Nepal is invariably held guilty because most of the rivers ... originate there before flowing into the Ganga. The Bihar Government maintains that Nepal's non-cooperation lies at the root of the annual cycle of human misery. ... This time the chorus of accusation reached fever pitch when Prime Minister Rajiv Gandhi ... demanded to know what preventative measures had been taken. ... Predictably, the [response] referred to the hill kingdom's lack of cooperation.
>
> The Nepal-bashers also scored a major victory at the second National Water Resources Council meeting in New Delhi last fortnight. State Irrigation and Power Minister Ramashray Prasad Singh managed to have the national water policy draft amended to say that the solution to Bihar's flood problems lay beyond its borders.
>
> (Farzend Ahmed, *India Today*, 15 October 1987: 77)

From the other side of the world, about a month later, *Newsweek* produced the dramatic prediction under the title: 'Trashing the Himalayas,' that 'a once fertile region could become a new desert.'

> Dense alpine forests once covered the lower slopes of Mount Everest, and the Khumbu Valley below the mountain used to blush dark green from its carpet of junipers. But that was the Everest of 1953, when Sir Edmund Hillary and Tenzing Norgay became the first men to conquer the highest peak on earth. Today the forest at Everest's base is 75 percent destroyed, replaced by a jumble of rocks interspersed with lonesome trees. All of the Khumbu's junipers have fallen to axes. . . .
>
> The degradation of the Himalayas is not confined to the tall peaks. In

> Pakistan, India, Nepal and Tibet, deforestation has eroded fertile top-soil from the hills, triggering landslides and clogging rivers and reservoirs with so much silt that they overflow when they reach the plains of the Ganges. ... At the rate trees are being felled for fuel and cropland, the Himalayas will be bald in 25 years....
>
> Although a significant fraction of the erosion stems from nature ... most of the damage is man-made.
>
> (S. Begley, R. Moreau, and S. Mazumdar, *Newsweek*, 9 November 1987)

While both of these articles tail off mildly to allow for the prospect that there may be other causes and that all may not be lost, they provide, through their emphasis, a thoughtless repetition of similar statements that have been reiterated for several decades. Considering the multi-million readership of these journals, their political and socio-economic impacts cause us grave concern.

Nevertheless, the most serious anxiety that faces us in our attempt to demonstrate that the Theory is an over-dramatization and distortion is that our position should not be interpreted as indicating that there is no problem and that, therefore, there is no need for re-thinking, action, or alarm. Our study is much more than an object of scholarly endeavour for its own sake. It is involved intimately with the well-being of several hundred million people and with the environmental conservation of a region that is a vital part of the world spiritual and cultural heritage. By focusing our approach in this way, however, we perforce enter a dangerous and labyrinthine arena composed of a complex of territories and imperatives that may well be defended by a host of vested interests. Nonetheless, we are convinced that the prevailing views on the Himalayan situation are indeed distorted and that a large part of the 'Problem' is due to the uncertainty that engulfs it. We must emphasize, moreover, that this uncertainty is not merely *technical*; that is, an absence of certainty. Rather it is technical *and structural* in the sense that, without their realizing it, some of the actors in the Himalayan debate have succeeded in imposing their desired certainties within it. It is these *unwarranted* certainties (or myths) that have evoked the Theory that we wish to dismantle. Our aim in doing this is two-fold: first to confront the full extent of the uncertainty (both technical and structural); second, to get to grips with it, both by reducing it, where this is possible, and by learning to live with it, where this is not possible. This theme will be deliberately reiterated throughout the text.

We recommend, therefore, that the Problem needs to be addressed at the highest level and that a course of action must be developed. Even in this final point, as academics, we are conscious of the hard-pressed decision maker's observation that, when the university professor does not know what to do, he collects more data, yet the decision maker does not have the time to wait for 'academic perfection.' Our counter observation, nevertheless, is that if a beginning had been made thirty years ago to collect relevant data on a systematic basis, a required course of action could have been much more readily defined today.

Similarly, the Himalayan region and its problem(s) will be with us for a long time, and the potential supercrisis, as we define it, is one of drift, or gradual development. Thus while urgent action tomorrow can still be perceived as vital, let us not waste the next thirty years arguing that more data are not the answer to formulation of a flexible response that can be adjusted as greater understanding is acquired. Let us set in progress a carefully constructed course of *relevant* data collection as soon as possible.

We hope that this book will constitute a scholarly contribution to the study of mountain environments in general, and to the Himalaya in particular. As such it should be of value to university upper-division undergraduates and graduate students in geography, in a variety of cognate disciplines including agriculture, forestry, economics, and the natural sciences, and in political science and related areas of environmental and Third World development studies. We also hope it will assist scholars who are involved, or who are contemplating research, in the region. However, because of the very complexity and uncertainty that we are anxious to emphasize, the book can hardly serve as a geographical treatise, nor can it be comprehensive. The primary intent is to provide enough background material, either directly, or indirectly through the references cited, so that the reader can focus on the conventional wisdom that has been built up over the past three or four decades and, we hope, be forced into reassessing both it and our own contribution. Thus our other, perhaps over-riding, purpose is to enter the political arena in the hope that we can influence the evolution of a new approach to development and environmental management. Thus we believe that there is a second readership, perhaps more numerous than the academic – resource managers, decision makers, the development and foreign aid establishment – as well as the general public concerned about what is, and what is perceived to be, happening in the mountain world.

The very fact that we have come to produce this book at all is due to an unusual degree of privilege that we have experienced in support from a number of institutions and many individuals and colleagues. This privilege also derives from our encounters with a large number of mountain people, in all their complexity, beauty, indigenous wisdom, and ethnic variety, into whose homes and fields we have been welcomed, and whose fears, perplexities, and aspirations we have learned to share. Because of the magnitude of our task, therefore, the remainder of this Preface is devoted to an explanation of how the book evolved.

We have worked together to establish the United Nations University's (UNU) project on Highland–Lowland Interactive Systems since its inception in 1977. We became involved in the Himalayan region in 1978 when reconnaissance journeys were made in Nepal, the Darjeeling Himalaya, and northwestern Thailand. While Thailand is not normally included in conventional definitions of the Himalaya, we intend to use the term 'Himalayan region' in a very broad sense.

The primary focus of these early journeys was to ascertain the viability and usefulness of transfer of Swiss Alpine and Colorado Rocky Mountain experience in mountain hazards mapping. The area of reconnaissance coverage was expanded to include parts of the Qinghai-Xizang (Tibet) Plateau, the Chinese Tien Shan, and the Hengduan Mountains of western Sichuan and northwestern Yunnan. The most sustained effort, however, was that associated with the Nepal Mountain Hazards Mapping Project.

These activities included the training of UNU fellows from the host countries, a series of small conferences, increasing linkages with other scholars, and extensive literature research. They led to a growing conviction that there was something seriously amiss with the conventional Himalayan environmental-scientific wisdom, upon which much foreign bilateral and international aid and development policy was based.

We believe that the UNU Highland–Lowland Interactive Systems Project was established at a critical time in the development of Himalayan research. Because it was staffed by a group of young and international scholars who had the opportunity of working for several years in a number of small field areas, it was possible to take a provocative and unconventional, and highly critical, look at some of the accepted scientific paradigms, and to begin to challenge them.

As fieldwork progressed, so a core of research results and new interpretations became available. In addition, progress was such that in 1984 UNU commissioned the preparation of a position paper on 'The Himalaya-Ganges Problem' from which this book directly evolved. With the formulation of the position paper strategy, a group of over a dozen scholars with Himalayan experience began producing draft reports of case studies and strengthening the access to existing, but sometimes obscure, literature on specific aspects of the Problem. In addition, contacts were established with Dr. David Pitt, who had begun a separate study of the Himalaya for UNO/UNEP, and with Dr. Michael Thompson, who was directing a study for IIASA (International Institute for Applied Systems Analysis)/UNEP. These contacts resulted in the formation of a steering committee, comprised of Ives, Messerli, Pitt, and Thompson, which met at Appenberg, Canton Bern, Switzerland, 16–27 November 1984, to refine further the strategy for the position paper (this committee was subsequently expanded to include Dr. Lawrence S. Hamilton, and Professor Sun Honglie).

During the Appenberg Workshop it became overwhelmingly clear that the task we were setting ourselves could easily swamp our available resources. In other words, as we learned more about the Himalaya-Ganges Problem, we came to understand that an exhaustive analysis was at least an order of magnitude beyond our combined capabilities. The descriptors, 'complexity' and 'uncertainty', came to be regular components of our discussions. This led to a degree of rationalization. Thus there followed a recommendation to the UNU that we move immediately to prepare for an international conference to address the Himalaya-Ganges Problem. This would provide the benefits of involving a wider range of participants, especially scholars from the region itself and experts from United Nations and national aid and development agencies. It would both focus *and* extend our efforts. These two, apparently contradictory, aims were viewed rather as being mutually supportive and interdependent. In the first place, we were responding to the perceived need to increase our available energies by extending the two-year period allotted for production of the position paper. In the second, we were substantially widening the scope of potential contributions.

The conference was eventually held, as the Mohonk Mountain Conference, 6–11 April 1986, at Mohonk Mountain House, New Paltz, New York State, USA, under the Honorary Chairmanship of Mr. Maurice F. Strong. A series of resolutions was passed and directed to senior administrators of the United Nations System and to the heads of state of the concerned governments of the region. In addition it was recommended that this book be produced, as well as the conference proceedings *per se* (see *Mountain Research and Development* 1987, Volume 7(3)), and a task force be established to carry our growing momentum further. These decisions, in turn, helped to limit the objectives of the UNU position paper and ensured its completion on schedule (June, 1986). They also became a stepping stone to a wider and more extensive endeavour (this book and additional new field studies) transcending the objectives of the position paper.

It was decided to edit and publish prior to the conference many of the position paper case studies as independent papers in a succession of issues of *Mountain Research and Development*, the quarterly journal of the International Mountain Society. This process also attracted responses from independent, and sometimes hitherto unknown to us, readers of the journal which, of course, further enlarged the effective size of the working group. Especially important amongst these contacts are:

- Shri Sunderlal Bahuguna, Messenger of the Chipko Movement;
- Drs. Jayanta Bandyopadhyay and Vandana Shiva who provided invaluable insights into the linkages between Indian Himalayan ecological research and the Chipko Movement;
- Professor David M. Griffin, Department of Forestry, Australian National University, and director of the Nepal–Australia Forestry Project;
- Dr. Lawrence S. Hamilton of the East-West Center, Hawaii, who became a member of the steering committee, and the East-West Center became a co-sponsor of the Conference;
- Dr. Janos Hrabovszky, senior agricultural consultant to FAO and to HMG, Nepal;
- Sir Robert Jackson, Senior Adviser to the Secretary-General, UNO;
- Dr. Rajni Kothari, director of the UNU project on Peace and Global Transformation, New Delhi;
- Dr. Corneille Jest, Centre National de la Recherche Scientifique, Paris, France;
- Dr. V. V. Dhruva Narayana, Director of the Central Soil and Water Conservation Research and Training Institute, Dehra Dun;
- Dr. Colin Rosser, Director, International Centre for Integrated

Mountain Development, Kathmandu;
- Dr. Hans Schreier, Department of Soil Science, University of British Columbia, Canada, and CIDA;
- Dr. C. K. Sharma, Executive Director, Water and Energy Commission Secretariat, HMG, Kathmandu;
- Mr. Maurice F. Strong, President, World Federation of United Nations Associations;
- Professor Sun Honglie, Vice President, Chinese Academy of Sciences.

This list is by no means complete. Nor is it intended as a formal acknowledgment, but rather as providing a few examples of the way in which the resources of the original working group were expanded. Thus the ensuing results are many times more extensive than the original funds allotted by UNU could possibly have commanded.

The first part of the book (to Chapter 6) represents an exposition and critical analysis of the Theory of Himalayan Environmental Degradation, primarily from a physical process point of view. Chapter 7 focuses on the 'human element' and the facts and pseudo-facts associated mainly with the indigenous subsistence farmer, the convenient scapegoat who is widely perceived as the destroyer of the mountain environment and the ogre of the plains. Chapter 8 presents two case studies in the form of prescriptions for averting disaster: the one approaches the issue from the village level as a fundamental self-help and self-directed undertaking, with governmental facilitation; the other, at the opposite extreme, is a large scale intervention and technological 'fix'. We do not necessarily support either but introduce them as extremes and as a basis for further emphasis of the complexity of the Problem. Chapter 9 takes us into the political arena and raises the question of 'crisis, pseudo-crisis, or supercrisis?' The final chapter (10) is drawn together as an outline for research – or rather as a fabric of overlapping designs.

Much of our presentation depends upon a large number of papers, many of which were prepared as background studies for the Mohonk Mountain Conference, together with others that they provoked or that reached us serendipitously as independent submissions to *Mountain Research and Development*. It should be noted that the deliberate emphasis on the Himalaya in the editorial policy of the journal has ensured that it has become not only a major repository of new information, new thinking, and debate, but also a motivating force in furthering this process of testing conventional wisdom and of searching for new insights. A second book is planned to follow this one, which will draw on *Mountain Research and Development*, and other publications, and serve as a major source volume for much of the present debate.

It must be admitted that, despite efforts to avoid it, a distinct bias in favour of Nepal will be apparent throughout this text. In large part this is due to the openness of Nepal and ease of access for both foreign and indigenous scholars, compared with many other parts of the region.

Our conclusions carry the conviction that it is essential to address, reduce, and/or learn to live with the Himalayan uncertainty. To do this we believe it is necessary to dismantle the Theory of Himalayan Environmental Degradation before we can begin to unravel the magnitude of the Problem, as a first step toward treating it. We strongly support the notion of multiple problem definition – multiple solution definition, and we emphasize this in the concluding chapter. It may seem trite to recommend that the Himalayan Problem is not environmental, but is socio-economic, and especially political. However, unless new political approaches can be devised, and unless quite profoundly different thinking can evolve – about access to resources, about what 'development' should be – and unless the overwhelming majority of the subsistence farmers in the region can be better accommodated (to anticipate: unless they can be regarded as part of the solution(s) rather than as part of the Problem), we believe that the Himalayan region *sensu lato* is destined to experience a supercrisis of unthinkable magnitude. This is the dilemma facing the Himalayan region.

The foregoing perspective of what we are attempting would appear to justify our fears that it is over-ambitious. This we have decided to risk. We feel humbled by the magnitude of the task and our inadequate resources to attempt it. However, as so eloquently argued at the Conference by Shri Sunderlal Bahuguna, Messenger of the Chipko Movement, we must all recognize the special role of the Himalaya as a unique part of the world cultural heritage and the importance of its spiritual contribution to the well-being of the world community. Some of the more serious threats to world civilization and environment are lethargy, complacency, despair, conviction that the problems are too great, the time too short, and that the damage already accomplished is irreversible. A great deal can be done given the inspiration, determination, and commitment, provided a proper focusing can be achieved. By tackling the uncertainties and misunderstandings that enshroud the Himalaya and the neighbouring plains, we hope that this book will make some contribution toward achievement of that necessary focus.

March 1988

Jack D. Ives

President, International Mountain Society
UNU Coordinator: Highland–Lowland Interactive Systems Project
Chairman, IGU Commission on Mountain Geoecology, 1988–92

Bruno Messerli

Chairman, IGU Commission on Mountain Geoecology, 1980–88
UNU Institutional Coordinator
Vice President, International Mountain Society

ACKNOWLEDGMENTS

This book is a product of the United Nations University (UNU) project on Highland–Lowland Interactive Systems. Centring on the project's Mountain Hazards Mapping study in Nepal, extensive reconnaissance and detailed field investigations were undertaken, not only in Nepal, but in several parts of the Indian Himalaya, in Tibet (Xizang Autonomous Region), Sichuan, Yunnan, and Xinjiang, of the People's Republic of China, and in northwestern Thailand. These various journeys, sometimes into remote mountain areas, afforded us an unusually extensive experience, often together, at other times with several colleagues, most of whom worked with us in the Kakani and Khumbu test areas of Nepal, or were members of the staffs of the Commission for Integrated Survey of Renewable Natural Resources (CISNAR) and the Geographical Institute of the Chinese Academy of Sciences. They included: Khagda Basnet, Markus Bichsel, Inger-Marie Bjønness, Barbara Brower, Alton Byers, Elizabeth Byers, Nel Caine, Chen Bao-lin, Chen Chuan-you, Sumitra M. Gurung, Heinrich Hafner, Hans Hurni, Corneille Jest, Kirsten Johnson, Narendra R. Khanal, Hans Kienholz, Li Wenhua, Liao Jungua, Liu Lan-hui, Lee MacDonald, Pradeep K. Mool, Elizabeth Ann Olson, Tjerk Peters, Guy Schneider, Kamal K. Shrestha, Ann Stettler, Rabindra M. Tamrakar, Colin E. Thorn, Daniel Vuichard, Yang Zhou-huai, Yao Zhiyun, Zhang Yongzu, Markus Zimmermann.

The fieldwork was largely financed by the UNU, with significant contributions of the Chinese Academy of Sciences, the Nepal National Committee for the Man and the Biosphere (MAB) Programme, and the UNESCO MAB Programme. Within UNESCO we owe special thanks to our colleague Dr. Gisbert Glaser. We were also greatly assisted by released time from our home institutions, including a Faculty Fellowship from the University of Colorado, and extensive material support from the Geographical Institute, University of Berne.

Many of the results of the field studies have been published in *Mountain Research and Development*, and elsewhere, or have appeared as doctoral, masters, and diploma dissertations. *Mountain Research and Development* has also been heavily supported by UNU.

None of this extensive body of work would have been possible without the total commitment, goodwill, abiding faith, and administrative skill of our guide and colleague, Professor Dr. Walther Manshard, both in his capacity as UNU Vice-Rector, and present position as UNU Senior Programme Director, ably assisted by Lee MacDonald and many of the UNU Tokyo headquarters staff. This support of UNU has been sustained by the current Vice-Rector Dr. Roland Fuchs; the special moral support of former UNU Rector Soedjatmoko was also of inestimable value.

The fieldwork, and extensive travels, much of it in restricted and politically sensitive areas, depended absolutely upon a number of national institutes and their staffs. In Nepal, the National Committee for MAB (UNESCO) served as our host institution, and we are especially indebted to National Committee Secretaries Dr. Cherunjivi Shrestha and Professor Suresh R. Chalise; we also appreciate support received from the National Planning Commission, and especially Dr. Ratna Rana and Professor Upendra Man Malla. Professors Kamal K. Shrestha and Suresh R. Chalise served as local UNU coordinators. In China we received vital administrative support and encouragement from Professor Sun Honglie, Vice-President, Chinese Academy of Sciences, and Professor Li Wenhua, Director, CISNAR. During the latter stages of the study in Nepal, Dr. Colin Rosser, Director, and staff of the International Centre for Integrated Mountain Development (ICIMOD) provided invaluable assistance, as did the Swiss Association for Technical Assistance (SATA).

Many of the ideas expressed in this book evolved from spirited informal discussions with all those so far individually acknowledged. However, the form and content of the book has depended heavily upon a group of colleagues whose tireless and selfless assistance has been indispensable and who really deserve credit as contributing authors. These include, amongst others, Sunderlal Bahuguna, Jayanta Bandyopadhyay, David Griffin, Lawrence S. Hamilton, Sun Honglie, Janos Hrabovszky, Donald Messerschmidt, David Pitt, and Michael Thompson. Maurice F. Strong warrants special thanks for serving as Honorary Chairman of the Mohonk Mountain Conference, for contributing the thought-provoking Foreword, and for several years of constant encouragement and moral support.

Andreas Lauterburg and Markus Wyss, Geographical Institute, University of Berne, contributed major reports that became vital parts of Chapters 2 and 6. Without the help of Professors A. B. Mukerji, Chandigarh, and R. B. Singh, New Delhi, the work of Andreas Lauterburg would not have been possible. The skill of Andreas Brodbeck, Martin Grosjean, Andreas Lauterburg, Susanne Wymann, and Markus Wyss, of the same institute, for their cartographic assistance is gratefully acknowledged. Professor Jean-François Dobremez kindly allowed us to borrow extensively from his book on Nepal to provide many of the figures used in Chapter 2; they are credited in the text. We are grateful to Dr. Barry C. Bishop, National Geographic Society, for making available prints of photographs taken in the 1920s and 1930s in Yunnan by Dr. Joseph F. Rock and for alerting us to the existence of this 'treasure' of old photographs in the archives of the Society.

This acknowledgment introduced the book as a product of UNU research;

equally, it evolved from the development of the International Mountain Society, and our thanks are due to all our IMS colleagues, but especially to Professor Frank P. Davidson, 'father' of the Society and its inveterate supporter and promoter. It is also a contribution of the Commission on Mountain Geoecology of the International Geographical Union and a tribute to its founder, the late Professor Dr. Carl Troll. We would like to record that it was Carl Troll who brought us together during a Commission symposium in the Canadian Rockies in 1972 and, as he has done so many times with his numerous colleagues, set us firmly on the path of mountain geoecology.

It remains to thank a group of skilled, loyal, and dedicated assistants, because, without their help, we would not have completed our task. Frau Florin, secretary of the Geographical Institute, University of Berne, better known to her many mountain friends as 'Flo', contributed innumerable services; Laura Koch, as loyal and astonishingly accurate secretary of the UNU-University of Colorado Mountain Project, 1979–87, provided immaculate typescripts of the several early drafts of the manuscript; Ann Underwood completed the final word processing with consummate skill.

Mary Ann Kernan guided us through many difficult moments with editorial aplomb and cheerful encouragement; the editorial assistance of Ruth Jeavons and Stephanie Horner was also of great value, and Pauline Ives provided extensive assistance in the preparation of the manuscript, as well as being a valuable sounding board for many of the ideas introduced. Without the indirect help and forebearance of both Beatrice Messerli and Pauline Ives, we ourselves would have entered a state of supercrisis!

With so much help it would seem that there can be no mistakes nor misinterpretations herein: there undoubtedly are and for these, acknowledgment is due only to ourselves.

We would like to dedicate this work to the wonderful, long-suffering, skilful, and hospitable subsistence farmers of the Himalaya.

1 THE THEORY OF HIMALAYAN ENVIRONMENTAL DEGRADATION: WHAT IS THE NATURE OF THE PERCEIVED CRISIS?

Conservationists, scientists, and administrators have expressed growing alarm about the rapid deterioration of the Himalayan environment over the past thirty years or so. This alarm has made itself felt in a large number of ways through the media. Television viewers, with striking regularity, are assailed with dramatic visions of deforestation, landsliding, and large-scale downstream flooding, coupled with statements about uncontrolled population growth, increasing poverty, and malnutrition. These processes – physical, human, socio-economic, and political – are frequently linked together into a gigantic cause-and-effect drama which is claimed to be pushing both the Himalaya and the northern plains of the Indian sub-continent to the brink of environmental and socio-economic collapse.

This pattern of thought, which can be divided into numerous sub-variants, has been widely accepted as established fact by large numbers of people who often lend their support to perpetuate it as a truism. In turn it seems to pervade the evolution of policy making in the areas of conservation, resource development, and foreign aid. In this context we feel we are justified in referring to it as the Theory of Himalayan Environmental Degradation.

While we are convinced that there is an enormous problem facing the Himalayan region, we believe that it is clouded in uncertainty and complexity. Much of the problem is contained within the overly simplistic view, as expressed in the Theory, with its assumed cause-and-effect relationships, that appears to have captured the imagination of so many people, both onlookers and actors.

During the early stages of our own Himalayan research efforts we also had accepted the Theory of Himalayan Environmental Degradation as self-evident. Fortunately, the nature of our research, initially concentrating on mountain hazards and the perceptions of, and the response to, such hazards by the local people, forced us to review critically the evidence for a number of the general assumptions upon which the Theory is based. This critical process was made the more rigorous by the mix of our co-workers (multi-disciplinary and multi-national). It was further assisted by the requirement of revisiting the intensive study sites at various times of the agricultural year and over a period of years.

As we began to realize that several of the widely accepted assumptions were either without factual support, or were demonstrably unsupportable, at least in a number of small field areas with which we were becoming familiar, our commitment widened to embrace an overall challenge of the Theory itself. This book, therefore, as indicated in the Preface, is the story of our increasing dissatisfaction with the Theory of Himalayan Environmental Degradation. It details our attempt to seek a fuller understanding of the physical and socio-economic dynamics of the Himalayan region. It is also intended to demonstrate the need for a much broader and deeper perspective of the problems facing the Himalayan region as a prerequisite for the development of more effective solutions.

The necessary first step is to provide a detailed exposition of the Theory of Himalayan Environmental Degradation itself. This is the purpose of the present chapter. What follows, therefore, has been abstracted from a large body of literature – and presented as a synthesis of the Theory that we are deliberately setting up for evaluation. We believe that this intellectually satisfying construct must be analysed, challenged, and dismantled before any real progress can be made toward solution of the Himalayan Problem. There must be a better attempt than hitherto available at defining the Problem before there can be hope for effective mitigation. Thus we will begin with a synoptic response to the question – what is the nature of the perceived crisis? While by no means acceptable to us, the response can be based upon a review of numerous reports in the news media, internal reports of aid and development agencies, and countless published books and papers in the scientific and conservationist literature.

We do not wish to imply that this body of literature is all inaccurate; we do believe, however, that the generalization, or accumulated perspective, is seriously distorted. Our primary objective, therefore, is to achieve a much more critical assessment of established thought as a necessary first step for advances in scholarly endeavour and for more effective aid and development policy formulation. This carries the implication that there is a pressing need for a much firmer linkage between scientific research, policy analysis, and policy making – in effect, we have become engaged in *science-for-public-policy*, whether we like it or not.

The most compelling and trend-setting characterization of the Himalayan region and its anticipated eco-disaster is that published by Erik Eckholm (1975, 1976), although he was exceeded by Claire Sterling (1976), amongst others; moreover, and more seriously, he is perpetuated by Norman Myers (1986) amongst many other environmental alarmists, the works of which Messerschmidt (personal communication, March 1987) describes as 'the Claire Sterling Effect.' The most startling visual presentation is contained in the superb movie, *The Fragile Mountain*, produced by Sandra Nichols (1982) with substantial financial support from the World Bank and other agencies. In addition, a spate of books and articles has been published, especially in India and Nepal. Some of the most prominent are Lall and Moddie (1981); Bandyopadhyay *et al.* (1985); J. S. Singh (1985); T. V. Singh and Kaur (1985); and Joshi (1986a).

THE THEORY

Any synthesis of this literature would include all, or most, of the following points although, strictly speaking, they apply to Nepal and have been extrapolated to characterize the much wider region:

1. Following the introduction of the modern health care, medicine, and malaria suppression in the Terai after 1950, an unprecedented wave of population growth occurred which does not yet appear to have peaked. For Nepal as a whole it appears to have reached 2.6 percent per annum for the 1971–81 census decade (Goldstein *et al.*, 1983) but in many areas it exceeds 3–3.5 percent per annum. Nepal's total population in 1988 is probably in excess of 16 million.
2. This veritable population explosion, with an overall doubling period of about 27 years, is augmented by uncounted and uncontrolled illegal immigration from India into the Nepalese Terai across the open frontier. Furthermore, over 90 percent of the 1981 population is rural and subsistence. This has led to rapidly increasing demands for fuelwood (more than 90 percent of Nepal's energy depends upon the combustion of biomass), construction timber, fodder (the domestic animal population has undergone a parallel, or even greater, increase to that of the human population), and agricultural land on which to grow food.
3. The next step in what has been described as a vicious circle, is that the needs of the burgeoning subsistence population are exerting increasing pressures on the forest cover. This has led to massive deforestation, amounting to a loss of half the forest reserves of Nepal within a 30-year period (1950–80) and a prediction that by AD 2000 no accessible forest cover will remain (while there are varying estimates of the rates of deforestation, a topic to be treated more fully in Chapter 3, it is widely assumed that the situation has reached crisis proportions).
4. The deforestation, which includes the cutting of agricultural terraces on steeper and more marginal mountain slopes, has led to a catastrophic increase in soil erosion and loss of productive land through accelerated landslide incidence, and to the disruption of the normal hydrological cycle.
5. This situation, in turn, has led to increased run-off during the summer monsoon and increases in disastrous flooding and massive siltation in the plains, and lower water levels and the drying up of springs and wells during the dry season. Related ills are: rapid siltation of reservoirs; abrupt changes in the courses of rivers; spread of barren sand and gravel across rich agricultural land on the plains; and increased incidence of disease in downstream areas.
6. The increased sediment load of the rivers emanating from the Himalayan system is extending the Ganges and Brahmaputra delta and causing islands to form in the Bay of Bengal. Amongst the evidence cited are extensive plumes of sediment that can be seen on LANDSAT imagery to extend several hundred kilometres into the bay.

7. The continued loss of agricultural land in the mountains leads to another round of deforestation to enable the construction of more terraces on which to grow subsistence crops. Yet, as the labour of walking greater distances from the village to fuelwood supplies increases with the receding forest perimeter, a critical threshold is reached whereby the available human energy (principally female) becomes progressively overtaxed and an increasing quantity of animal dung is used for fuel.
8. Consequently, another vicious circle is linked to the first one: terraced soils are deprived of natural fertilizer – the animal dung now being used for fuel, thus depriving the agricultural terraces, in many instances, of their only source of fertilizer. This lowers crop yields. Also, the ensuing weakened soil structure further augments the incidence of landslides. Even more trees are cut on more marginal and steeper slopes to make room for more agricultural terraces to feed the ever-growing subsistence population.

Many other facets can be added to the eight-point scenario. These include the pressures generated by the subdivision of the finite amount of agricultural land as the population continues to double every twenty-seven years; at present it is calculated that there is less than 1 ha of land per family. Similarly, the added pressures of collecting and carrying fuelwood and fodder and fetching water falls predominantly on the women. They, in turn, become progressively overworked and undernourished and their next generation of children begins life more and more deficient in essential nutrients, so that the situation worsens further. Domestic animals, essential to the Middle Mountain subsistence mixed-farming system as suppliers of fertilizer and draught energy, depend heavily on fodder from the depleted forests, so that their capacity also diminishes.

It follows from this brief exposition of the Theory of Himalayan Environmental Degradation that a series of linked vicious circles is envisaged as operating inexorably to drive a downward spiral. The apparent impossibility of breaking any of these circles thus leads to the prediction of widespread environmental and socio-economic ruin in the near future. There is perceived to be a progressive and accelerating shift from *potential* instability to massive *actual* instability. This includes: mountain slopes, from a physical point of view; hill-village subsistence agriculture; breakdown in traditional mountain culture; disruption of the regional or national economy. All these gathering tragedies will put increasing pressure on the already fragile political balances of the wider Himalayan region.

In the face of the irreversible destructive processes in the Middle Mountains, for instance, out-migration increases. This in turn not only deprives the source areas – the villages and hamlets of the Middle Mountains – of a proportion of their youngest, fittest, and most creative members, but it adds to the already existing heavy population pressure on the resource base of the Nepal Terai. Population growth in the Terai, for example, including both natural increase and in-migration, is calculated at over 4 percent per annum

(1971–81 census; cf. Goldstein *et al.*, 1983; Hrabovszky and Miyan, 1987). Fifteen to twenty years ago it was believed that the opening up to settlement of the previously malaria-infested jungle of the Terai would provide a breathing space by enabling the absorption of excess population of the Middle Mountains, Nepal's most densely populated area. Today it can be seen that the 'new land' in the Terai is virtually used up, a process accelerated by the surge of illegal immigration across the open border with India. And yet, despite extensive out-migration, the population increase in the Middle Mountains continues at an unacceptable, or unmanageable, rate in excess of 2 percent per annum. Goldstein *et al.* (1983), for instance, characterize Nepal as being in transition, demographically speaking, from a highland, rural country to a lowland, urban one.

The net results of the various destabilizing processes in the Middle Mountains are perceived as absolute deforestation, lowered crop productivity (both in terms of total national production and as yield per unit area), increase in absolute numbers and percentage of the subsistence farming population with nutrient intake below a minimum acceptable level, and progressive mountain desertification. Since the mountain desertification is assumed to be occurring on steep slopes, the associated processes of gullying, soil erosion, and landsliding are cited as having calamitous downstream effects. Thus are envisaged the rapid siltation of reservoirs, excessive shortening of the useful life of major hydroelectric and irrigation projects, increased flooding on the plains (already an annual disaster for India and Bangladesh), increases in the levels of river beds, and destruction of rich lowland farmland by the spread of sand and gravel as rivers break their banks and change their courses. In short, the worst-case scenario foresees that the terrain of Nepal and that of adjacent areas of the Himalaya, and certainly the very basis of life, the topsoil, will virtually flow down the Ganges and Brahmaputra rivers by the year AD 2000. It has even been suggested that, in preparation for such an event, His Majesty's Government of Nepal should transfer its patronage of the Swiss technical-aid system (SATA) to that of the Dutch. In this manner Nepal can begin the struggle to reclaim (and legally claim) land below sea level and establish polders in the Bay of Bengal, the product of its own topsoil (Indian and Bangladesh gunboats are already rumoured to be patrolling extensive new islands that are being added to the outer Sundarbans, the outer delta of the distributaries of the Ganges and Brahmaputra).

The last sentence of the preceding paragraph, while an extreme interpretation that may be criticized as a macabre joke, has been introduced to demonstrate both the seriousness and the science-fiction attributes of this powerful Theory of Himalayan Environmental Degradation. More than a decade ago Eckholm (1976) wrote eloquently of the process whereby Nepal was exporting to India the commodity that it could least afford to part with, namely topsoil, and in the form that India could least afford to receive it – as silt that clogged reservoirs, turbines, and irrigation works. The broad theory, nevertheless, is an intellectually satisfying concept which seems so reasonable that it is hardly surprising that it is widely accepted as fact. And, of course, there are further ramifications, such as claims by environmentalists that the deforestation, in turn, is affecting the climate in such a way as to reduce normal annual rainfall amounts. This, of course, would set up yet another vicious circle to accentuate the effects of the others.

The eight-point scenario and brief discussion presented above lead to a number of critical implications which further enlarge what can be described loosely as the *perceived* Himalaya–Ganges Problem. It infers that a few million Nepalese hill farmers are responsible for the massive landscape (and climatic) changes that are affecting the lives and property of several hundred million people in Gangetic India and Bangladesh. This raises two related points: (1) that the downstream countries, as victims of this unwarranted and irresponsible environmental disruption, could justify reprisals in economic, political, or military terms; and (2) that Nepalese interests are served well (assuming no reprisals are actually taken) by this perceived image of helpless drift into environmental and socio-economic chaos, since it may account for its disproportionate amount of international and bilateral development aid in relation to its total size and population.

A further point is applicable within Nepal itself, and within several of the Indian states that have a plains and a Himalayan component, such as Uttar Pradesh. This is that the popular image of the hill farmer as the cause of the growing environmental disaster makes him a convenient scapegoat; it has been claimed that the relatively few mountain farmers are holding hostage the very many on the plains. Once more, effect is taken for cause, and corrective measures are misdirected.

Whether or not the eight-point scenario of disaster for Nepal can be extended along the entire Himalayan system will not be discussed here except for emphasis of several related points. The Kumaun and Garhwal Himalaya appear to fall within this framework, with two additional components. One is the excessive commercial cutting of mountain forest stands to meet the timber demands of the lowland population centres (until recently checked by the Chipko Movement - see below page 67). The other, associated with it, is the extensive development of mountain roads, especially as a military response on the part of India resulting from the border war of 1962 with China (see p. 119 below). Much of the road construction is substandard and has caused a great increase in landslide incidence; the roads also opened up extensive mountain forests to commercial clear-felling and to large-scale movement of people.

As we proceed westward into Himachal Pradesh, Jammu and Kashmir, and the Karakorum and Hindu Kush, we enter a series of mountain and highland landscapes with very different climatic regimes. Increasing aridity with distance from the influence of the summer monsoon greatly reduces the value of comparison with the Nepal and Central Indian Himalaya. Conversely, eastward from Nepal, into Sikkim and the Darjeeling (West Bengal) Himalaya, and Bhutan we enter mountain areas with increasing amounts of summer monsoon precipitation. Bhutan, and probably the even less accessible and less well-known Arunachal Pradesh, should be accepted as unique cases where assumed deforestation and environmental disturbance are modest or insignificant and where fuelwood supplies are reported as abundant. However, even these regions can be assumed to be poised to follow the same road to disaster along which Sikkim, Nepal, and the Central Himalaya of India are travelling.

There are several additional contributing problems that need to be introduced. The first is the political processes that beset the region: the border dispute between India and China; the three Indo–Pakistan wars of 1947, 1966, and 1971 and the continued border tensions, especially along the Kashmir cease-fire line; the frictions between India and Bangladesh, and especially the very slow progress in negotiations over management of the flow of the Ganges and Brahmaputra. There are also the political tensions within India – Punjab, Assam, West Bengal – generated, in part, by the competing demands for access to natural resources by different ethnic groups.

The 1959/60 exodus of approximately 120,000 refugees from Tibet, and their impacts on the natural resource base of Khumbu Himal and Mustang, for instance, and the consequences of China's closing of the frontier with Nepal, must be seen as unprecedented disruptive events. The much more extensive disruption in Afghanistan, and the impacts of over three million refugees on the Pakistan Hindu Kush and adjacent northern areas, is a problem of world magnitude in itself, notwithstanding the widespread political implications.

The rapid growth in tourism is also an important contributing factor. While most data, again, are available for Nepal, popular access to the Garhwal Himalaya, especially the Valley of Flowers and the Nanda Devi Sanctuary, is causing a large increase in environmental pressure of an entirely different kind, the result of demands for recreation and adventure from people outside the mountain area. Fifteen years ago the growth of tourism in Nepal was perceived as a panacea for that country's balance-of-payment deficits. Today that perception has changed. Certainly in specific areas, such as the Khumbu Himal and the Kali Gandaki and Annapurna circuit, the numbers of tourist visitors each year now exceed the total indigenous populations. And their demands are seen as increasing the threat to local forests as well as being disruptive of the way of life of the local people.

It follows from the foregoing discussion that, regardless of whether or not we accept the cause-and-effect linkages that together make up the Theory of Himalayan Environmental Degradation, the broader Himalayan region must be viewed as dynamic in the extreme – geophysically, climatically, and in the socio-economic and political senses. It is also useful to extend our consideration beyond the limits of even the broadest conventional depiction of the Himalayan region since similar catastrophic processes are presumed to be operating. For instance, the perceived linkage between mountain deforestation and downstream effects also has been applied to the Hengduan Mountain–Chengdu Basin system, far to the east of the Himalaya proper.

Here extensive deforestation in the 'River Gorge Country' of western Sichuan and northwestern Yunnan, and its impact on the hydrology and sediment load of the Jinsha Jiang (Yangtze), is credited with increased flooding, siltation, and damage to heavily populated and rich agricultural land downstream (the Chengdu Basin). As with the Himalaya–Ganges system, population growth, deforestation, and soil erosion have been perceived as post-1950 phenomena, in this case with the added overtones of mismanagement during the Mao Tse-tung years with the Great Leap Forward, and Cultural Revolution, and the ensuing chaos created by the 'Gang-of-Four.'

It must be said that the 'save-the-world's-forests' lobby has had a powerful influence on large sections of public and scientific opinion. It has been widely agreed that serious problems will arise from loss of the tropical rain forests (and also from mountain deforestation in the Himalaya and elsewhere). This we accept and strongly support: our intention is not to challenge the importance of the world's forest resources but to examine critically some of the claims that are made in terms of the assumed effects of forest clearance, and their facile extension to other, especially mountain, regions. We believe that such a critical assessment is a vital undertaking, a prerequisite to rational corrective policy development.

LINKAGES OF THE THEORY AND THEIR IMPLICATIONS

To sum up our discussion of the perceived Himalaya–Ganges Problem, it is necessary to point out a number of basic assumptions about and salient linkages between the component parts of the eight-point scenario. There is also a basic philosophical issue – what should be our attitude to the ignorant subsistence farmer who is seen, unthinkingly, to produce swarms of children, and irresponsibly to devastate the mountain forest cover and so to accelerate landslide occurrence on his poorly constructed and badly maintained agricultural terraces, or in some areas, by his catastrophic slash-and-burn (swidden) agriculture? To take the basic assumptions and salient linkages first:

1. That a population explosion was initiated shortly after World War II due to the introduction of modern health care and medicine and the reduction of malaria and other diseases;
2. That increased population in subsistence mountain societies has led to:
   (a) reduced amount of land per family
   (b) deepening poverty
   (c) massive deforestation;
3. That mountain deforestation, on such a scale, will result in total loss of all accessible forest cover in a country such as Nepal by AD 2000, and is the cause of accelerating soil erosion and increased incidence of landsliding;
4. That destabilized mountain slopes resulting from points 1, 2, and 3 above cause:

(a) increased flooding on the Ganges and Brahmaputra plains,
(b) extension of the delta and formation of islands in the Bay of Bengal,
(c) drying up of wells and springs in the hills and lower dry-season river levels downstream,
(d) massive siltation and drastic reduction in the useful life of highly expensive water resource projects;

5. That deforestation also leads to climatic change in general and reduced rainfall amounts in particular.

It is not our intention to dispute the facts, wherever reliable information exists, but the assumptions that so frequently are not based upon facts. Nevertheless, throughout this attempt to dissect the Theory of Himalayan Environmental Degradation the causal relationships between timing and degree of population growth, deforestation, loss of agricultural land, and downstream effects are paramount. We will attempt to demonstrate that most of these linkages and assumptions are founded upon latter-day myth, or falsely based intuition, or are not supported by rigorous, replicable, and reliable data. They are the 'sacred cows' of the perceived Himalayan Problem, and we will seek to dismantle them, in part or in whole. However, our claim of lack of reliable data cuts both ways – we cannot demonstrate unequivocably that *all* the linkages are inoperable in *all*, or even in most, cases. Nevertheless, we do believe that we can dispose of enough of the 'sacred cows' and damage others sufficiently to support our claim that the overall Theory of Himalayan Environmental Degradation is untenable and that the Himalayan Problem needs to be much more rigorously defined.

The perceived problem, in our view, is in the minds of the vested interests – whether the World Bank, the Chipko Movement, different national governments, or the scientists. It is likened to a kaleidoscope, which will change its pattern depending upon the way in which it is tilted, or upon the angle of view. This is the essence of Thompson and Warburton's (1985a) 'Uncertainty on a Himalayan Scale.' The uncertainty is a large element of the Problem. Thus the present claim that we expect to demolish most of the underpinnings of the Theory itself must be qualified by the very nature of the uncertainty. There must be the qualification that *in certain instances*, and *in specific areas*, we believe we can show that many of the widely preferred assumptions are untenable. The widespread temptation to extrapolate, or generalize, must be resisted or else we ourselves would commit the error that we are seeking to expose – unwarranted generalization. The single and obvious generalization that we do make, however, is that the Himalayan region is so varied and so complex that generalization is counter-productive. Hence, the application of broad panaceas by aid and development agencies in most, if not all, instances will not succeed; in some instances they may well exacerbate the problem.

But what of the ignorant and fecund subsistence farmer whose well-being lies at the crux of the Problem? He has indeed become a convenient scapegoat. We will demonstrate this by a single observation, illustrative of many others.

Figure 1.1 Middle Mountains, Nepal. View of intensely terraced landscape a few kilometres from Kathmandu; from the Trisuli Road below Kakani. The bare, outward-sloping terraces on the right are rainfed (*bari*) and are fallow during the dry winter season. A crop of winter wheat emphasizes the irrigated terraces (*khet*) down the centre of the small valley which gives relatively easy access to water.

In an otherwise impressive review of the Nepal Agricultural Sector, the following quotation from a report by the Asian Development Bank (ADB 1982: (II) 34) indicates the extent of the misunderstanding of the subsistent farmer's role: 'Terraces, especially on rainfed land, are often poorly constructed; they are outward rather than inward sloping and do not have a grassed bund on the edge.' The fact is that *bari*, or rainfed terraces, in Nepal mostly support maize, millet, buckwheat, and other crops. They are constructed usually on the upper, steeper slopes in the Middle Mountains which are inaccessible to irrigation systems. They slope outward from the hillside so that these crops are not damaged by waterlogging. In our Kakani field area (Johnson *et al.*, 1982; Gurung, 1988) the local farmers are well aware that an increased accumulation of water on terraces (such as would

Figure 1.2 Maintenance of terraces in the Middle Mountains on slopes of up to 45 degrees requires constant labour. Here *bari* terraces are being prepared for early summer planting of maize or millet. Note the careful trimming back of terrace fronts in the lower right. The soil from this is then worked into the next terrace step below, with manure and crop residue added.

result from inward-sloping forms) would greatly exacerbate the problem of landsliding by increasing the degree of soil saturation and adding the weight of the ponded water itself. Furthermore, annual repair of the terraces would require a much larger labour input if they sloped inward. The summer monsoon rain is intended to run off the outward-sloping terraces.

Although there are undoubtedly poorly maintained terraces in Nepal, and in other areas, many are very well maintained and have vegetated steps; absence of a bund on the *bari* terraces (in contrast to the *khet*, or irrigated, terraces) is deliberate and *ensures* rainwater run-off. It can be argued that both *bari* and *khet* terraces are, for the most part, superbly engineered in Nepal. Admittedly, during heavy monsoon downpours available human energy is concentrated on repairing damage to the *khet* and irrigation

Figure 1.3 A mixture of *khet*, *bari*, and rough grazing on the northwestern edge of the Kathmandu Valley. Young rice seedlings stand a few centimetres above the irrigation water in the foreground. The rainfed terraces above the house will be used for maize or millet, or a variety of vegetables.

systems, and the *bari* terraces may have to be left to collapse; this is because the much higher-yielding *khet* terraces, usually under paddy rice, are consequently more vital to the survival of the subsistence family. Any apparent neglect of the terraces may be due to shortage of available labour at the particular moment that they were observed by the visiting 'expert' rather than a reflection of the ignorance of the farmer.

The famous English mountaineer, H. W. Tilman, more sensitive to the hill farmer than many latter-day experts, made a poignant observation many years ago when the Himalaya was still a distant fantasy land of 'Shangri La' to most of us:

Whether it takes place little by little or in one swift calamity, soil erosion is

Figure 1.4 Sindhu Palchok District, Nepal Middle Mountains, near Chautara, headquarters of the Nepal-Australia Forestry Project. Note the relatively dense forest cover on the steeper, more distant slopes, the large number of individual trees on private land near the farm houses, and the precise maintenance of the *khet* terraces in the fore- and middleground.

> generally attributed to man's careless greed, his idleness or neglect. It would not, I think, be fair to blame the people of these valleys on the Himalayan fringe for the frequent landslides which occur here. In turning the steep slopes into fruitful fields they have neither been lazy nor neglectful.
>
> (Tilman, 1952: 126–7)

The date of Tilman's writing is significant in that the Theory of Himalayan Environmental Degradation had not then been formulated.

Frequently the subsistence farmer can be shown to be a highly knowledgeable and intelligent land manager with a wealth of accumulated, traditional wisdom of great potential value to the 'educated' elites, if only they would listen (Whiteman, 1985). This leads to the claim that there is a need for gift exchange in contradistinction to *charity* - a synonym for international and bilateral aid (Hatley and Thompson, 1985).[1] Nevertheless, we do not wish to imply that all subsistence farmers are intelligent indigenous scientists, nor that even the most gifted amongst them can necessarily control the change which is sweeping them along; and there are ignorant and foolish farmers, just as there are ignorant and foolish factory workers, tradesmen, scientists, and decision makers.

Before concluding this chapter two further points must be made. First, there are no claims to be established for any individual's academic or scientific precedence, despite the intellectual satisfaction of having the opportunity to fault a widely accepted paradigm. Members of the United Nations University/Nepal MAB-Mountain Hazards Mapping Project began to suspect the reliability of some of the pre-existing claims of the Theory of Himalayan Environmental Degradation. They began to doubt that deforestation and increased landsliding were linked in a simple cause-and-effect relationship. They also began to understand, as fieldwork progressed over several years and during different parts of the annual agricultural cycle, that the human impacts, principally those of the subsistence farmer, were not all negative. Part of the farmers' coping strategy was to re-terrace landslide scars and stabilize slopes. They also responded to prospects of immediate landslide initiation by such acts as agricultural de-intensification (Johnson *et al.*, 1982; Messerschmidt, 1987). Similarly, reconnaissance of the Qinghai-Xizang (Tibet) Plateau, and in the Hengduan Mountains of western Sichuan and northwestern Yunnan (Ives, 1981, 1985; Messerli and Ives, 1984) led us to suspect that the assumptions of post-1950 massive deforestation were also over-simplifications, and that the actual history of deforestation was a very much longer and more complex process. This gradual growth in understanding of the complex nature of the region and the processes operating therein led to this questioning of conventional wisdom. Our doubts about recent mountain deforestation brought us into contact with the work of forest historians Richard Tucker (1986, 1987) and John Richards (1987), with ecologists and Chipko activists Vandana Shiva and Jayanta Bandyopadhyay (1986a and b), with Tej Mahat, David Griffin, and Kenneth Shepherd (Mahat *et al.*, 1986a and b; 1987a and b), with Michael Thompson, Michael Warburton, and Tom Hatley (Thompson *et al.*, 1986), with David Pitt (1986), Lawrence Hamilton (1987), and Deepak Bajracharya (1983a), and many others, together with the spiritual leadership of Chipko Messenger, Sunderlal Bahuguna. We discovered from these contacts that simultaneous doubts and challenges had been developing along similar lines (cf. Carson, 1985). It is the coming together of this group, facilitated by the United Nations University's support of the Highland–Lowland Interactive Systems project, that has led to this concerted effort to challenge the Theory of Himalayan Environmental Degradation.

The second point is equally important. All members of this now considerably enlarged working group do not necessarily agree on all points, or even on any one particular point; nor do we, nor can we, all have the same perspective. But we all do agree that a major arena of enquiry has been opened up that is fraught with an unusual level of uncertainty. We also wish to stress that the enquiry has been encouraged by many individuals within several major agencies of the United Nations Organization, despite the occasional criticisms that appear to be levelled against them. We believe, however, that the enquiry has an important potential bearing on the well-being of several hundred million people and on the socio-economic and political stability of a pivotal region of the world. Thus further exhaustive pursuit of the enquiry should become a major endeavour, not only for the United Nations University, but also for other relevant UN agencies, bilateral aid and development agencies, and the governments of the region.

To conclude this chapter we wish to emphasize again that it is not our intention to dispute the validity of established facts, nor to imply that there is no Problem facing the Himalayan region. We believe that there is a most serious problem; that it has been exacerbated by the very tendency to generalize, to accept uncritically a large number of inter-related assumptions, and to precondition policy making by rigidly defended perceptions. One of the more destructive of these perceptions, for instance, is that deforestation is necessarily bad (cf. Hamilton, 1987); another, related to the first, is the habit of using the term *human impact* invariably in a negative sense (cf. Messerschmidt, 1987).

Now that we have set forth a synthesis of the Theory of Himalayan Environmental Degradation and begun the process of questioning the validity of some of its component parts, following our geographical overview (Chapter 2), we will devote the next four chapters to a more exhaustive examination of the linkages that form the vital fabric of the physical basis of the Theory.

NOTE

[1] Hatley and Thompson maintain that, in its present form, foreign aid is perceived as charity: only when a clearly perceived two-way sharing of benefits can be established will development aid produce more convincing results. The 'ignorant' subsistence farmer has much to offer; from various parts of the mountain world he has already provided the world community, amongst many other 'gifts,' potatoes, maize, quiñoa, and intricate sustainable farming systems. Much more remains to be discovered, including a wealth of indigenous knowledge.

2 THE HIMALAYAN REGION: A GEOGRAPHICAL OVERVIEW

INTRODUCTION

The purpose of this chapter is to provide a very brief overview of the general geography of the Himalayan region. This is a necessary but rather unsatisfactory task, in part because of the constant outpouring of publications on practically every conceivable aspect, with a heavy emphasis on applied topics – resource development, environmental degradation, foreign aid – and in part because of the very complexity of the topic. Regrettably, there is no recent systematic treatment. It is even difficult to produce a justified regional break-down. We have added a short section at the end of the chapter which is a synthesis of the most recent attempt to tackle this difficult problem of regionalization. This is the result of a graduate diploma study of Markus Wyss, Geographical Institute, University of Berne (1988). We will largely limit ourselves, nevertheless, to a brief indication of the region's complexity in the process of sketching some of the major components of the topography, climate, vegetation, and human geography. We are not attempting a geography *per se*; rather we are providing selected background material and additional literature citations to support the discussion that forms the *core* of the book. First we will define the area under review.

The traditional definition of the Himalaya, *sensu stricto*, is that great range of mountains that separates India, along its north-central and northeastern frontier, from China (Tibet), and extends between latitudes 26°20′ and 35°40′ North, and between longitudes 74°50′ and 95°40′ East. In this sense the Himalaya extend from the Indus Trench below Nanga Parbat (8,125 m) in the west to the Yarlungtsangpo–Brahmaputra gorge below Namche Barwa (7,756 m) in the east, a west-northwest to east-southeast distance of about 2,500 km. This definition includes, politically, the independent kingdoms of Nepal and Bhutan, a small part of Pakistan, parts of China (Xizang Autonomous Region), as well as the western, central, and eastern sections of the Indian Himalaya (see Figure 2.1): sections 6 (Kashmir Himalaya), 7 (Central Himalaya), and 9 (Assam Himalaya), together with portions of the Plateau (8) and the Plains, as shown on Figure 2.2.

Also, traditionally, a north–south topographical transect across the Himalaya would include the whole, or parts, of several aligned physiographic

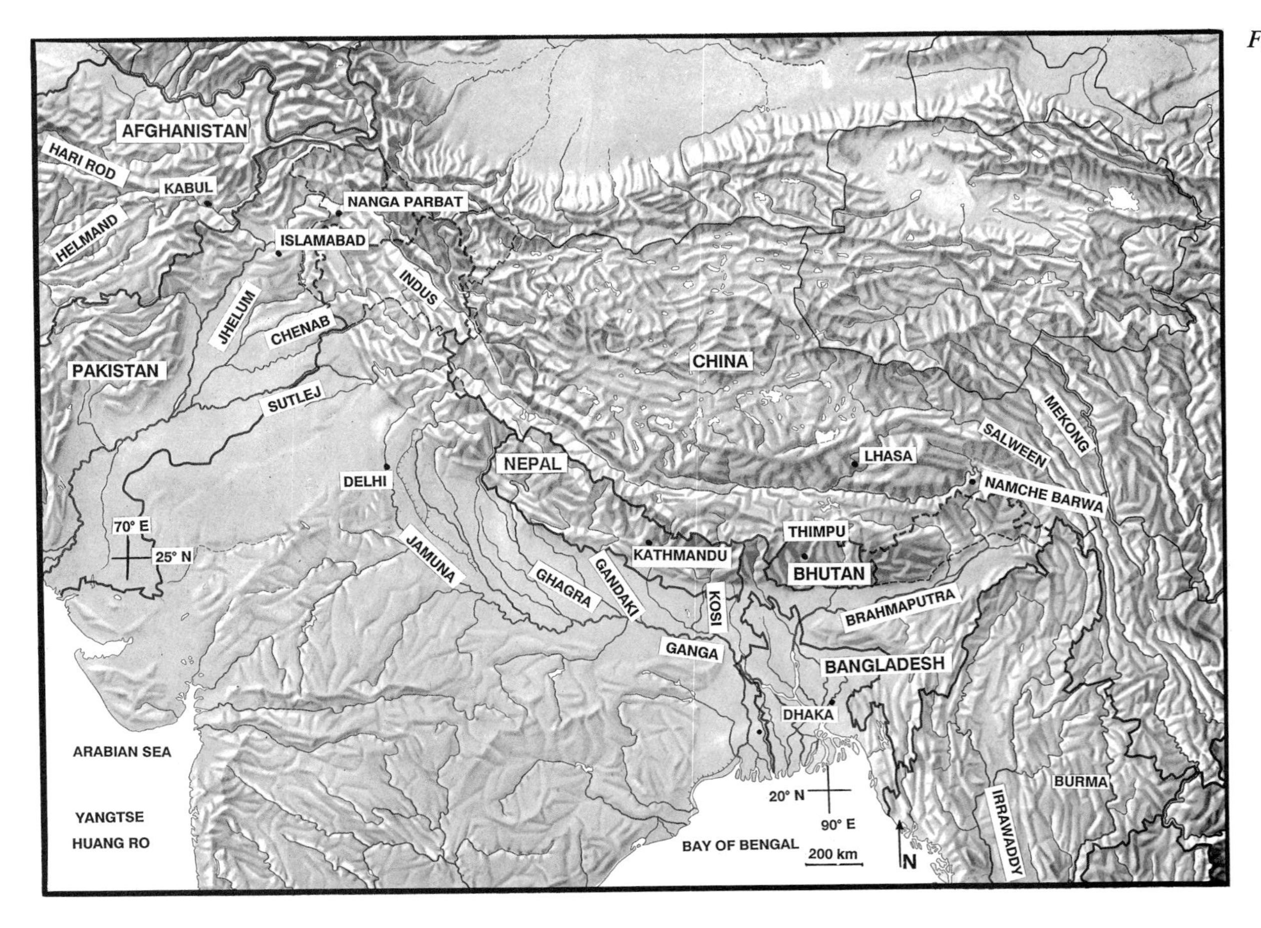

Figure 2.1 The Himalayan region, southern Tibet and northern India. The Himalaya *sensu stricto* extend from Nanga Parbat, above the Indus Gorge, to Namche Barwa, above the Brahmaputra. The wider region takes in the Hindu Kush and the Hengduan Mountains. Topography: copyright, Swiss *High School Atlas*, 1988.

Figure 2.2 Main topographical subdivisions of the Himalayan region, *sensu lato*. Topography: copyright, Swiss *High School Atlas*, 1988.

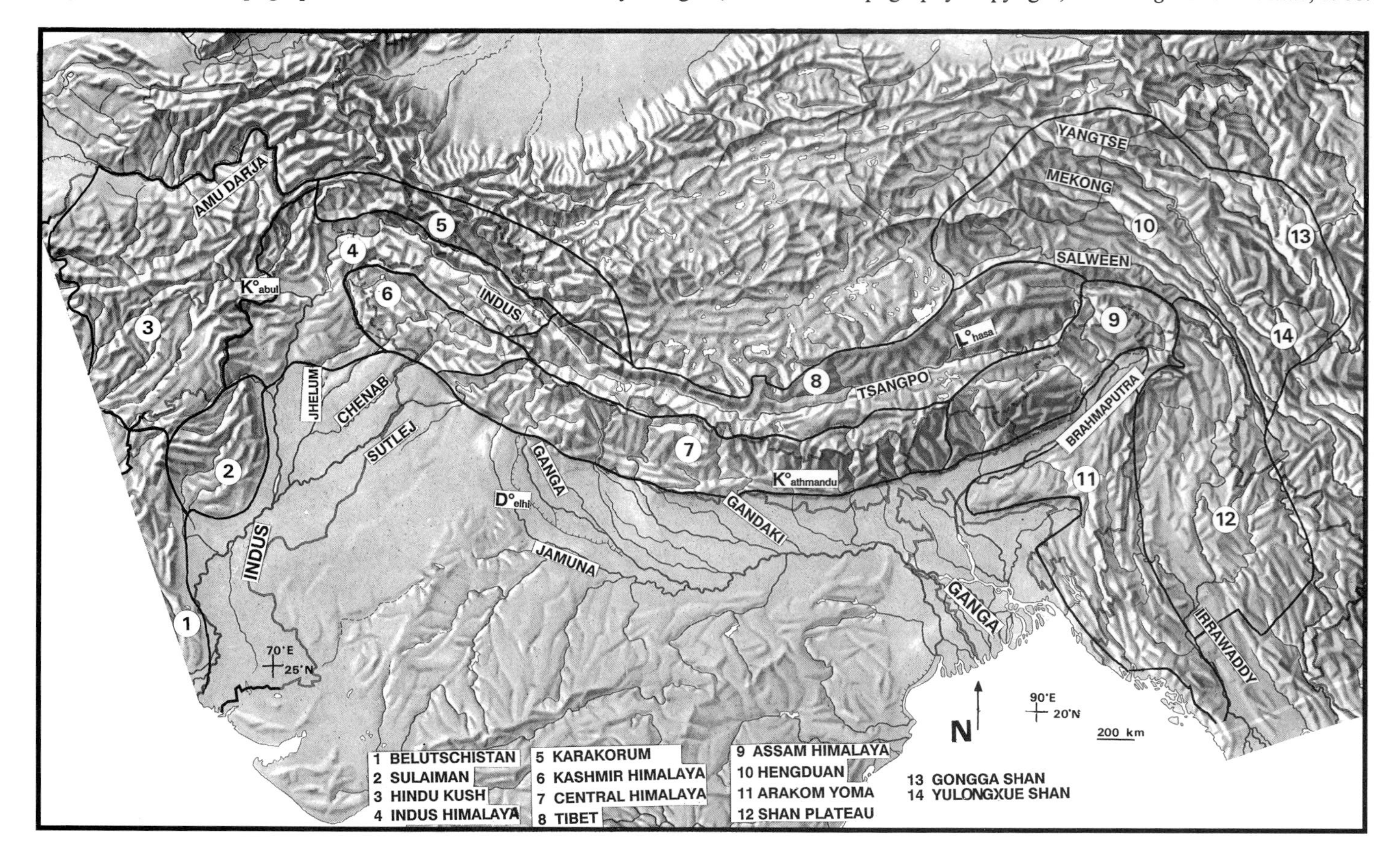

provinces: a small slice of the Qinghai-Xizang (Tibet) Plateau; the trans-Himalayan ranges and intervening valleys; the Greater Himalaya; the foothills (locally referred to as the Siwaliks or Churia Hills); the Middle Mountains, the Lesser Himalaya (including the Mahabharat Lekh); and the Terai. The Terai can best be described as the upper section of the Ganges Plain; it is also convenient to subdivide it into the *barbar* (porous place), which immediately abuts the Siwalik front and is composed of a series of giant coalescing fans that have been laid down by numerous torrents and major tributaries of the Ganges, and the Ganges flood plain proper, into which the *barbar* grades imperceptibly.

Figure 2.3 is a schematic north–south transect characterizing the Central, or Nepal, Himalaya. Similar transects, drawn to represent sections further east or west, would show the same major physiographic divisions, but with many local variations. The terminology that we have adopted for the main physiographic divisions is not universally accepted, as will be illustrated in the next section (pp. 22–24).

The Ganges includes the catchment of the enormous Ganga (Hindu spelling) river system; this embraces much of the Himalaya as defined above, as far east as Sikkim, a large part of densely populated northern peninsula India, and part of Bangladesh. The western-most section of the Himalaya is drained by the Jhelum, Chenab, and Sutlej, major tributaries of the Indus, as well as by the upper Indus itself. The Brahmaputra is the Hindu name for that section of the other main trans-Himalayan river (Yarlungtsangpo–Brahmaputra) from the point where it enters Indian territory. It rises high on the Tibetan Plateau in longitude 82° East, remarkably close to the main headstream of the Indus, and flows eastward for more than 1,200 km north of the Himalayan crest-line before making its spectacular turn to cut through the mountains in one of the world's most impressive gorges. It enters India through Arunachal Pradesh and flows roughly westward across Assam and Bangladesh before turning south again to produce a maze of distributaries that merge with those of the Ganges, and eventually enters the Bay of Bengal through the great delta.

The term Himalayan region is being used in a very broad sense, and we will draw no precise boundaries. It will include the Hindu Kush, Karakorum, and the Hengduan mountain ranges, and a large slice of the Qinghai-Xizang (Tibet) plateau (see Figures 2.1 and 2.2). We will also include the Ganges and Indus Plains. In other words, our discussion will draw freely on available information from that vast tract of mountain, plateau, river gorge, and plains that can perhaps best be described as the south-central Asian mountain, plateau, and plains region. Thus, Bangladesh is included, as well as part of Afghanistan, northern Pakistan, Xizang Autonomous Region of China, western Sichuan, and northwestern Yunnan, and the border regions of northern Burma and Thailand. Nevertheless, we will concentrate heavily upon the Himalaya, *sensu stricto*, and especially on Nepal, and the immediately subjacent sections of the Ganges, Teesta, and Brahmaputra plains.

The region, broadly defined, provides the life-support base for about 50

Figure 2.3 Schematic cross-section of the Nepal Himalaya: geology, Daniel Vuichard, Institute of Mineralogy, University of Berne; topography, modified after W. J. H. Ramsay.

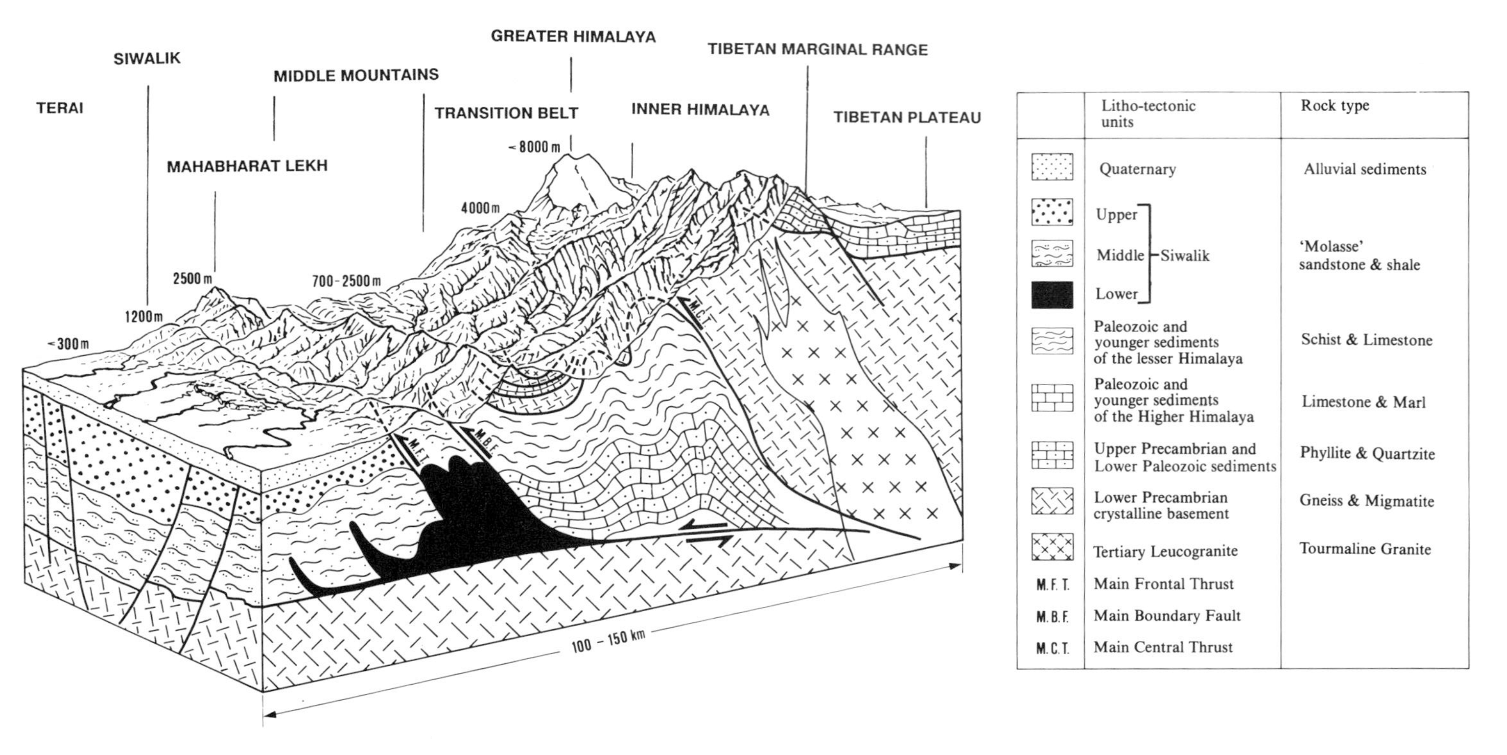

million mountain people and probably in excess of 450 million people of the plains – the very densely populated areas of the Indus, Ganges, Brahmaputra, and upper and middle Jinsha Jiang (Sichuan Basin) – a significant proportion of humankind. While this book is principally a Himalayan story, the mountains and plateaus are the source of Central and South Asia's great rivers. As we seek to show, erosion of the mountains, over geological time, is the prime reason for the existence of the plains; and the lowlanders tend to perceive many of the 'natural' catastrophes with which they must contend as the result of landscape changes in the mountains brought about by the mountain people. The reverse of this antagonism, albeit over-simplified, is that many of the political and economic forces that are disrupting life in the mountains have a lowland base. Thus, the Himalaya–Ganges system, in the narrow sense, or the south-central Asian mountains-plateaus-plains system, in the broad sense, can be considered as one of the world's largest 'highland–lowland interactive systems.' It also embraces one of the world's greatest accumulations of poverty, malnutrition, and accelerating population growth.

It is recommended that the reader consult a good world atlas.[1]

THE PHYSICAL BASIS

Figure 2.3 has been introduced as a general model for describing the basic landscape elements of the Nepal Himalaya. For the purpose of our discussion it is necessary to emphasize that the 2,500 km west-northwest to east-southeast alignment of the Himalayan crest-line, reaching altitudes in excess of 8,000 m in a series of distinct mountain massifs separated by areas of lower altitude, often the locus of major river gorges, is paralleled by lower and less spectacular physiographic divisions to the south. This large-scale pattern is essentially the result of the Indian tectonic plate thrusting beneath the Central Asian plate. Despite the rapid uplift that this plate convergence has caused throughout the past several million years, some of the main rivers have been able to maintain their courses – as antecedent streams. The entire river network, however, displays a rectilinear pattern, with long sections of the mainstreams having developed courses along the structures at right angles to the gorge sections that cut across the structure. These patterns, the extreme altitudes, and also the very limited north–south extent (less than 150 km in places) reflects the enormous crustal shortening, over-thrusting, faulting, and folding that has occurred, and is occurring. The reader is referred to Hagen (1960, 1969) for a general account of the geology and structure of the Nepal sector. Of special importance, however, in terms of tectonics, are the absolute heights, with a roughly east–west crest-line, the continued rapid uplift (in many areas in excess of the estimated rates of down-wasting – see Chapter 5), and the great accumulations of sediments in the foredeep, which underlie the Ganges Plain and neighbouring plains. The close proximity of maximum altitudes in excess of 8,000 m, and maximum depths (of sediment) to more than 5,000 m below sea level, is dramatized by three major tectonic discontinuities. These are: the Main Central Thrust Zone (MCT); the Main

Boundary Fault (MBF); and the Main Frontal Thrust (MFT), which, in essence, parallel the entire Himalayan system. As indicated in Figure 2.3 (and discussed in more detail, from the perspective of susceptibility to erosion, in Chapter 5) this 'standard' north–south transect includes nine distinct physiographic divisions. Gurung and Khanal (1987) discuss the names of these units from a Nepalese point of view and, while we have not adhered closely to them, their discussion is introduced here for basic reference.

Gurung and Khanal (1987) proceed from Hagen's (1960) division into seven zones, from north to south:

Tibetan marginal Mountains
Inner Himalaya
Himalaya
Nepal Midlands
Mahabharat Lekh
Siwalik Hills
Terai

Hagen (1969) later added three more: the Dun Valleys ('Mid-terai'); the Fore Himalaya (Lesser Himalaya); and the Tibetan Plateau. Gurung and Khanal (1987) explain that for Nepal there exists a long-standing traditional terminology: thus, terai, or *madhes* for the plain; *pahar* for the hills; and *himal* for the mountains. They differentiate the plains into the terai and the inner terai (*dun*), the latter being tectonic depressions in the *Chure* range (or Churia = Siwaliks), or between the Siwaliks and the Mahabharat Lekh. Similarly, the limit of the sub-tropical middle hills (Hagen, Nepal Midlands; Ives and Messerli, Middle Mountains) is the higher and temperate highlands – the Mahabharat Lekh – on the south. Finally, the trans-Himalayan valleys (*bhot*: hence Bhotia people = of Tibetan or Mongol origin) are enclosed by the *himal* (Great Himalaya) and the Tibetan border ranges. Gurung and Khanal maintain that the distinction between 'Himalaya' and 'Mountain' is spurious as they are synonymous terms, and that the epithet 'high' in this context is superfluous. The 'Middle Mountains,' a term we retain, actually relates to the conventionally recognized 'hill' region. Also, the native terms differentiate between the hills (*pahar*) which have no snow, the highland (we prefer 'higher mountains') or *lekh*, which only experiences snow in winter, and the mountain or *himal*, with permanent snow. Siwalik, according to these authors, is a geological term and is a composite of the Chure range (Churia) and the intermontane valleys and depressions, known as *dun* in the west (for example, Dehra Dun; also Doon Valley), *marhi* in the centre, and *khonch* in the eastern inner terai. Terai is often spelled 'Tarai,' as it is in Gurung and Khanal (1987).

Undoubtedly, there will prove to be many more local names; an examination of this topic is both beyond our competence and the limits of this study. Such a review would be of great future value and interest. For our purpose, however, we have adopted an amalgam that will not likely meet with universal approval. For much of this discussion we are indebted to Dr. Harka

Table 2.1 *Nepal: Physical divisions – a comparison between several sets of terminology from different sources (after Gurung and Khanal, 1987).*

Geographic regions (native term)	*Hagan physical features (7)*	*Gurung physical features (9)*	*FAO/HMG/UNDP ecological zones (5)*	*LRMP[1] physiographic divisions (5)*
	Tibetan Marginal Mountains	Border Range		
Mountain (Himal)	Inner Himalaya	Trans-Himalayan Valleys (bhot)	High Himalaya	High Himalaya
	Himalaya	Himalaya (himal)		
		Temperate (lekh)	Transition Zone	High Mountain
Hill (Pahar)	Midlands	Sub-tropical (pahar)	Middle Mountain	Middle Mountain
	Mahabharat Lekh	Mahabharat Lekh		
Plain (Tarai or Madhes)	Siwalik Zone	Inner Tarai (dun) Chure Range	Siwalik	Siwalik
	Tarai	Tarai	Tarai	Tarai
Source	Hagan (1960)	Gurung (1971)	Nelson (1980)	HMG (1986) (Kenting Surveys)

[1]LRMP Land Resources Mapping Project (Canada/HMG) (from Gurung and Khanal, 1987)

Gurung whose documents reached us after the main body of this book had been completed (Dr. H. Gurung, personal communication, December 1987). Table 2.1 indicates some of the variety of terminological usages.

Given the enormous range of altitude in such a short north–south horizontal distance, it follows that the 'normal' climate for the latitude, a sub-tropical monsoon type, is strongly modified by the presence of the extremely high east–west trending Himalaya which, by reducing outbursts of cold air from Central Asia, ensures for northern peninsula India warmer winters than would otherwise be the case; in addition, the regional climates are modified with increasing elevation. This is best portrayed by a reconstruction of the natural vegetation belts that range from tropical monsoon rain forest (*Shorea robusta* = sal forests) in the south, through a series of forest belts, to the upper timberline at approximately 4,000–4,500 m. Above this a rhododendron-shrub belt gives out onto alpine meadows, a sub-nival belt of extensive bare ground and scattered dwarf plants, mosses, and lichens, and finally, at 5,000–5,000 m, permanent ice and snow with steep rock outcrops. Joshi (1986b), following Numata (1981) gives the following generalized pattern of vegetation belts for the central part of Nepal, chosen to relate to our schematic transect shown in Figure 2.3:

Nival belt	above 5,500 metres
Alpine belt	4,500–5,500
Rhododendron–Juniperus belt	3,700–4,500
Betula–Abies belt	2,900–3,700
Acer belt	2,500–2,900
Quercus belt	1,900–2,500
Schima belt	1,000–1,900
Shorea robusta belt	0–1,000

Dobremez (1976), in a major study of the plant ecology of Nepal, provides a wealth of detailed information, and Figures 2.4 and 2.5 have been taken from his work. Figure 2.4 shows the extreme complexity of vegetation types in Nepal, using a strictly schematic approach based upon his own fieldwork (Dobremez, 1976: 244). This provides a combined altitudinal (latitudinal) transect and a moisture gradient (decreasing moisture with decreasing longitude). Figure 2.5 provides a comparison of two altitudinal gradients that summarize the work of Schweinfurth (1957) and Troll (1959), and four deriving from the extensive Japanese research in the Central and Eastern Himalaya (Nakao, 1957; Kawakita, 1956; Kanai, 1966; and Numata, 1966).

So far we have restricted ourselves to the simplest possible phytogeographic description using basically south-facing slopes in central Nepal. To take the next, more complicating, step, it must be emphasized that each east–west crest-line separates a wetter, south-facing, and a drier, north-facing slope. Precipitation from the summer monsoon, as the moist air is forced to rise against each successively higher ridge, is generally much heavier to the south

Figure 2.4 Schematic representation of major vegetation communities as a function of altitude in Nepal. The stippled boundaries correspond to the four phytogeographic regions of Nepal: from left to right, Northwest Nepalese, West Nepalese, Central Nepalese, and East Nepalese. From J-F. Dobremez, 1976: 244, Figure 169.

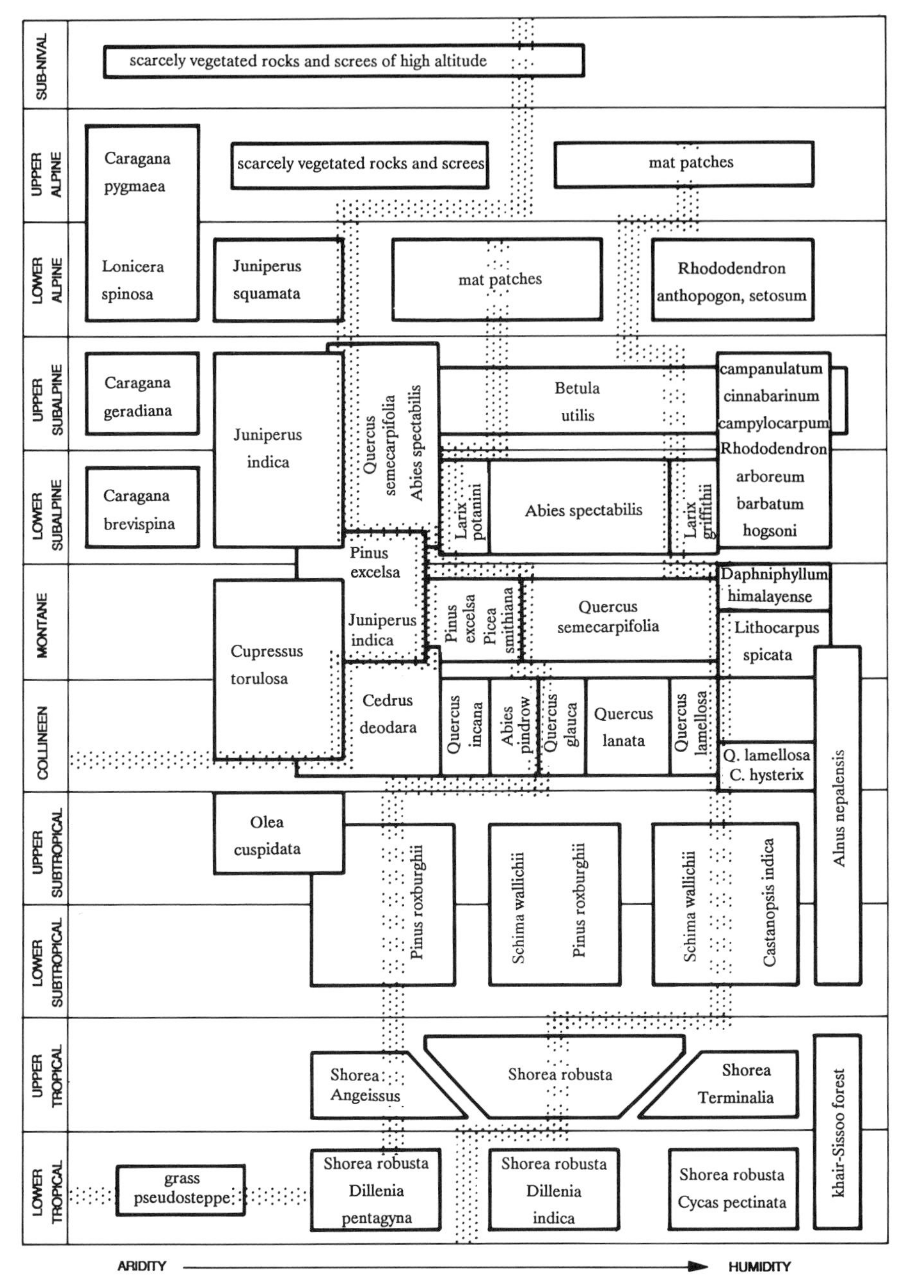

Figure 2.5 Schematic representation of climatic zones and vegetation belts: a) after Schweinfurth (1957) and Troll (1959); b) 1 after Nakao (1957), 2 after Kawakita (1956), 3 after Kanai (1966), and 4 after Numata (1966). Reproduced from J-F. Dobremez, 1976: 248, Figures 172, 174 and 249.

a)

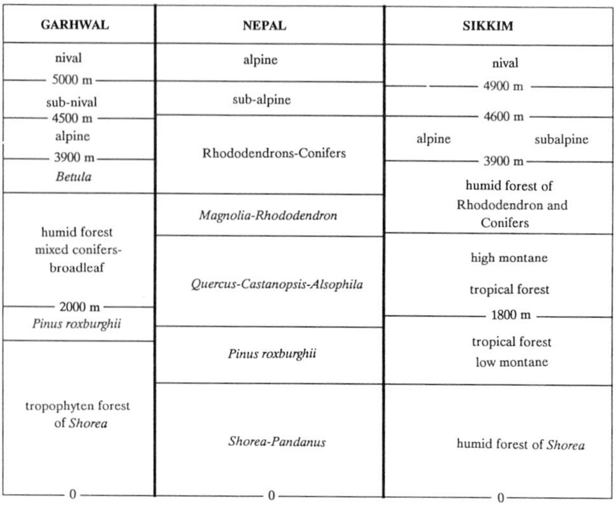

GARHWAL	NEPAL	SIKKIM
nival	alpine	nival
5000 m		4900 m
sub-nival	sub-alpine	
4500 m		4600 m
alpine	Rhododendrons-Conifers	alpine subalpine
3900 m		3900 m
Betula		humid forest of Rhododendron and Conifers
humid forest mixed conifers-broadleaf	*Magnolia-Rhododendron*	high montane tropical forest
	Quercus-Castanopsis-Alsophila	
2000 m		1800 m
Pinus roxburghii	*Pinus roxburghii*	tropical forest low montane
tropophyten forest of *Shorea*	*Shorea-Pandanus*	humid forest of *Shorea*
0	0	0

b)

1	2 CLIMATE	2 VEGETATION	3	4 CLIMATE	4 VEGETATION
nival					
5000 m	nival	cold desert	5000 m	5000 m	
alpine zone	4600 m		alpine	arctic	*Juniperus Rhododendron*
	alpine	alpine	4000 m		
	3900 m		*Rhododen drons*	3900 m	
Abies Betula Tsuga-Picea	subalpine	*Abies*	*Conifers*	sub-arctic	*Abies*
	2900 m			3000 m	
	cold tempered	*Picea-Quercus*	2600 m	cold tempered	*Tsuga*
	2500 m			2500 m	
Oakforest	tempered	*Quercus*	*Quercus Castanopsis*	tempered	*Quercus*
	1900 m			1900 m	
mixed forest *Castanopsis indica*	warm temperate	*Schima-Castanopsis*	mixed broadleaf	warm temperate	*Schima*
	1200 m			1200 m	
1000 m	subtropical tropical	*Shorea*	800 m	subtropical	*Schorea*
Shorea			rainforest		
0	0		0	0	

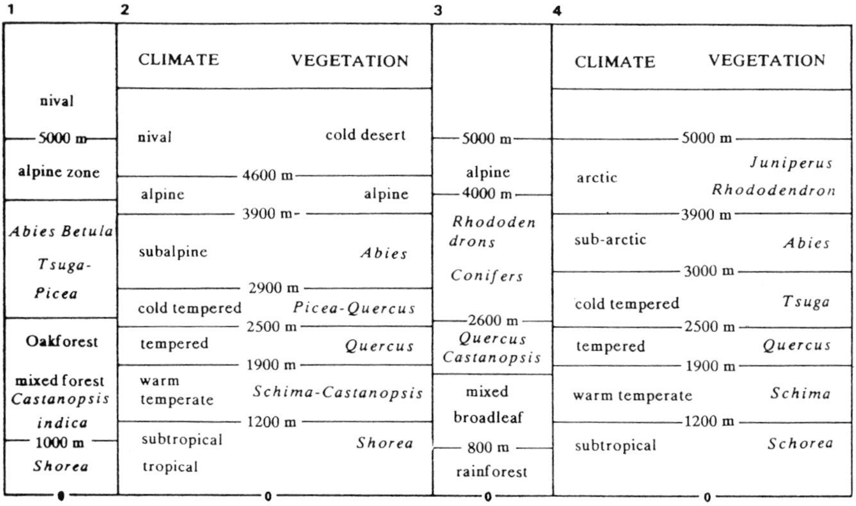

of the major topographic alignments. In contrast, the succession of north-facing slopes experience rain-shadow effects to varying degrees.

Annual rainfall totals appear to increase with increasing altitude in the Himalaya of central Nepal to about 3,000 m; thereafter, with increasing altitude (and increasing northerly latitude) annual totals diminish. Above about 5,500 m all precipitation is in the form of snow. The most complete topographical barrier, the Greater Himalaya, produces a major rain shadow on its northern side. Thus, annual precipitation totals show marked differences over very short horizontal distances. One of the most dramatic examples is provided by a pair of climatic stations (Lumle, altitude 1,642 m, and Jomosom, altitude 2,650 m) to the south and north of the Annapurna massif respectively and only 50 km apart. The former recorded 5,964 mm in 1961 and a five-year average (1971–75) of 5,551 mm; the latter has a mean annual precipitation of 255 mm.

Even in a *relatively* simple transect through the central Nepal Himalaya, however, extremely sparse data, especially long time series, render generalizations rather dangerous. The increasing aridity with increasing altitude and latitude, especially on north-facing slopes, reaches its most pronounced development north of the Greater Himalaya on the Tibetan Plateau and in the trans-Himalayan valleys. Thus Dobremez has added a 'steppe belt' to the simplified altitudinal vegetation–climate belts outlined above. Until recently much of our generalized climatic knowledge of the Himalayan region was based upon inferences drawn from the mapping of vegetation (Schweinfurth, 1957; Ohsawa *et al.*, 1986). Recently, data from newly established climatic stations are beginning to indicate the presence of 'pockets' of extremely high annual precipitation totals. One such pocket has been identified north of Pokhara in Nepal on the lower flanks of the Annapurna massif, which includes the Lumle record over five years (5,000+ mm) introduced above (Dhar and Mandal, 1986). The results of Dhar and Mandal's appraisal of Nepal's precipitation pattern is provided in Figure 2.6. Equally important, however, are the pockets of very low precipitation receipts. These also have long been recognized from vegetation mapping (Schweinfurth, 1957), and also from the travels of F. Kingdon Ward in the eastern Himalaya and in the Hengduan Mountains that resulted in the description of the 'dry valleys' (Wissman, 1959; Schweinfurth, 1985; Zhang Yongzu, personal communication, May 1985). Thus the large number of examples of precipitation data that can be found in the standard climatological references can only be regarded as providing a rough scale of the range of variation. Other important climatic parameters, usually noticeable by unavailability of relevant data, are annual variability, short-term rainfall intensities, and pronounced variations over very short horizontal and vertical distances.

The scatter, and the great range of climatic data, can best be viewed in the context of the general rhythm of the monsoon climate at large. A good general account is given by Das (1983) who provides a breakdown of the seasonal rhythm as:

Pre-monsoon	(April–May)
Monsoon	(June–September)
Post-monsoon	(October–December)
Winter	(January–March)

April–May provides the highest temperatures, with maxima exceeding 40°C at many lowland stations. Of the total annual precipitation, 70–85 percent falls during June–September, depending upon location. The highest totals, according to available statistics, occur between the Annapurna massif and the eastern end of the Himalaya *sensu stricto*, with the highest amounts being recorded north of the Bay of Bengal and in Assam-Arunachal Pradesh (Cherrapungi: altitude 1,326 m, annual precipitation 11,615 mm). With increasing distance along the Himalayan front toward the west-northwest, total annual precipitation decreases and the occasional winter westerly disturbances become more important. At the western end of the Himalaya, *sensu stricto*, and more particularly in the Karakorum and Hindu Kush, a winter maximum regime predominates with most precipitation occurring in the form of snow. The lower valleys and gorges are very dry and local agriculture is dependent upon snow-melt and glacial-melt irrigation. Summer monsoon influences here are slight or absent. This overall trend was noted by Troll (1938, 1939) while mapping altitudinal vegetation transects in the Nanga Parbat area (see also Troll's north–south reconstruction of the altitude of the regional snowline – Figure 2.7). Troll's work ws greatly expanded by Schweinfurth (1957), who produced the first composite vegetation map for the entire Himalaya. This map, a fundamental research resource to this day, reflects the parallelism between the climate and vegetation trends from north to south and from east to west.

This same pattern is also important in terms of the region's glacio-hydrology and the associated variation in water source and availability throughout the year (Young, 1982; Hewitt, 1985). The eastern Himalaya, in general, provide moisture surpluses from direct runoff of the abundant summer monsoon rainfall; the snow-melt contribution is comparatively insignificant. With increasing distance toward the west-northwest meltwater becomes critically important. A particularly heavy summer monsoon, for instance, which produces excess water (and flooding) in the eastern half of the region, may only serve to lower the summer flow of the western rivers since the increased cloud cover (with little or no rain/snow) will serve to reduce incoming solar radiation and thus limit meltwater production. Additionally, summer snow at high altitudes in the northwest, by greatly increasing surface reflectance (albedo), will curtail melting. A general lack of systematic studies in glacio-hydrology is a serious deficiency in terms of the great importance attached to hydroelectric and irrigation schemes by several governments of the region as well as United Nations and bilateral aid agencies (cf. the Canadian–Pakistan collaboration: Hewitt, 1986, 1987).

So far we have referred mainly to the impacts of the summer monsoon as a mass of warm moist air flowing northward from the Bay of Bengal and

Figure 2.6 Mean annual precipitation for Nepal, showing 500mm isopleths. The 'pocket' of very high precipitation near Pokhara (5,000mm) is based upon recent data from newly established climatic stations (after Dhar and Mandal in Joshi, 1986a).

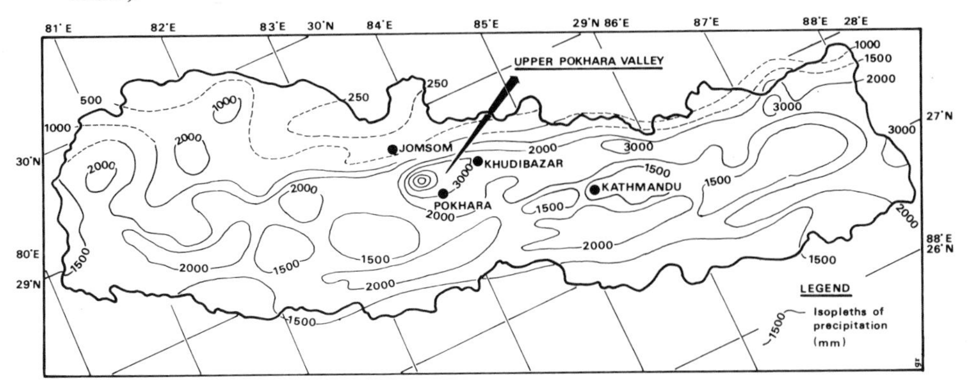

moving predominantly westward along the mountain front. While the mountains and valleys of the northwest are influenced mainly by winter westerlies, the easternmost ranges, the Hengduan Mountains, have their own predominating monsoon regime. In this case, however, the influence of the Bay of Bengal monsoon flow (southwesterly) and the southeasterly monsoon flow from the South China Sea are both important. But again, sparsity of climatic stations, and especially lack of data from higher altitudes, in this equally complex mountain system, inhibits any level of detail. Figure 2.8 provides some information for the Gongga Shan and Yulongxue Shan areas of western Sichuan and northwestern Yunnan respectively (Messerli and Ives,

Figure 2.7 Topographic and climatic profile from south to north across the Himalaya and the Qinghai-Xizang (Tibet) Plateau. The broken line shows the average altitude of the regional snowline (after Troll, 1960).

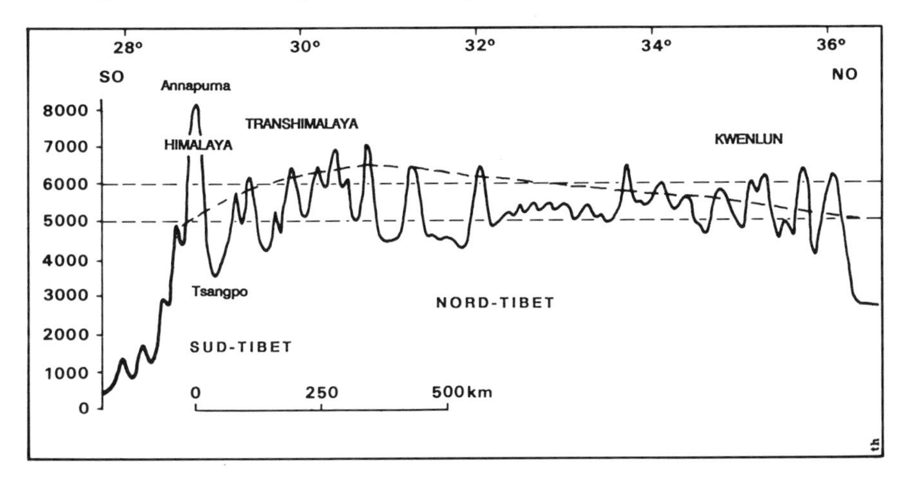

Figure 2.8 The pattern of vertical belts: forest, vegetation, and land use, Gongga Shan and Yulongxue Shan, southwestern China (after Messerli and Ives in Lauer, *Natural Environment and Man in Tropical Mountain Ecosystems*, 1984: 63, Figure 5).

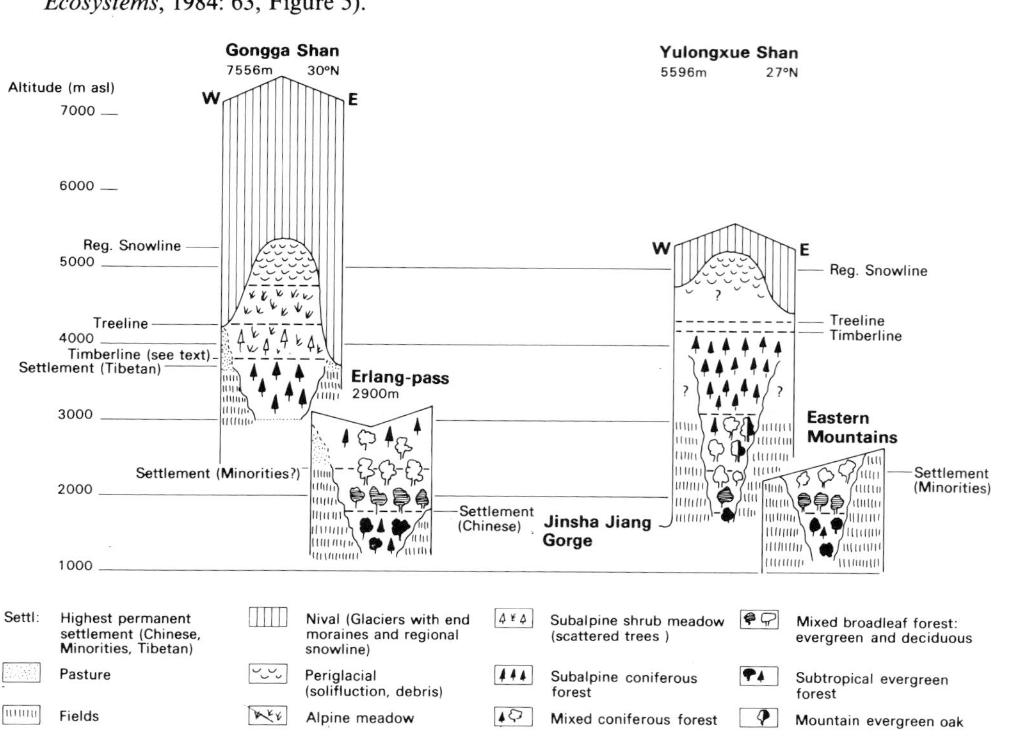

1984). The Yulongxue Shan (Jade Dragon Mountain) is of special climatic interest as it is the site of the most southerly glaciers in Asia, and some of the highest known rice culture is to be found in its vicinity (Uhlig, 1978; Messerli and Ives, 1984). For a more complete discussion of the complexities of mountain weather and climate the reader is referred to Barry (1981).

PEOPLE AND POPULATION

Complicated as the physical geography of our region is, the present-day cultural, economic, and ethnic patterns, not to mention the many rivalries at the state level, defy easy description. This in turn is partly due to the influence of the physical base and partly a result of the very long and complicated history.

Our knowledge of the pattern and extent of prehistoric settlement in the Himalaya, based upon archaeological evidence, remains very sparse. There are only a very few palaeolithic and neolithic sites scattered, for instance, throughout the eastern basin of the Brahmaputra. Consequently, attempts to reconstruct the social and intellectual forces that have shaped contemporary society in northern India and the Himalaya rely heavily upon the great body of Sanskrit epic literature, the earliest fragments of which date from 1200 BC. Much of this literature is derived from an oral tradition that had been continually modified over several thousand years before being committed to writing (O'Flaherty, 1975). This, and other evidence, indicates a continuous political and cultural relationship between the Himalayan foothills and the Ganges plains from at least the fourth century BC to the reign of Ashoka Maurya, the great patron of Buddhism (Rose, 1971).

These introductory remarks are taken from English (1985: 65), who proceeds to outline the cultural and economic history of the Himalaya with emphasis on state formation and the impact of British Rule during the nineteenth century. English describes three patterns of settlement. The western Himalaya was widely settled from 1500 BC onward by a population of nomadic warriors called Khas who were part of a succession of waves of Aryan migration into India from the northwest. The Khas are believed to have subjugated the indigenous inhabitants and to have relegated them to a rigidly inferior social status. The Khas gradually became acculturated to the predominant Hindu influences of the northern plains.

The developing pattern of settlement of the central and eastern Himalaya appears as a migration of Tibeto–Burman tribal peoples from Southeast Asia who moved westward along the mountains north of the Brahmaputra in the early millenia BC. They were reputed to be great hunters, skilled in the arts of magic, and are believed to have practised cannibalism. They were referred to in Sanskrit writings as *Kirata*.

The third pattern is the settlement of the Bhotias and related peoples in the high Himalayan valleys dating from the early centuries AD. Successive waves of these nomads appear to have occurred, transforming in the process to a more settled life combining agriculture, pastoralism, and trade, and coming to occupy what is now northern and central Nepal.

The emergence of a powerful centralized Tibetan monarchy patronizing Buddhist monasticism dates from the seventh century AD (Beckwith, 1987). During the course of two hundred years this monarchy expanded the frontiers of Tibet from western China to Kashmir and northward into Central Asia. Unseated nobles and adherents of the pre-Buddhist Bon religion migrated southward to take refuge in the Himalaya where they established semi-autonomous kingdoms. Those kingdoms south of the Sutlej River were eventually absorbed by the Khas Malla dynasty which, following upon the eclipse of Tibetan power, unified the Garhwal–Kumaun region and much of western Nepal into an integrated Hindu polity based upon trade and agricultural revenues (Pant, 1935). The kingdoms of Mustang and Dolpo, in northern Nepal, and the later Sherpa entity, remained virtually independent of Hindu domination until well into the present century.

The Muslim conquest of northern India during the fourteenth century led to widespread movements of Hindu and Buddhist populations into the northern mountains in the wake of massive religious persecution. The high-caste refugees, who migrated into the western foothills, claimed descent from the Rajputs, who were legendary for their resistance to a succession of Turko–Afghan invasions. Over the next three centuries, these Rajput nobles displaced the ruling Khas lineages in the hills and extended their control of the Himalaya from Kashmir to the eastern Terai of present-day Nepal. The fortunes of many princely states waxed and waned and the political landscape remained extremely fragmented. At the time of the Gorkhali conquest, late in the eighteenth century, there were about eighty separate principalities in this section of the Himalaya, *sensu stricto*, alone (Stiller, 1975).

Sikkim was settled in the thirteenth century by herders from eastern Tibet, and Bhutan became unified in the early decades of the seventeenth century. The expansion of the hill state of Gorkha culminated with the conquest of the Newar city-kingdoms of the Kathmandu Valley in 1769. King Prithivi Narayan Shah extended his empire to embrace a 1,500 km area of the Himalaya from the Sutlej to the Teesta. At its short-lived maximum extent it included Sikkim, and the Darjeeling area of present-day West Bengal, the southern tracts of Tibet, and a large section of the Ganges Plain (Figure 2.9) before coming into violent contact with the expanding territorialism of the British East India Company.

This somewhat fragmentary 'history' of settlement can be summed up with the statement that the northwestern part of the region evolved under Muslim influence, the southern flanks of the centre and east under Hindu influence, and the northern fringe under Buddhist influence. This broadly sweeping overlay conceals innumerable small ethnic groupings and says little about the independent entities such as Hunza, the great complexity of settlement of the Arunachal Pradesh Himalaya, and the several dozen distinct ethnic and linguistic groups of the Hengduan Mountains, which were eventually infiltrated along the main valleys by Han agricultural settlers and traders. Karan (1987b) has sought to generalize the major religious–ethnic patterns of the Himalaya, *sensu stricto* (Figure 2.10), while Figure 2.11 provides greater

Figure 2.9 Nepal at the maximum extent of the Gorkhali Empire in 1814 (after English, 1985).

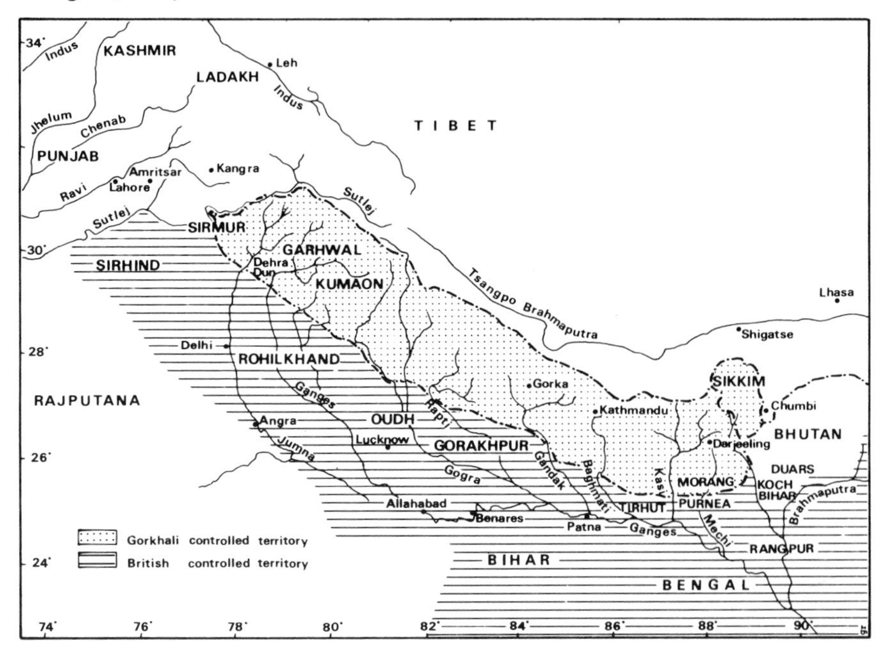

detail for Nepal (after Dobremez, 1976 (Figure 117): 108).

Karan (1987b) claims a total 1981 population of about 33 million for the Himalaya, *sensu stricto*. This includes the entire population of Nepal (15 million) so that, strictly speaking, this takes in a significant number of lowlanders, since about 7 million inhabit the Nepalese Terai (including the inner terai: Goldstein *et al.*, 1983). Our overall estimate of about 50 million

Figure 2.10 'Cultural regions' of the Himalaya, *sensu stricto* (after Karan 1987b).

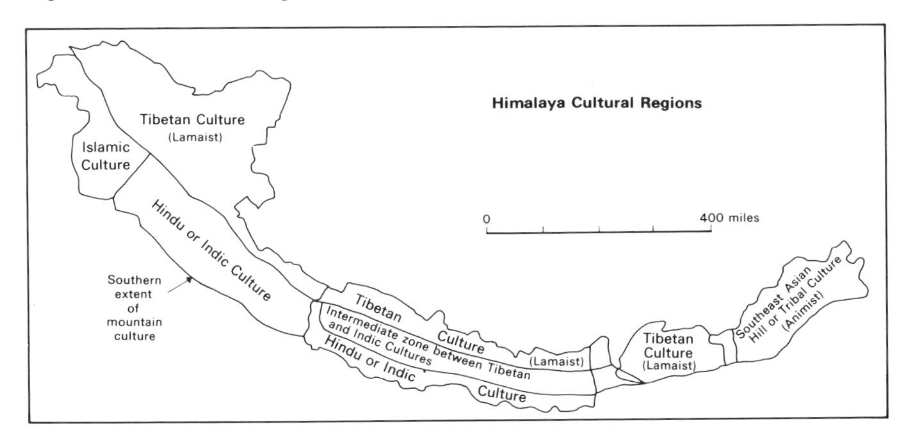

Figure 2.11 Ethnic groups of Nepal. Wide diagonal lines show the extent of ethnic groups under the Nepalese caste system; close diagonal lines indicate the extent of groups of Tibetan origin; between these and within the heavy black lines are shown the Tibeto-Burman groups (after Dobremez, 1976: 108, Figure 117).

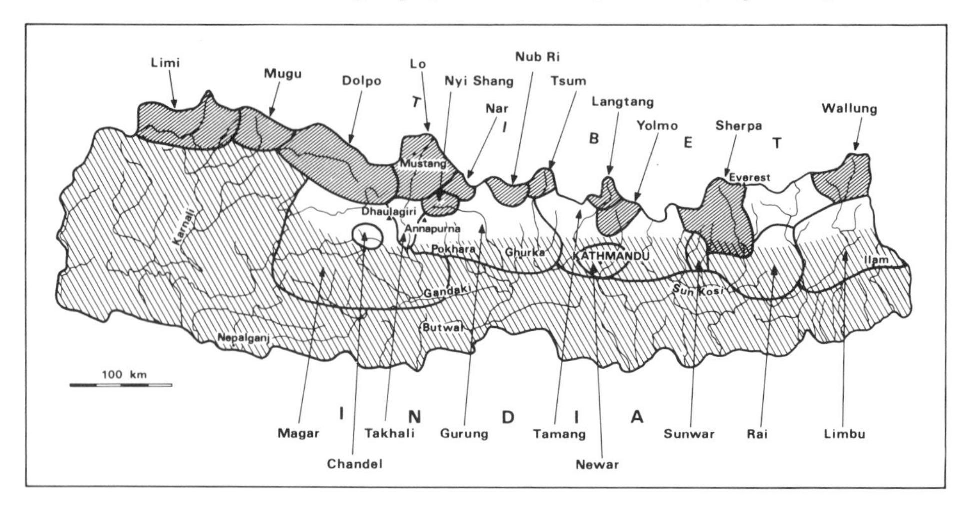

for the mountain (plus Terai) region, *sensu lato*, is probably of the correct order of magnitude, although presumably this has increased at a rate of about 2 percent per annum, or more, since 1981.

Karan's (1987b) study indicates that the population of the Himalaya, *sensu stricto*, has trebled between 1901 and 1981 (from 11 million to over 33 million: Table 2.2). He provides an annual growth rate of 1.26 percent for the region as a whole, for the first five decades of this century, and a rate of 2.7 percent between 1951 and 1981. Broken down into different regions, this gives especially high annual rates of growth (1901–51) for Sikkim (5.8 percent), Darjeeling (3 percent), and Kumaun (2.6 percent). Since 1951 high rates have characterized Sikkim and the Eastern Himalaya (4.3 percent), Kashmir and Kumaun (2.7 percent), and the Punjab and Nepal Himalaya (2.5 percent). In recent years it would appear that the annual growth rate has exceeded 2 percent in all areas of the Himalaya, except Bhutan. It seems reasonable to extend this estimate to the wider region, with the exception of Afghanistan, where 'normal' population trends will have been obliterated by the massive flow of refugees into northern Pakistan and Iran.

When these growth estimates and absolute 1981 population numbers are converted into densities, allowance must be made for the large proportion of land that cannot be used for agriculture or pastoralism (high mountains and land under permanent snow and ice) and the probably equally large amount of marginal land (on account of a combination of altitude and steep slopes). Thus, as has been repeated in many publications, it is necessary to convert population densities into numbers of people per hectare of arable land, or even irrigated *khet* land. While it is difficult to do this with any degree of

Table 2.2 *Himalaya*, sensu stricto: *area and population of different sub-regions (1901, 1951, 1981) and percentage growth (1901–51, 1951–81) (after Karan, 1987:272).*

	Area (km²)	*Population 1901*	*Population 1951*	*Population 1981*	*Percentage growth 1901–51*	*Percentage growth 1951–81*
Kashmir Himalaya	222,797	2,139,362	3,253,852	5,981,600	52.09	83.83
Punjab Himalaya	55,500	1,920,294	2,385,981	4,237,569	24.25	77.60
Kumaun U.P. Himalaya	51,100	1,207,030	3,106,356[a]	4,815,326	157.35[b]	55.01[c]
Nepal Himalaya	142,124	5,638,749[d]	8,473,478[e]	15,020,451	50.27[f]	77.26[g]
Sikkim Himalaya	7,100	30,458[h]	137,725	315,682	352,18[i]	129.21
Darjeeling Himalaya	3,200	249,117	624,879	1,006,434	150.83	61.06
Bhutan Himalaya	46,500	n.d.	n.d.	1,162,000	—	—
Eastern Himalaya Arunachal	83,700	n.d.	336,558[j]	628,050	—	86.60[k]
Himalaya (total)	612,021	11,185,010	18,318,829	33,167,112	63.78	81.05
Density (km²)		18.27	29.93	54.19		

Sources: Census of India, 1901, 1951, and 1981; Central Bureau of Statistics, Kathmandu, Nepal; Planning Commission, Royal Government of Bhutan, Thimphu.

Notes: Kumaun–a. 1961 data; b. 1901–61 growth; c. 1961–81 growth. Nepal–d. 1911 data; e. 1952/54 data; f. 1911–52/54 growth; g. 1952/54–81 growth. Sikkim–h. 1891 data; i. 1891–1951 growth. Arunachal Pradesh–j. 1961 data; k. 1961–81 growth. n.d. = no data.

accuracy, it is reasonable to conclude that such 'real' densities would greatly exceed those of the rich farmlands of the plains. This comparison is made the more striking if we bear in mind that triple cropping has become a widespread practice in the plains, a form of intensification that is not practised at higher altitudes (most of the Middle Mountains) because of lower temperatures.

As Karan (1987b: 272) points out, despite its rugged terrain, the Himalaya is also a region of extensive population movement. Migration is constantly occurring, from one rural area to another, from urban to urban areas, and from rural to urban areas. Goldstein *et al.*, (1983), for instance, have characterized Nepal as showing a distinct trend in transformation from a highland rural society to a lowland urban one; although the total urban population remains very small it is beginning to grow with increasing rapidity. Nor do these remarks take into account the massive amount of seasonal migration and the day-to-day movement of young males from rural areas seeking wage labour in the nearest towns, and on the plains. This is as characteristic of the Indian Himalaya as it is of Nepal. Karan (1987b) indicates that the two major areas of permanent out-migration are Nepal and Kumaun. There are indications of extensive reverse-flow migration into the mountains – from Assam into Arunachal Pradesh (Goswami, personal communication, October 1987).

BREAKDOWN INTO REGIONS

Different attempts have been made over the past three decades to provide a regional breakdown of the Himalaya and contiguous areas. Schweinfurth (1957) has demonstrated in his seminal treatise the great variability of the natural vegetation cover. His approach is based upon a massive literature review without any first-hand experience in the field. Troll (1967) subsequently synthesized his own extensive experience gained during his scientific expeditions to different parts of the Himalaya and produced a climo-ecological regionalization. Uhlig (1978), as a third example of the long tradition of German research, compares the natural altitudinal differentiations with socio-cultural altitudinal systems. The approach introduced here is a new attempt undertaken by the Geographical Institute of the University of Berne. Figure 2.12 provides a summary of the steps used in this procedure.

By extrapolating the annual precipitation amounts for more than 200 climatic stations, the mean annual precipitation pattern for the region (Figure 2.13) was produced with the aid of a computer programme. Next, using a method developed by FAO (1982) to calculate humidity available (moisture) for vegetation, or agricultural production, the monthly precipitation amounts were compared with the potential evapotranspiration amounts (Step 1 in Figure 2.12). To delineate the actual thermic situation three different isotherm maps were superimposed (Voeikow, 1981). These were: mean annual temperature; mean temperature of the coldest month; mean temperature of the warmest month (Step 2 in Figure 2.12).

The temperature classes for the map (Figure 2.14) were made to correspond

Figure 2.12 Schematic procedure for the regionalization process, showing Steps 1 to 4 as discussed in the text (prepared by Markus Wyss, Geographical Institute, University of Berne, 1988).

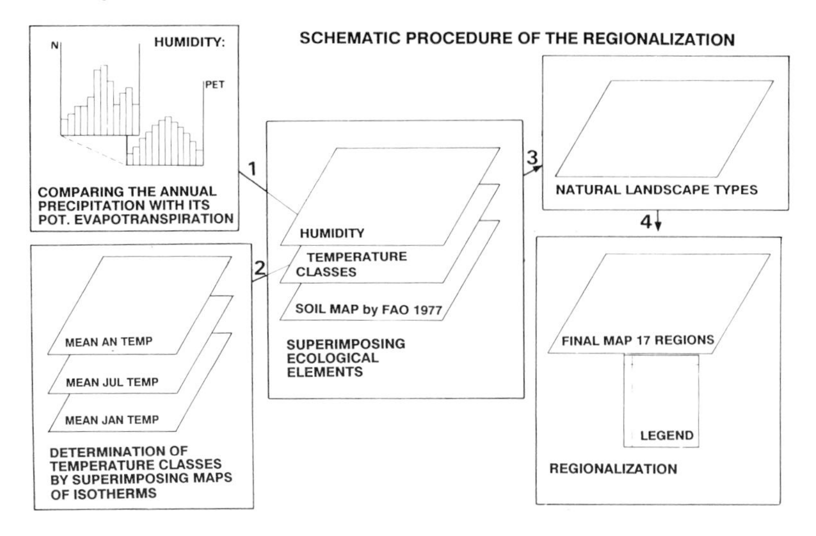

with the altitudinal ranges of different types of vegetation. Next, the definition of soil units (Figure 2.15) follows in a generalized way the FAO soil map of the world (FAO, 1977). In taking the world soil map as an overview of the edaphic situation, however, two limitations need to be considered: (1) the

Figure 2.13 Mean annual precipitation (prepared by Markus Wyss, Geographical Institute, University of Berne, 1988).

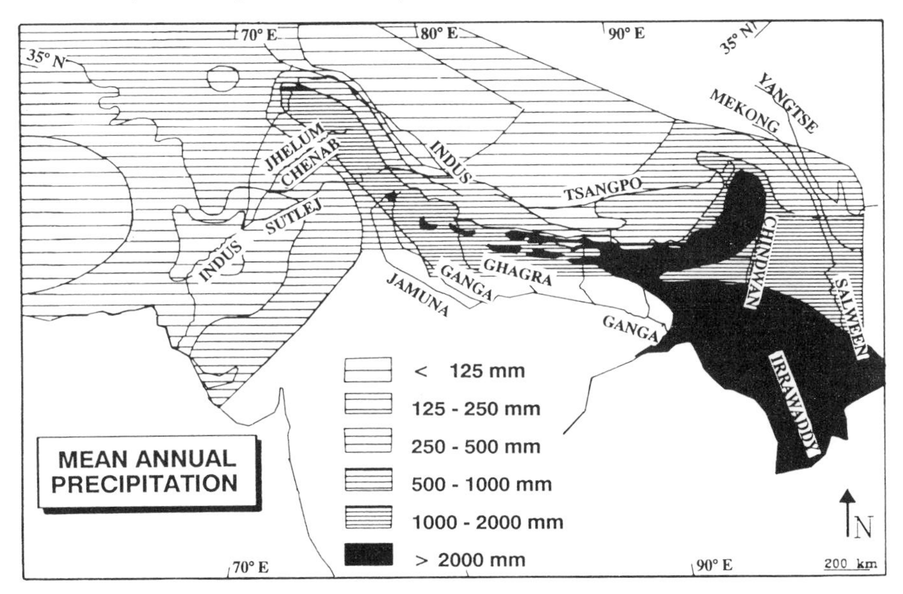

Figure 2.14 Temperature classes (prepared by Markus Wyss, Geographical Institute, University of Berne, 1988).

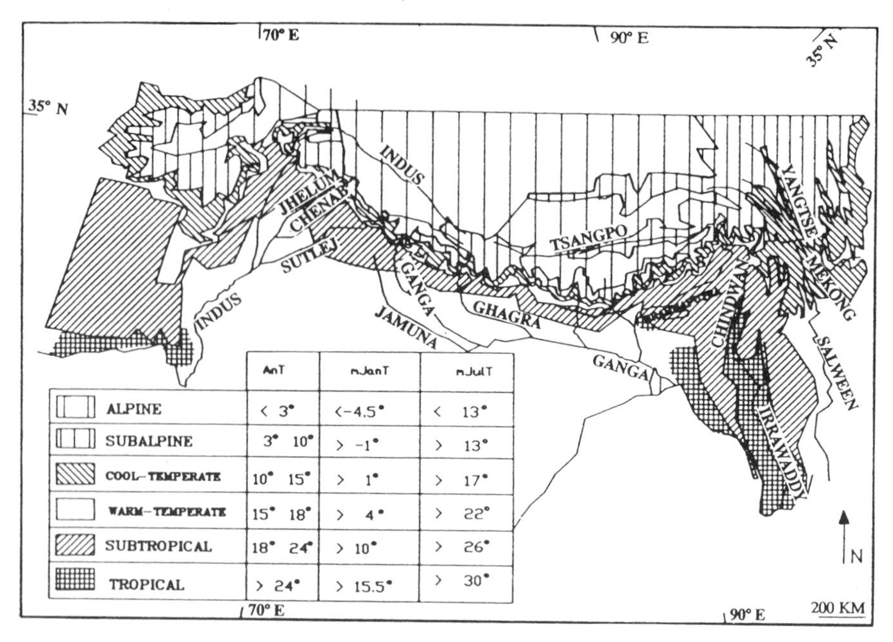

FAO information on soils is based upon very limited field data; and (2) the map projection for the soils of the Himalayan Region is not compatible with those of other maps that were used. To overcome, at least in part, this latter limitation the soil map was redrawn using a more compatible scale and projection. This was accomplished by computer-aided design (Auto CAD).

The three maps showing precipitation (Figure 2.13), temperature classes (Figure 2.14), and soil units (Figure 2.15) provide a general overview of the physical base of the Himalayan Region and serve at the same time as intermediate steps for the regional breakdown. However, the several limitations introduced by this approach must be indicated. First, accuracy and detail of information that can be shown are obviously restricted by the small scale of the maps and by the fact that data from some parts of the area are missing. This is especially critical in high mountain areas where site differences can be much more important than sub-regional differences. In particular, the maps cannot take into account the great variability that occurs within short horizontal and vertical distances in mountainous regions.

The next step (3 in Figure 2.12) involved the delineation of natural landscape types by superimposing the three base maps (Figures 2.13–15). For ease of interpretation, this procedural step was taken at the enlarged scale of 1:2,000,000 and, in turn, was used for the final step (4 in Figure 2.12): delineation of the seventeen regions. Figure 2.16 depicts the seventeen regions, and the corresponding legend provides a summary of the ecological

Figure 2.15 Soil units (prepared by Markus Wyss, Geographical Institute, University of Berne, 1988, generalized from FAO, 1977).

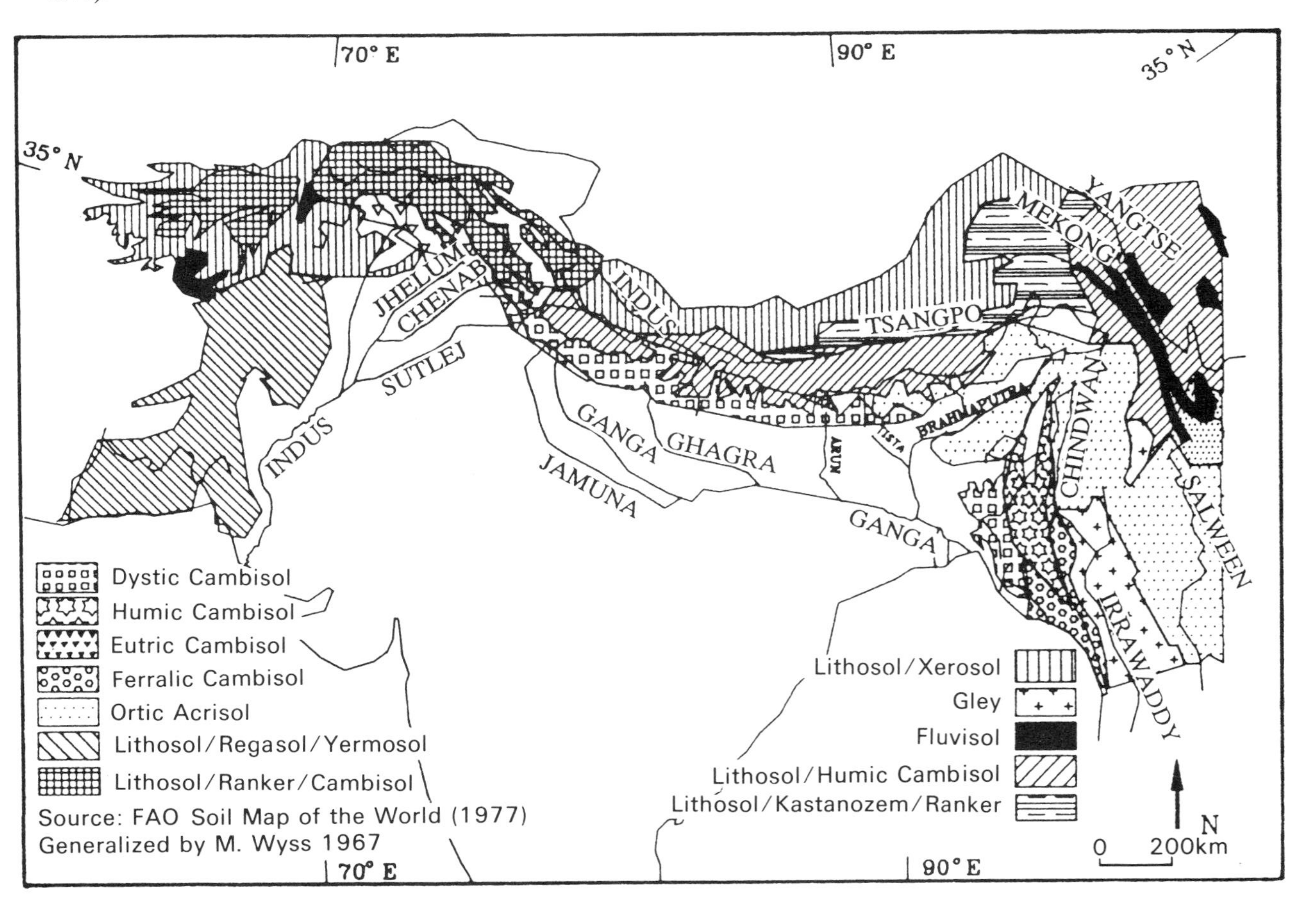

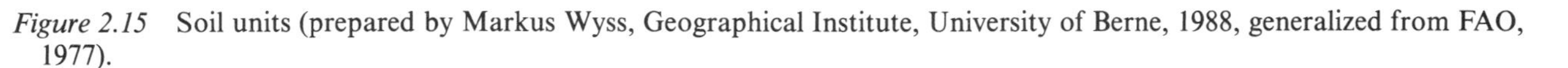

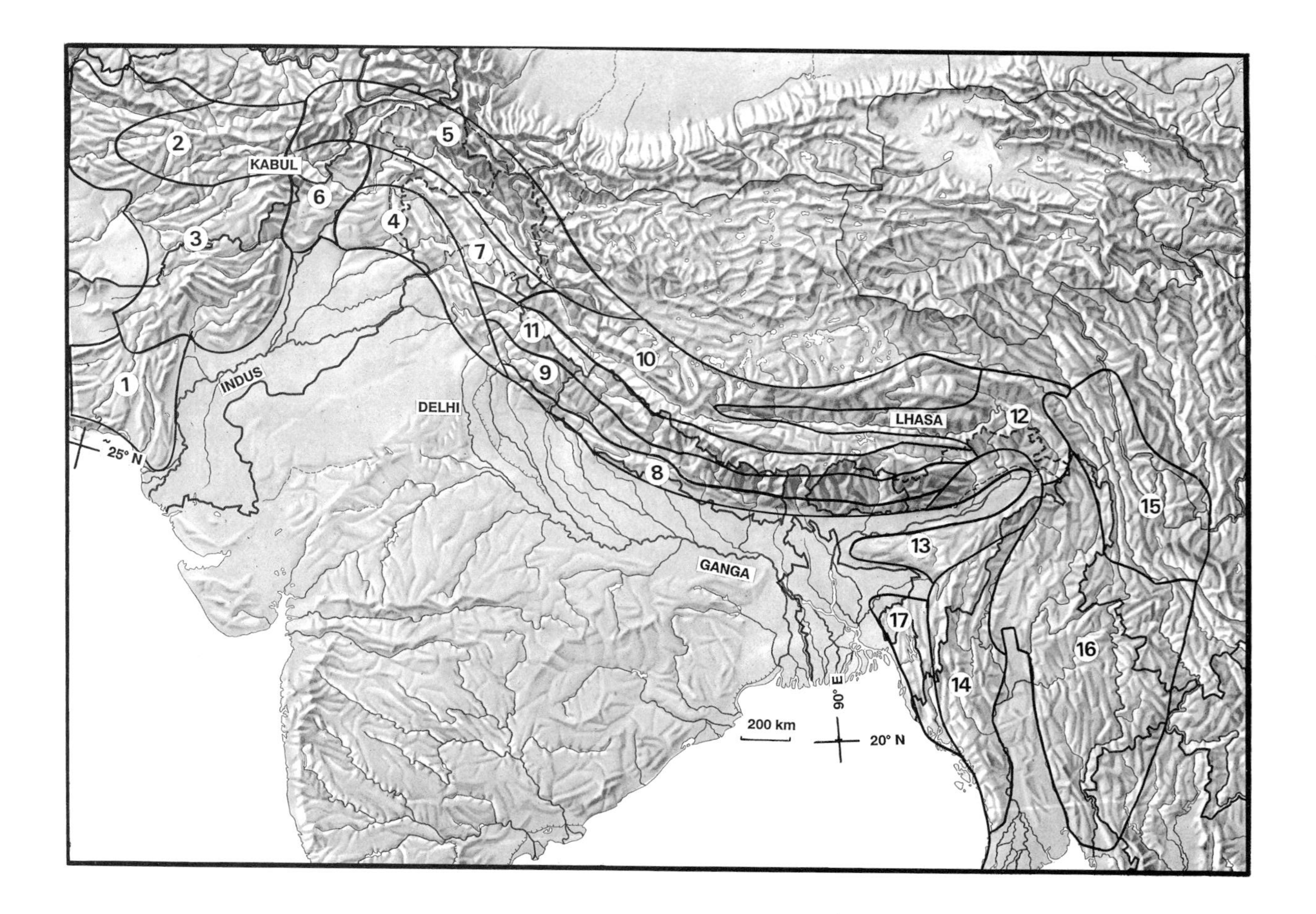
KABUL
INDUS
DELHI
LHASA
GANGA
~25° N
90° E
20° N
200 km
1
2
3
4
5
6
7
8
9
10
11
12
13
14
15
16
17

Figure 2.16 A proposed regional breakdown for the Himalayan region, *sensu lato* (prepared by Markus Wyss, Geographical Institute, University of Berne, 1987).

Region	Humidity			Temperature				Soil
	P mm	Distribution	L.G.P.	Temp-Types	An.T	Jan T	Jul T	Units
1	<250		0	Subtropical	18-24	>7	>26	Yermosols
2	250-500	Mediter	<90	Subalpine	3-10	>-13	>13	Lithosols + associations
3	+/− 250	Mediter	<90	Cool temperate	10-15	>-3	>17	Xerosol Lithosols
4	500-1000	Bimodal	>90	Warm temperate	15-18	>0·4	>22	Cambisols
5	<125	Mediter		Alpine	<3	<-13	<13	Lithosols + associations
6	500-1000	Bimodal		Warm temperate	15-18	>0·4	>22	Lithosols + associations
7	250-500	Bimodal		Subalpine	3-10	>-13	>13	Lithosols
8	>1000	Monsoon	>210	Subtropical	18-24	>7	>26	Dystic Cambisols
9	>1000	Monsoon	>210	Cool temperate	10-15	>-3	>17	Lithosols + associations
10	>250		<90	Alpine	<3	<-13	<13	Xerosol associations
11	?			Nival				
12	250-500	Monsoon		Subalpine	3-10	>-13	>13	Kastanozem/ Ranker - Lithosol
13	>2000	Monsoon	>270	Subtropical	18-24	>7	>26	Acrisols
14	>1000	Monsoon	240	Subtropical	18-24	>7	>26	Acrisols
15	250-1000	Monsoon		Cool temperate	10-15	>-13	>17	Fluvisols
16	>1500	Monsoon	>270	Subtropical	18-14	>7	>26	Ferralic Cambisols
17	>2000	Monsoon	>300	Tropical	>24	>9	>30	Dystic Cambisols

P mm — Annual amount of precipitation in millimetres
Distribution.... of maxima of precipitation: Monsoon: Max in summer; winter max: Mediterranean; Bimodal distribution: twice a maximum
L.G.P. — Length of growing period. (FAO, 1982: 9) in days
An. T — Mean annual temperature in degrees celsius
Jan T — Mean temperature of the coldest month in degrees celcius
Jul T — Mean temperature of the warmest month in degrees celcius
Units — Soil units defined by FAO in: Soil Map of the World (1977)

situation of each. In considering humidity (available moisture) we differentiate between the annual precipitation, the season of maximum rainfall, and the length of the period suitable for plant growth. The effects of vertical gradients and different aspects are particularly important in this exercise and are probably under-estimated. Nevertheless, the final product (Figure 2.16) provides a coherent and basically rational overview of the major subregions of the Himalayan region, *sensu lato*. It should serve as a framework for the discussion that follows and, regardless of its limitations, enables us to focus our attention on some of the aspects of the physical diversity of the Himalaya. We have not attempted to superimpose the cultural and historical diversity since that would carry the burden of complexity beyond our cartographic competence.

NOTE

[1]As an example we cite *The Times Atlas of the World* (The Times, 1985 (7th edition): Plates 23, 24, 27, 28, 29, 30, 31).

3 WHEN DID DEFORESTATION OCCUR? A HISTORICAL PERSPECTIVE ON HIMALAYAN FOREST-COVER CHANGES

INTRODUCTION

This chapter and the following three chapters are devoted to an examination of a *selection* of the salient linkages within the Theory of Himalayan Environmental Degradation. The selection, in part, is a reflection of the unevenness of the available information and, in part, is related to the fact that some of the primary linkages – for example, the relationship between population growth and poverty: which is cause, which effect? – are topics that are too complex to be handled exhaustively and competently within the format of this book. Nevertheless, by introducing them, however briefly, we hope to reinforce our concern about the scale of the uncertainty that pervades the entire Theory of Himalayan Environmental Degradation.

We will include here, therefore, a synthesis of material produced as background papers and case studies for the Mohonk Mountain Conference, held on 6–11 April 1986 (Ives, J. D., and Ives, P. 1987), and also material derived from a general, but by no means exhaustive, survey of the literature. We will also draw on personal experiences in the field, and group discussions at all levels.

WIDELY HELD PERCEPTIONS ABOUT THE MOUNTAIN FORESTS OF THE HIMALAYAN REGION

Climax forest cover, or even extensive secondary growth, or plantation forest, is perceived the world over as one of society's most important natural resources. This perception is based not only upon an appreciation of the directly available forest products, such as construction timber and fuelwood, but also upon the aesthetic and recreational importance of forests and the protective role that forests are presumed to play in terms of reducing soil erosion, modifying runoff of precipitation, and providing security against avalanche and other catastrophic processes. As the world's natural, or wilderness, areas continue to shrink in the face of excessive exploitation by large-scale industry for short-term monetary gain, or through extension of cultivated land, accelerated fuelwood collection, and overgrazing by the animals of the rapidly growing subsistence populations, widespread concern is being expressed over the likely consequences of progressive worldwide deforestation. Depletion of the gene pool and loss of wildlife habitat, both implying extinction of species, many as yet unknown to science, are additional major concerns. The recent anxiety about the relationship between forest cover and climate, while still imperfectly understood, is perhaps one of the most critical facets of the deforestation debate.

All of these concerns are shared by us. It is perfectly understandable that deforestation has come to be regarded as inherently 'bad' – an emotional response. We are highly sensitive, therefore, to placing ourselves in a position in which we appear to be swimming against the current of conventional thought.

We began our research efforts in the Himalaya in that delectable state of drifting with the current of scholarly and populist conviction. In this, and subsequent chapters, we will appear to have reversed our stand. Certainly, many of the popular beliefs about the role of forests, and deforestation *per se*, seem to be based upon emotion rather than fact in so far as the specific region with which we are dealing is concerned – the Himalayan region, *sensu lato*. Thus we must insist that our argument is not necessarily intended to apply to the question of deforestation and the function of forests at large. We do hope, however, that our region-specific questioning of cherished beliefs will provide some food for thought in terms of the broader issues.

The role of the forests, the history of their progressive removal and conversion to arable and grazing land, and the pressures inducing that removal permeate most of the eight-point scenario of the Theory of Himalayan Environmental Degradation (see pp. 3–4). We will examine the claim that massive recent deforestation has occurred in Nepal, and throughout the wider mountain region, and that it was triggered by a population explosion, itself induced by the introduction of modern health care and medicine after 1950. The emphasis will be first on forest history, or the history of deforestation. Perhaps central to this emphasis is the oft-repeated statement that Nepal has lost half its forest cover within a thirty-year period (1950–80) and that by AD 2000 no accessible forests will remain (World Bank, 1979).

To appreciate the importance of the role of the forests in the life of the Himalayan mountain societies it is necessary to understand the general pattern of subsistence agriculture. While hill agriculture takes on many forms according to altitude, latitude, longitude, and social group, it is basically a form of mixed farming, including arable cultivation (of cereals and tubers), animal husbandry, and intensive use of forests. The importance of this three-fold component system and the interdependence of each of the three parts cannot be over-stressed. The basic energy intake of the subsistence villagers is satisfied by food crops, usually grown on irrigated (*khet*) or rainfed (*bari*) terraces, supplemented by some animal products. At lower altitudes, where irrigation is feasible, the *khet* produce a winter, or dry-season, crop in addition to the preferred summer monsoon crop of rice. At increasingly higher altitudes, the proportion of *bari* to *khet* increases, due to cooler dry-season conditions, cooler summers, and increasing slope angle and difficulty of access to water. Livestock provide draught power and serve as the primary, and often the only, source of fertilizer. Animal dung is usually mixed with forest green, animal bedding materials, and crop residues. The population explosion, over the longer term, must also be viewed as a livestock increase as well as an increase in the number of human beings. Thus, over the past several centuries, as forests have been converted to agricultural terraces to feed a growing population, so the livestock numbers have risen to maintain a viable farming system. It is only recently that increase in livestock numbers, at least in some areas, has been reversed, possibly due to decrease in fodder availability.

This close relationship between human population growth, expanding area under subsistence crops, and increase in livestock numbers is equally closely tied to intensifying demands on the forests, therefore, to supply animal fodder and fuelwood and land for farming. Additional demands on the forests' resources include the need for materials for making wooden implements, house thatch, and so on. The fodder is derived from unrestricted grazing by livestock in the forests and the actual lopping of trees for green fodder. Forest leaf litter provides bedding material for animals, which is eventually mixed with dung and crop residue to provide nutrient replacement for the agricultural terraces. While there are serious problems in determining total available forest cover, biomass productivity, biomass demands, and actual consumption (discussed in more detail in Chapter 4), there are also conflicting estimates of the amount of forested land needed to 'support' one hectare of arable land. While actual figures are expected to vary with altitude and with each specific annual agricultural cycle, it is reasonable to conclude that sustainable use of one hectare of arable land will require between one and four hectares of forest. The primary demands on the forest are first for fodder, second for fuelwood, third for other subsistence requirements (the demands for timber, charcoal, and commercial uses are important but can be set aside from the present discussion, which is concerned primarily with the inter-relationships between subsistence farming, forest cover, and slope stability). Thus the vital role of the forests can be appreciated. Any complete conversion of forest land to agricultural land would require the introduction of an entirely different farming system; if access to forest land were to disappear by AD 2000, as predicted, the existing forms of subsistence farming would collapse totally. In this sense, estimates of forest cover, in terms of quantity, quality, and rates of change, are therefore essential to any prediction of environmental or socio-economic (and presumably political) stability or, as is usually the case, degradation and collapse, bearing in mind that these conditions are also tied to population growth estimates.

From the foregoing brief discussion it can be seen that claims that 50 percent of Nepal's forest cover has been lost in as little as thirty years and a prediction of total deforestation by the end of this century are dramatic indeed: they are central to the prediction that 'eco-disaster' for the Himalaya is imminent. Furthermore, the assumed relationship between forest cover, rain-drop impact and soil erosion, gullying and landslides, downstream impacts, and climatic change imply equally serious consequences.

Nevertheless, we believe that the data supporting the claims for catastrophic deforestation are inaccurate and totally unreliable on several counts. The studies of Bajracharya (1983a and b), Thompson and Warburton (1985a and b), Mahat *et al.* (1986a and b, 1987a and b); Griffin *et al.* (1988), and others, support this statement, as do our own observations in several widely scattered field areas.

The various surveys of forest cover in Nepal, using air photography and satellite imagery, are inconsistent. The Nepal Water and Energy Commission study (HMG Nepal, 1983) presents the most comprehensive picture available to date on the extent of forested area and the loss of forest cover in the different geographic regions of Nepal since 1964. Comparison of air photographs dating from 1964 and 1977 suggests a loss of 47,200 ha (1.5 percent of the original cover) of forest in the Middle Mountains. This represents an annual loss of 0.11 percent over thirteen years, a figure that is not statistically significant for any of the major watersheds of the Middle Mountain region, first, because it is a minute rate of loss in itself, and second, because it is smaller than the limits of accuracy of the methods used. For the Terai and Siwaliks losses of 250,000 ha and 148,500 ha respectively (27.5 percent and 10.2 percent of the forest area) are indicated. This gives annual loss rates over the thirteen-year period of 2.1 percent in the Terai and 0.8 percent in the Siwaliks, and these figures are statistically significant. Mahat *et al.* (1987a) contend that it has been this loss of forest in the Terai and Siwalik regions of Nepal that has given rise to the view that the entire forest cover is disappearing. Moreover, deforestation in the Terai, while of great importance in terms of supply of, and demands for, forest products, is of much less significance in the context of assumed acceleration of soil erosion and landsliding on steep slopes.

Other attempts to determine loss of forest cover have been confounded by the fact that different surveys did not use the same definitions for 'forest' and 'grazing land.' Also, in many instances, areas designated as forest were in fact mere shrubberies or, at a further stage of degeneration, rough grazing land. In mapping exercises, it appears that the distinction between shrubberies and much middle-altitude grazing land is still arbitrary and many areas mapped as shrubberies are actually almost devoid of shrubs and used entirely for grazing. There have also been problems with incomplete or not strictly comparable image coverage, variable quality and scale, which add to the difficulties of interpretation. In general, it must be concluded that in recent decades there has been little loss of forest area in the Middle Mountains, as determined from conventional air photography and analysis of satellite imagery. However, while the rate of loss of forest area may have been slight and in some areas forest cover may have expanded, the density, and hence the quality, of the forests is seen to have been seriously reduced (HMG Nepal, 1983). Thus, crown cover has been reduced, suggesting a deterioration within the forest itself. Many of these points, while relating specifically to Nepal, are also relevant to the Indian Himalaya and elsewhere.

The conclusion so far, therefore, is that the linkage: population growth→deforestation→increased soil erosion and landslide activity, specifically a Middle Mountain process driving the vicious circle concept, is open to challenge on the first link alone. That massive deforestation has occurred since the 1950s in the Terai and Siwaliks is not in question. The issue that is central to the present discussion is the extent of *recent* deforestation in the Middle Mountains. If the fundamental question is now asked – if forest cover in the Middle Mountains has not changed significantly since the 1950s, what was the previous history of forest conversion to arable land? – we can come closer to a more complete renunciation of this linkage.

History of Deforestation of the Nepal Middle Mountains

Documentation for detailing the history of deforestation of the Middle Mountains is very fragmentary. However, important insights can be obtained from the work of Tej B. S. Mahat, David M. Griffin, Kenneth R. Shepherd, and their co-workers. Mahat not only served for seven years as District Forest Officer for the districts of Sindhu Palchok and Kabhre Palanchok, which lie immediately east of the Kathmandu Valley (Figure 3.1), but he worked as a staff member of the Nepal–Australia Forestry Project and completed his doctoral dissertation (Mahat, 1985) on problems of forestry in these two districts. Much of the detailed information from his own and related studies is published in a series of papers in *Mountain Research and Development* (Mahat *et al.*, 1986a and b, 1987a and b; Griffin *et al.*, 1988). The two districts provide a representative cross-section of the *Pahar* region of Nepal (defined

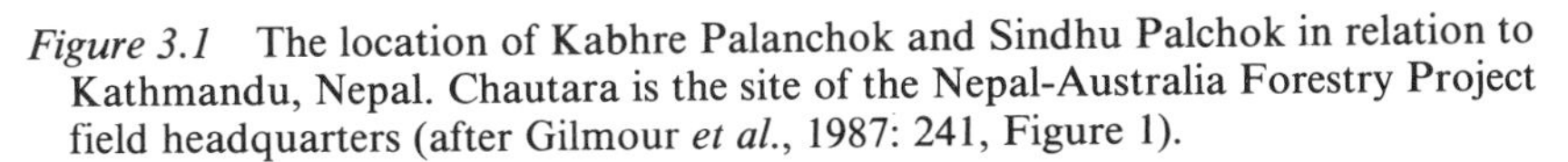

Figure 3.1 The location of Kabhre Palanchok and Sindhu Palchok in relation to Kathmandu, Nepal. Chautara is the site of the Nepal-Australia Forestry Project field headquarters (after Gilmour *et al.*, 1987: 241, Figure 1).

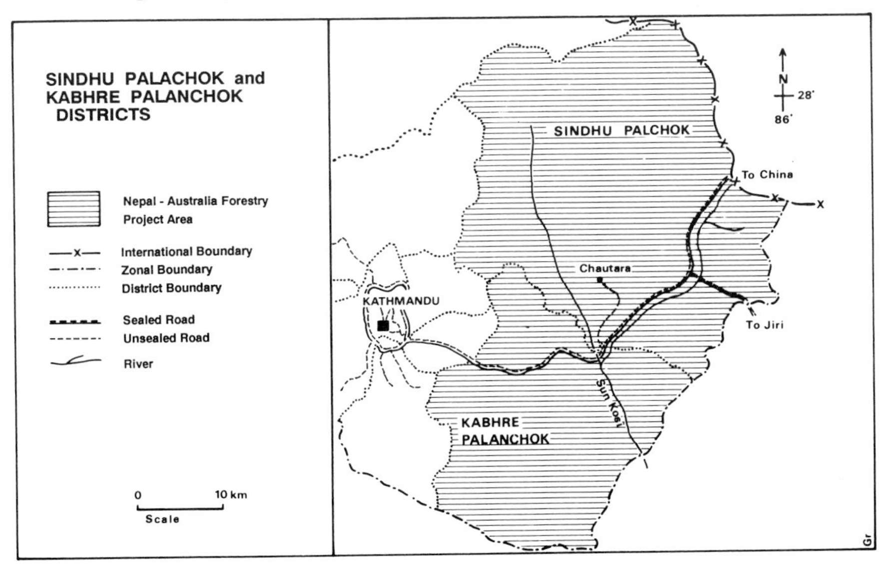

as including the Middle Mountains, the Mahabharat Lekh to the south, and the high montane forests up to timberline at about 4,000 m on the flanks of the Greater Himalaya to the north). The single exception to the claim for representativeness of this particular cross-section of the *pahar* is its proximity to Kathmandu, which will be shown below to have had an important influence on forest use.

Mahat, Griffin, and Shepherd describe the process of deforestation as one that extended over several hundred years. Acceleration in the rate of deforestation occurred following King Prithivi Narayan Shah's (1743–75) militant unification of the numerous petty principalities and the founding of the State of Nepal under the House of Gorkha (1769). It must be understood that land, specifically arable land, provided the prime source of income for the State. Various forms of land tenure and taxation policies ensured the progressive conversion of forest to arable land. Arable land was taxed in such a way that half the crop produced was forfeit either to the State, or to the landholder who had been allotted the land in recompense for services to the State. To ensure an acceleration in the rate of conversion from forest to arable land the King declared that newly cleared forest land could be farmed tax free for three years. The rent of half of the produce of a peasant's land, even during periods of good harvest, was very high and reclamation of forest lands was a vital relief. It is also necessary to explain that while *deforestation* was rampant in the Middle Mountains, the opposite was occurring in at least parts of the Terai. The policy of the House of Gorkha after 1769 was to protect the southern lowland forests of Nepal, and to encourage natural regrowth in formerly cleared areas so as to develop a dense, impenetrable and malaria-infested jungle as a defence against British expansion (Michael Thompson, personal communication, April 1987). A detailed account of the forms of taxation, land tenure, and compensation to military and civil servants is presented in Ludwig Stiller's histories of the *Rise of the House of Ghorkha* (Stiller, 1975). M. C. Regmi has also produced an invaluable series of books on the political economy of Nepal (Regmi, 1964, 1963–68, 1978).

Another important form of taxation was a requirement for provision of iron. Local low-grade ores were extracted, smelted, and carried to Kathmandu to supply the armouries of the rapidly expanding militant state which, by 1815, extended from the River Teesta in the east for some 2,100 km westward to the Sutlej River in what is today Himachal Pradesh (see Figure 2.9). The smelting of iron ore required large amounts of charcoal and, since the annual demand was heavy and the iron-smelting process very inefficient, depredations on the forests of Sindhu Palchok and Kabhre Palanchok were extensive. Remnant evidence of iron smelting and charcoal making is widespread throughout the two districts today.

These pressures that encouraged deforestation continued throughout the Rana Period from 1845 until its collapse in 1950. In addition, the great temple- and palace-building episode in the Kathmandu Valley, lasting from the early nineteenth century into the 1930s, placed another heavy burden on the forest resources.

Mahat *et al.* (1986a) conclude that most conversion of forest land to arable land had occurred by the early 1900s. Since this process had resulted in the conversion of all the better-quality forest land to arable land and had extended onto marginal slopes with poor soils, change in forest cover in the Middle Mountains over the past half century has been largely reduced to fluctuations of the forest margins brought about by cycles of deforestation and spontaneous natural, secondary, reforestation. However, the reforestation process has been retarded in areas where livestock grazing has been intensive.

From the foregoing discussion it can be concluded that in Sindhu Palchok and Kabhre Palanchok districts deforestation has a long history, that it became very intensive after about 1769 (at the latest), and that it climaxed between about 1890 and 1930. Little subsequent deforestation has occurred, although some areas of forest under excessive utilization have deteriorated to shrubberies or even, in limited areas, to grassland (the 'nibble effect' of Moench and Bandyopadhyay, 1986). As noted above, some abandoned farmland has reverted to secondary forest. In recent years the pressures of charcoal making for iron smelting and of palace and temple building have been eliminated. Nevertheless, it must be concluded that by the middle of the twentieth century the total remaining area of forest cover in the Middle Mountains had been reduced to a dangerously low proportion. The subsistence population, which had been expanding throughout the preceding two centuries, was faced with a reduced ratio of forest to arable land upon which the various systems of mixed farming depended. This, and the heavy demands of the State and landowners on the peasantry, also induced extensive migration to Sikkim, the Darjeeling area, and Assam, a process that is not without serious repercussions today. The progressively decreasing quality of the remaining forest is critical, therefore, to the survival of subsistence farming. But this is not central to the present line of argument, where our intent, in the present context, is to raise two questions: (1) how accurate (or inaccurate) is the claim for extensive deforestation of the Middle Mountains during the past thirty to fifty years? and (2) if, as we believe, we can show that much of the deforestation actually occurred during the eighteenth and nineteenth centuries, how can assumed recent increases in soil erosion and landsliding be accredited to deforestation?

Another relevant question is: how representative are Sindhu Palchok and Kabhre Palanchok districts of the rest of the *pahar*, bearing in mind the effects, especially during the period 1769 to 1950, of their proximity to the Kathmandu Valley? Bajracharya (1983a), following a detailed study of the fuelwood–food nexus of Pangma Panchayat in the eastern *pahar* of Nepal, came to conclusions similar to those of Mahat, Griffin, and Shepherd, as far as fuelwood pressures are concerned. While the emphasis of Bajracharya's study was to challenge the claim that demand for fuelwood was the essential driving force behind deforestation, he was also able to demonstrate a long history of deforestation; nevertheless, he does conclude that deforestation has continued up to the present. In the Karnali Zone of far western Nepal a

Table 3.1 *Changes in vegetation cover of Dehra Dun district, 1880–1980. Vegetation categories in left column have suffered losses, those in right column, gains (from Richards, 1987:300).*

	Hectares		*Hectares*
Forest woodlands	–27,421	Arable	+22,729
Major wetlands	– 4,607	Human use/transition woods	+ 4,982
Grass/shrub	– 2,269	Woods/shrub	+ 7,382
Semi-desert shrub	– 442		
Total	–34,739	Total	+35,093

similar pattern of deforestation over a very long period has been established (Bishop, 1978, 1986, personal communication). Nevertheless, neither Bishop nor Bajracharya underestimates the current precarious balance between a depauperate remnant forest and the sustainability of traditional subsistence agriculture. This theme is also central to much of the work of Mahat, Griffin, and Shepherd, and the Nepal–Australia Forestry Project will be discussed in more detail in Chapter 4. More recent work by Gilmour (1989, in press) in Sindhu Palchok and Kabhre Palanchok districts, and by Gurung (1988) in the Kakani area close to Kathmandu, provide extensive indications of an actual improvement in the extent and quality of tree and forest cover in certain localities over the past twenty years or so. That this process is the result of a direct response of the local people both to their growing awareness of forest depauperization and progressive government decentralization, rather than due to 'outside' aid and development activity, is highly significant. It is also strongly supportive of our defence of the 'ignorant' subsistence farmer who has so often served as a convenient scapegoat. The next logical question is: can this line of reasoning be extended beyond the limits of Nepal? A partial answer can be provided through examination of a few case studies – the Dehra Dun–Mussoorie area of the Garhwal Himalaya, Sikkim, Bhutan, and two areas in the Hengduan Mountains of southwestern China. In addition, to provide some insights into the high mountain forest situation, as

Table 3.2 *Dehra Dun district 1877–1971. Changes in area devoted to officially designated reserved and protected forests (increase of 96%) (from Richards, 1987:301).*

			Hectares	
Date	*Dehra Dun Tahsil*	*Chakrata Tahsil*	*Total*	*% of Area*
1877	77,000	–	77,000	24.9
1901	77,002	36,778	113,780	35.2
1950	77,287	36,518	113,805	35.2
1961	70,082	45,089	115,171	37.3
1971	–	–	150,803	48.8

distinct from that of the Middle Mountains, the case of Khumbu Himal is introduced.

Garhwal Himalaya: Dehra Dun-Mussoorie

Richards (1987), drawing upon a rich source of government documents, has reconstructed the pattern of land-use changes and population growth (1880–1980) for the Dehra Dun administrative district. The official extent of the district is 308,800 ha which, in addition to Dehra Dun Valley, includes Chakrata administrative district or *tahsil*, the much less densely populated subdistrict to the north of the valley. With an area of 136,000 ha it comprises 44 percent of the total district.

During the hundred-year period of the survey the population rose by 426.3 percent, from 144,070 to 758,241 (1981 census). The most dramatic growth has occurred since 1940 with a 185 percent increase from 229,850 to its 1981 level. Population densities have increased from 47/km^2 in 1880 to 246/km^2 in 1980. Much of this growth has been driven by the rapid development of the Dehra Dun metropolitan area, with an urban population of 293,628 in 1981.

This development is reflected in extensive changes in land use and vegetation cover with the main categories shown in Table 3.1. Thus, while the growth of a major metropolitan area renders this district not representative of the Central Indian Himalaya in general, the early establishment of the central offices and institutes of the Forestry Department in Dehra Dun partially offset what otherwise would have been an unusual degree of forest destruction. This ensured that a large portion of the district's forests and woodlands came under official management early. The area devoted to officially designated reserved and protected forests has grown by 90 percent between about 1880 and 1970 (Table 3.2), and today they occupy 48.8 percent of the total area.

The impacts of forest management, therefore, on agricultural expansion have been severe. Also, government control may have limited, but certainly did not halt, the deterioration of forests, woodlands, and wetlands in the district. Richards documents that much unrestrained timber cutting had already taken place between the assumption of British control in the 1850s and the demarcation of the reserved and protected forests in 1880. Subsequently, there followed some sixty years of British control with a forest policy designed to ensure a substantial financial return for the Crown. For instance, in the 1900s Dehra Dun's average annual timber export to the plains was 6,333 m^3, mostly sold for railway sleepers and building timbers. During the same period the Forestry Department licensed the extraction of 27,088 m^3 for fuelwood and charcoal. After the early 1900s illicit cutting has grown progressively, with a rapid increase after 1947, to the effect that by 1980, for many peasants, subsistence rather than merely supplemental income may be at stake in the forests. This is an indication of the serious deterioration of forest quality, if not of actual area of designated forest cover. Of course, without reliable biomass estimates precise figures for rate of forest loss cannot be derived.

Attendant problems facing Dehra Dun district are rapid loss of wildlife, including tigers, leopards, hyena – the last *Khedah*, indicating a catastrophic reduction in the number of elephants, was held in 1905 – while the growth in limestone quarrying, and its mode of operation, and the concentration of cement works have served to turn one of the 'gems' of the Himalaya into a *cause célèbre* of air and noise pollution and serious disruption of water resources and surface cover. At least this has been checked in recent years by India's first Supreme Court environmental ruling (Shiva and Bandyopadhyay, 1985). It must also be borne in mind that all of those processes together have brought about a decline in available arable land per person from about 0.3 ha in 1880 to a little over 0.1 ha in 1980. Similarly, per capita access to natural vegetation has declined from about 1.8 ha/person in 1880 to under 0.4 ha/person in 1980. There can be no doubt that the district's population has been extensively and progressively impoverished over the hundred-year period, a trend that also implies an acceleration in the human pressures on the vegetation of the district.

A very different situation can be seen to prevail within a few tens of kilometres north of Dehra Dun. Moench and Bandyopadhyay (1986) have conducted a fascinating study of the forest ecology and forest use by inhabitants of the small village of Munglori. It is situated at an altitude of 1,700 m on the slopes of Nag Tibba (3,022 m) in the partial rain shadow of the Mussoorie ridge (2,200 m). Moench and Bandyopadhyay have estimated, from a careful survey of the types of forest cover and land use in conjunction with a study of total village biomass consumption, that while total biomass productivity exceeds human consumption by a wide margin, there is a progressive loss in forest cover. This they refer to as peripheral degradation or the 'nibble effect.'

In practice, villagers, usually women, are collecting firewood and fodder along the margins of the forest. This occurs on south-facing slopes, at the lower elevations, and in areas where the demand is high. As lopping begins in a forested area and opening of the canopy occurs, grasses and forbs enter in large quantities. This in turn creates increased competition for oak seedlings, encourages grazing, and accentuates risk of fire. Hence the 'nibble effect,' whereby the forest margins are progressively rolled back. Moench and Bandyopadhyay (1986) go on to demonstrate many other aspects of the villager–forest balance and to suggest a number of critical forestry policy implications. Most important amongst these is the need for 'participatory management' whereby the local people are fully incorporated into official programmes. The latter so far have not taken seriously into account either the creative or the consumptive roles of the local people.

Sikkim and Bhutan

Sikkim is generally regarded as having incurred extensive deforestation. Pressures of land tenure and taxation in Nepal brought about by the demands of the House of Gorkha after 1769 resulted in the out-migration of large numbers of Nepalese subsistence farmers to what is currently the Indian State of Sikkim, the Darjeeling Himalaya, Assam, and Arunachal Pradesh. Extensive increases in subsistence agriculture and the widespread development of tea plantations certainly reduced the forest cover in the Sikkim and Darjeeling outer ranges and hills. While this process of deforestation continues today (Karan, 1984, 1987a), as in Sindhu Palchok and Kabhre Palanchok districts in Nepal, it also has a much longer history than the Theory of Himalayan Environmental Degradation would demand.

Karan (1987a) notes that Sikkim has an overall density of population of 44/km^2, a total population of 315,682 (1981 census), and a growth rate of 5 percent per annum. Only 36 percent of Sikkim's total area remains under forest, and extensive clear-cutting continues to occur, especially in the more accessible southern districts. While the northern forests at higher elevations are reported to be in good condition, lack of survey, demarcation, and an inventory of forest land has hindered the formulation of a forest management plan.

In addition, over the past twenty years, Sikkim has acquired one of the highest road network densities in the Himalaya (12 km/100 km^2 compared with 0.44 km/100 km^2 in Bhutan). This in turn has caused accelerated landsliding and further loss of forest cover (see Chapter 5).

Bhutan, in contrast, has a much more extensive forest cover, amounting to 60 percent of total area, a much lower population density and a much lower population growth rate (2 percent per annum). In Bhutan, unlike many areas throughout the Himalaya, there is a need for an increase in the extent of logging roads to reduce loss of forest products through more effective harvesting. One of the special features of Bhutan is the beneficent effects of government control, in terms of stringent protection of the country's forest and in many other respects (Karan, 1987a).

The Hengduan Mountains

Reconnaissance expeditions into the Gongga Shan area of western Sichuan and the Yulongxue Shan–upper Jinsha Jiang gorge area of northwestern Yunnan (Messerli and Ives, 1984) led to the preliminary conclusion that in these areas the claim for massive deforestation since 1950 was a serious over-statement. First, extensive high montane areas in western Sichuan were seen to be reforested due to natural processes, with a strong growth of immature or sub-mature trees depending upon the time span since the reduction of human impacts. In specific areas that were photographed or painted in watercolours by Eduard Imhof (1974) during the 1929–30 Swiss expedition to Gongga Shan (7,556 m) virtually no change had occurred. Where replicate photography could be obtained it could be demonstrated that no significant change in vegetation cover has occurred in the past half century. In addition, where the Imhof documents indicated extensive deforestation, this had likely occurred long prior to his visit. In other areas, especially those close to main roads, there is also recent deforestation.

Figure 3.2 Hengduan Mountains, northwestern Yunnan, People's Republic of China. With the recent (post–1979) improvement in housing, additional pressures have been put on the mountain forests closest to the villages. This view in Lijiang autonomous county looks across the Jinsha Jiang River on to heavily impacted mountain slopes. Note the numerous skid trails, some converting into gullies, with the more inaccessible high slopes retaining a much fuller forest cover.

This line of enquiry was pursued more thoroughly in the Yulongxue Shan (5,596 m) area of northwestern Yunnan during a ten-week field study in 1985 (Ives, 1985). Here it was possible to compare present-day vegetation cover with that shown on high-quality photographs taken by Joseph F. Rock between 1923 and 1947. In some areas extensive deforestation has occurred and the impact of heavy monsoon downpours on log skid trails has produced significant loss of top soil and gullying (Figure 3.2). Also, an extensive expansion of village housing following the introduction of the 1979 liberal 'local responsibility in rural areas' policy of Mr. Deng Xiaoping has had a noticeable impact on forest cover adjacent to many Naxi villages. Naxi house construction consumes a large amount of timber (Figure 3.3). Local government authorities in Lijiang autonomous county explained that at least

Figure 3.3 A traditional Naxi farmhouse under construction, Lijiang autonomous county, Yunnan. Note the extensive use of timber. The Yulongxue Shan (Jade Dragon Mountains) loom over the houses.

a third of the county's houses in existence in 1985 had been built since 1979. In addition, the methods of harvesting timber, including squaring of the tree trunks with hand axes, heavy destruction of saplings and seedlings during felling operations, and abandonment of logs with a spiral grain, resulted in at least 50 percent wastage (Ives, unpub.). However, many other areas show a much improved, denser and more vigorous forest cover when compared with that shown on Rock's photographs taken as long as sixty years ago (Figures 3.4 and 3.5). Additional large areas, which have gone through at least one cycle of clear cutting and regrowth since the mid-1920s, currently have a forest cover strictly comparable to that shown on Rock's photographs (Figures 3.6 and 3.7). Finally, travel along the main highway between Lijiang, Dali, and Kunming provided evidence that vast areas of the Yunnan Plateau had not only been stripped of forest cover, but that subsequent soil erosion has removed the soil cover completely over wide areas so that the multi-

Figure 3.4 Photograph taken in 1928 by Dr. Joseph F. Rock, Lijiang autonomous county, northwestern Yunnan. The time is post-monsoon and the doline is partly filled with water forming a shallow lake; this both empties subterraneously and evaporates by late winter or spring. Note the rather poor condition of the forests, both in the foreground, and especially on the steep hillslope to the right, and on the opposite side of the lake. (Courtesy, Dr. Barry C. Bishop, National Geographic Society.)

coloured bedrock of slates, shales, and limestones displays the underlying geology most vividly (Figure 3.8). The important point here, however, is not that deforestation has occurred, but that it occurred hundreds, if not more than a thousand, years ago.

To sum up the Sichuan and Yunnan reconnaissance, it is concluded that deforestation has been a very long-term process, that it has been interspersed with cycles of natural reforestation, that many areas have experienced formation of badlands, but that the overall picture is one of extreme variation from one specific site to another and certainly not a simple process of extensive deforestation in recent decades. Additionally, the extensive areas of the Yunnan Plateau that have been stripped of forest cover over the past thousand years or so have been a source of sediment over the same length of time, casting further doubt on the claims of massive increases in soil erosion and sediment transfer during the past four decades. This does not imply that there are no problems deriving from forest management in Sichuan and Yunnan. It does imply, however, that the emotional claims for sudden catastrophic destruction occurring in recent decades must be viewed with scepticism; a much longer historical perspective is required.

Figure 3.5 Replicate of photograph in Figure 3.4 taken in April 1985 by Jack D. Ives. The lake is now dry. Of special interest, the foreground forests are quite luxuriant. Note the closed-crown cover on the steep hillslope to the right. Also, despite the considerable distance, it is clear that the forests on the far side of the lake are more extensive and more luxuriant than they were in 1928.

The Southeastern Tibetan Plateau (Xizang)

The Tibetan Plateau has been regarded traditionally as a high-altitude mountain or plateau desert, or semi-arid land devoid of trees, on account of both altitude and sparse precipitation. While little detailed historical information relating to vegetation cover change is available, members of the first scientific group invited to Lhasa by the Government of the People's Republic of China, and to traverse the Plateau and Himalaya to Kathmandu, reacted to signs of much more extensive forest cover in the past than

Figure 3.6 The east face of Yulongxue Shan, northwestern Yunnan, as seen by Dr. Joseph F. Rock in 1928. Rock's camp is situated on the far side of the karst lake. The forest cover of the middle ground is somewhat open and immature, being cut over and fired from time to time. (Courtesy, Dr. Barry C. Bishop, National Geographic Society.)

previously supposed. Trees surviving today at very high altitudes (in excess of 4,000 m) at the sites of former small temples and monasteries suggest that much of the lower valley slopes may have supported extensive forests prior to the establishment of the Tibetan state. Supporting indirect evidence was located in the form of thick peat deposits, and the very success of the current Chinese programme of afforestation (Ives, 1981). Reiter (1981) goes much further and raises the question of how a population of between one and four million (cf. Goldstein, 1981, for discussion on population size), supporting significant cities and large religious structures, could have evolved and continued to thrive for more than a thousand years if there had been a total dependency on importing timber and fuelwood over great distances and across very rugged terrain. Reiter even raises the fascinating idea that widespread deforestation of the Tibetan Plateau, and the consequent increase in albedo, particularly during periods of snow cover, may have had a marked effect on the mechanism and strength of the Indian monsoon.

The case for former extensive forest cover in parts of Tibet, and deforestation over the past 1,200 years or so, is by no means proven. It is perhaps appropriate to introduce the possibility here as an example of the type of direction in which scholarly enquiry could move once the comfortable

Figure 3.7 The same scene as that shown in Figure 3.6 but taken in April 1985 by Jack D. Ives. The lake is now dry. The conspicuous rock in the foreground greatly aided exact relocation of Rock's camera position. The forest in the middle ground is in approximately the same condition although most of the trees are only twenty-five to forty years old (based upon tree ring count); the area is being actively harvested today. Most of the trees growing when Rock visited the area were subsequently felled.

assumptions upon which conventional wisdom is based are questioned. The tentative hypothesis for Tibetan deforestation provides at least a similar chronological framework to that of the much more defensible account of deforestation in neighbouring western Sichuan and Yunnan.

Khumbu Himal, Nepal

The problem of deforestation in Khumbu Himal is introduced here for two reasons. The first is that it provides a useful case study of the condition of the

Figure 3.8 Large sections of the Yunnan Plateau have been totally deforested and soil erosion has proceeded to such an extent that the C-horizon has disappeared exposing the bedrock. This scene is traversed by the main road between Kunming and Dali, which is lined by eucalypts. Only on sheltered valley floors has enough soil been preserved to permit agriculture. Natural climax vegetation would be closed-crown monsoon rain forest. This deforestation, however, occurred hundreds of years ago, if not more than a thousand.

forests of a widely known high montane Himalayan region, with world-wide attention being focused on the Sagarmatha (Mt. Everest) National Park, home of the Sherpas, and a World Heritage Centre. The second reason is that it is an ideal example of Thompson and Warburton's (1985a) 'uncertainty on a Himalayan scale,' and of the confusion being created by totally contradictory observations that have been widely reported and discussed.

First, within the framework of the eight-point scenario of the Theory of Himalayan Environmental Degradation, the forests of Khumbu Himal have been depicted as being in the process of rapid destruction (Fürer-Haimendorf,

1975; Eckholm, 1976; Jeffries, 1982; Andrews, 1983; Coburn, 1983). Various reasons have been given for this dramatic state of affairs and the main points are as follows:

1. The rapid increase in mountaineering expeditions and in trekking tourism since the 1960s has brought into Khumbu Himal large numbers of visitors. By the late 1970s the total number of annual visits, of an average fourteen-day per person, exceeded 5,000, double the total indigenous Sherpa population. This is claimed to have greatly increased pressures on the local forests in several ways: fuelwood for camp fires, both for cooking and for the warmth and enjoyment of the large numbers of trekkers and mountaineers, and their larger numbers of porters; timber for housing, in part to provide lodges, small hotels, and tea houses for tourists, and in part to enable the increasingly affluent Sherpas to build larger private houses. Additional requirements have been fuel for heating water to provide hot showers for tourists, wood fires for the Japanese-built Mount Everest View hotel, and construction timber for better bridges across the Dudh Kosi.
2. In 1957 the Nepalese government nationalized the forests in a misguided effort to protect them. Campbell (1979), Messerschmidt (1981, 1987), and others have discussed this point in some detail. It was claimed that the Sherpas had a superb system of forest management, as had many indigenous groups in Nepal. This flourished on the basis of community recognition of the importance of the forests to their continued well-being. Forest guards (*shinga naua*), appointed by the village headmen, controlled forest use. Fuelwood collection, timing, and location of felling of large trees for construction and maintenance of sacred forests was controlled and policed by the forest guards who had power to levy fines on offenders. The nationalization of the forests destroyed this efficient traditional system and replaced it with an ineffective alternative. For instance, when a Sherpa was faced with a three-day walk to obtain permission to cut a tree from a disinterested government agent who was not even a Sherpa, he simply went into the forest and helped himself. Thus a wave of indiscriminate felling occurred.
3. The establishment of Sagarmatha National Park in 1977 had been heralded by discussion of the prospects for implementation of total forest protection. In fact, the possibility of forcibly ejecting the entire Sherpa community from their traditional homeland had been raised as essential to the development of a national park, a policy that had been put into effect when the Lake Rara National Park was established in western Nepal. In this instance, several hundred high-altitude Chhetris were forcibly expelled from their traditional homeland and abandoned without compensation. Fürer-Haimendorf (1975) reported that the threat of restricted access to the forest products set off a further wave of forest destruction whereby the local people sought to obtain all they could of *their* forests before formal establishment of the park deprived them of

their traditional rights.

4. The influx of several hundred Tibetan refugees into Khumbu Himal after 1959 placed further demands on the forests for construction timber and fuelwood.

These various pressures on the forests of Khumbu Himal were perceived as causing devastation to the montane forests and the forest-alpine ecotone shrub juniper and rhododendron cover. To provide a few examples:

- Fürer-Haimendorf (1975:97–8) claimed that 'forests in the vicinity of the [Khumbu] villages have already been seriously depleted, and particularly near Namche Bazar whole hillsides which were densely forested in 1957 are now bare of tree growth and villagers have to go further and further to collect dry firewood.'
- Blower (1972, cited in Mishra, 1973:2), as a justification for the establishment of the Sagarmatha National Park, emphasizes the need to conserve the 'depleting forests of the Khumbu ... since destruction would result in disastrous erosion leading to enormous economic and aesthetic loss to the country.'
- Speechly (1976:2) explained that the 'forest areas in the proposed Sagarmatha National Park are, as a result of a combination of influences, in a depleted state, such that if present pressure of use is continued, severe environmental damage will result.'
- Hinrichsen *et al.* (1983:204) suggested that 'more deforestation [has occurred in Khumbu Himal] during the past two decades than during the preceding 200 years.'

Thus Khumbu Himal came to have its own mini-version of the Theory of Himalayan Environmental Degradation. Deforestation is assumed to have occurred, accompanied by changing patterns of yak and cross-breed grazing, as a result of the impacts of tourism on the traditional life style, leading to overgrazing near the villages (Bjønness, 1980a and b). This is believed to have accentuated soil erosion, landslides, and gullying and to have caused a variety of downstream impacts. To this was added trampling of vegetation by tourists and the spread of refuse so that the world-famous trekking route to the Mount Everest Base Camp became known as 'The Garbage Trail'.

It may come as a pleasant surprise, therefore, to find that the above combined descriptions and predictions of portending disaster for one small, but special, part of Nepal are totally turned on their head by the thirty-year comparison of a sympathetic and careful observer. A former US Peace Corps Director in India, Dr. Charles Houston, revisited Khumbu Himal in 1981, more than thirty years after being a member of the first mountaineering reconnaissance team ever allowed into the southern approaches to Mount Everest. He wrote that, with the exception of a thicket of dwarf juniper at Pheriche, in 1981 there was 'as much or more forest cover than there was in 1950 and I have the pictures to prove it' (Houston, 1982). Thompson and Warburton (1985a) use this statement to argue that Houston has questioned the very existence of a possible crisis facing Khumbu Himal, or even Nepal. While we do not think that Houston intended his remarks to be taken in this way (cf. Houston, 1987), it is convenient for our argument to introduce his quotation because, to a large degree, it coincides with more recent deliberate examination of the Khumbu Himal forests by Byers (1987a and b) and Stevens (1986). Thompson and Warburton (1985a) themselves talk about the 'vast tracts of forest some distance from the villages and trekking routes [that] are still intact. Indeed, large fallen trees lie rotting within them' (but even here we must ask: how far from the villages?).

In 1984 Byers spent ten months in Khumbu Himal attempting to reduce some of this uncertainty by collecting field data from carefully selected study sites. While the primary purpose was to quantify rates of soil erosion and degree of slope instability relative to vegetation cover, slope angle and altitude, and rainfall amount and intensity (see Chapter 6), important secondary tasks were to assess the changes in forest cover and to determine rates of fuelwood usage. His observations on forest cover will be introduced for comparison with the claims of Fürer-Haimendorf, Jeffries, Coburn, and Hinrichsen, and others.

One of the methods employed by Byers was repeat photography using Erwin Schneider's terrestrial photogrammetic surveys of 1956–63 that were the base material for the production of the 1:50,000 topographic map 'Khumbu Himal'. Byers also undertook numerous studies of forest stands and made many spot observations. His conclusions are that the claims for forest disaster in Khumbu Himal are grossly over-dramatized, especially where applied to the montane forests. It is likely that, prior to the arrival of the Sherpas approximately 400 years ago, the forests of Khumbu Himal were much more extensive than today; that deforestation has progressively occurred over this long period of four centuries of human occupation.

The most recent findings of Byers (1987c) even raise the question of a longer period of progressive deforestation, possibly ante-dating the traditionally accepted time of the initial movement of Sherpas into the area. Pollen analyses and radiocarbon dating of buried soil horizons and charcoal fragments are providing the first indications that complex changes in vegetation cover due to human impacts over a millenium may have occurred. According to these preliminary analyses, the valleys between about 3,500 and 3,900 m were occupied by open birch/fir/alder forest from several thousand years ago to a period represented by the relative age of the 15-cm level in one of the sampled soil profiles (it has not yet been possible to provide an 'absolute' date for this level due to lack of datable material). At the 15-cm level (i.e., upward in the profile) there occurred an abrupt transition to a grasslands formation on the drier, south-facing slopes, such as we see today. Byers (1987c:200) makes the tentative estimate that the replacement of the open birch/fir/alder forest by grassland communities occurred between 400 and 800 years ago. What is even more interesting, however, is the speculation by the palynologist (Markgraf in Byers, 1987c:199) that the open forest itself shows indications of human alteration in the form of burning and grazing.

Furthermore, grains of cereal pollen were identified at each of the four soil levels analysed, and fragments of charcoal from 31-cm and 41-cm depths give dates of 1,480 ± 360 and 2,170 ± 330 radiocarbon years BP, respectively.

It is interesting to speculate that significant landscape changes may have been effected by other ethnic groups (from the south?) or that Sherpa groups arrived in the Khumbu much earlier than is generally supposed on the basis of historic and linguistic evidence and analysis of folk-lore (i.e., about 400 years ago).

What is important in the present context is that, as this relatively new phase of more rigorous research begins to produce results, it becomes apparent that the broad assumptions, to which we refer as the 'conventional wisdom,' that have led to the emergence of the Theory of Himalayan Environmental Degradation, become more and more inadequate.

To return to the general claims of recent widespread deforestation in the Khumbu, some of the specific contentions, such as that of Fürer-Haimendorf to the effect that the formerly forested slopes above Namche Bazar and above the villages of Khunde and Khumjung had been clear-cut within his own memory, demand an equally specific response. Byers (1987a) replicated photographs taken in the 1950s and 1960s by Erwin Schneider. An example, the Imja Khola–Namche Bazar–Everest 1962 panorama, was taken from an altitude of 4,488 m on the northwest slope of Tamserku (6,608 m) (Byers, 1987a:78, 79). Despite difficulties in lighting, angle of view, scale, and differences in optical qualities of the lenses used, Byers was able to conclude:

1. that most forested areas in the Namche–Khunde–Khumjung region appear to be relatively unchanged;
2. considerable thinning of certain juniper woodlands has occurred;
3. little change of a medium- to large-scale geomorphic nature is discernible;
4. several distinctive tourist- and National Park-related structures are very evident.

Byers comments: 'In general the photographic evidence does not support the hypothesis of widespread deforestation, nor the assumed linkages between tree removal, grazing, and geomorphic damage within the specific geographic areas under discussion' (Byers, 1987a: 80). The open and rigorous discussion about details of the deforestation history of the Khumbu, and other areas, that is now rapidly emerging is an effective vindication of our initial decision to question the validity of much of the conventional wisdom. In terms of the Khumbu, therefore, we must conclude that the claims of many of the supporters of recent catastrophic forest and soil degradation are simply false.

Byers (1986, 1987c) does point out that in the higher altitude fuel-scarce areas, specifically within and above the juniper and rhododendron shrub belt in the upper part of the forest–alpine meadow ecotone, serious damage has occurred and continues to occur. Here Sherpas and trekking groups are pulling up even the roots of shrub juniper for cooking fires and thereby exposing sandy soils to extensive wind and rill-wash erosion. His two soil-study plot losses in this belt were twenty five times greater than losses from the typical lower altitude shrub/grassland plots, and forty-two times greater than those from the forest plots. Two points arise from Byers's observations: one, that individual scientists have seriously misunderstood, or misreported, local conditions; and two, that the furore raised by over-dramatized claims of deforestation is resulting in attention being focused on the main montane forest belts and the development of new forest plantations. While this is by no means a bad thing on its own account, attention is being diverted away from the really critical areas at higher altitudes. In this sense, Houston has performed an invaluable service in prompting the resolve of the new wave of scientists to make a critical assessment of the actual conditions. This point is argued on the conviction that good policy cannot be based upon bad information, however 'sympathetic' it may be to a worthy cause. Houston and Byers also add to the findings of Mahat, Tucker, Richards, Bishop, Messerli, and Ives that deforestation in many specific areas within the broader Himalayan region is a process that has a very long history, and is not, *at least in these specific areas*, a recent, post-1950, catastrophic phenomenon.

Stevens (1986) has taken several aspects of the discussion on forest use, protection, and destruction in Khumbu Himal a stage further. His report, following more than fourteen months of resident study, concludes that a picture emerges of traditional Sherpa forest-use that is far different from the one presented by Fürer-Haimendorf, and others. 'Instead of a well-regulated traditional system geared to sustainable use of forests, there appears to have been a pattern of unregulated use which has led to substantial' changes in forest composition, density, and total forest area throughout the 400 or so years of Sherpa occupation of Khumbu Himal. Sherpa subsistence use of the forests seems to have consisted of gleaning the floors of protected forests for fuelwood and fodder and then outflanking the forest guard and other forms of local control to supply their needs for fuel, construction timber, and tools. Their use of the forests went to the extent of grubbing out the tree roots so that nothing remained. This progressive assault on the forests resulted in the clearing of areas near the villages first and then moved further and further afield. Lack of any tradition of reforestation, together with the heavy grazing pressures, prevented any significant recuperation of forest cover. These conclusions by Stevens have been anticipated by Messerschmidt (1981, 1987) who has also documented a recent shift to *re-emphasis* on traditional management and control systems.

To sum up for the Khumbu Himal forests, it is apparent that conservationist claims for rapidly approaching disaster stemming from a change in forest use over the past three or four decades is a gross exaggeration. It must be concluded that the history of deforestation in this area, at least in a general way, parallels that of the other areas – Sindhu Palchok and Kabhre Palanchok districts, parts of Karnali Zone, Pangma Panchayat in the eastern *pahar* of Nepal, and further afield, in the Dehra Dun–Mussoorie, Yulongxue Shan, and Gongga Shan areas. In terms of the viability of present-day subsistence agriculture in these areas, however, this does not mean that there is no crisis. In fact, with continued population growth in many of these areas, and the apparent accompaniment of forest depauperization, shortages of fodder, and other forest products, and eventually of fuelwood, cast a serious shadow over the future well-being of the mountain people. What we wish to emphasize here, however, is that if significant deforestation since 1950 has not occurred in the areas for which reasonably reliable data is available, and if these results are applicable, at least to adjacent areas, the claim that post-1950 deforestation in the mountains has led to increased flooding, siltation, and other deleterious impacts on the Ganges and Brahmaputra (and Chengdu) plains is tenuous at best. This conclusion appears reasonably well validated without considering whether or not deforestation accomplishes the damage that has been asserted for it. This issue, the linkage between forests, precipitation, and soil erosion, will be considered in Chapter 5. First, however, it is necessary to analyse the actual pressures that are affecting the remaining forests and to discuss why so much confusion has arisen concerning the presumed rates of forest product consumption. Chapter 4 is devoted to these issues.

4 PERCEIVED PRESSURES ON THE HIMALAYAN FORESTS AND THEIR ROLE AS ENVIRONMENTAL SHIELD

The previous chapter has led us to the conclusion that the popular claims about catastrophic post-1950 deforestation of the Middle Mountain belt and areas of the high mountains of the Himalaya are much exaggerated, if not inaccurate. This does not imply, however, that deforestation is not occurring in the wider Himalayan region, nor that damage to the remaining forests should not be a matter of serious concern.

Similarly, there is available and reliable data for extensive deforestation of recent vintage in specific areas of the mountains, as distinct from the Terai. As an example, Messerschmidt (personal communication, March 1987), on the basis of personal experience between 1964 and 1983, cites serious loss of forest cover at Ghora Pani, a major tourist-trail stopover in Myagoli/ Parbat district near the Annapurna Sanctuary en route to Jomosom. The 'massive destruction of the forests' has been caused in part, according to Messerschmidt, to satisfy perceived tourist desires for huge camp fires. More recently some lodge owners at Ghora Pani have installed efficient wood-burning stoves, and some tourists prefer to frequent such lodges in order to reduce what they themselves perceive as extensive and wasteful forest cutting.

As already indicated, extensive planned and spontaneous deforestation has occurred in the Nepal Terai, and extensive commercial felling in the Kumaun and Garhwal Himalaya and other areas, has caused widespread destruction and social unrest. This latter development led to the formation of the Chipko Movement (a grass-roots movement, initiated by village women in the Kumaun Himalaya to prevent 'outside' exploitation of mountain forest resources) and, in 1984, to a fifteen-year moratorium on green felling decreed by the late Mrs. Indira Gandhi, former Prime Minister of India. This chapter, however, will be devoted to an overview of the perceived and actual pressures on the remaining mountain forests and an examination of the responses to the perceived trends and of the reasons for the enormous range in estimates of per capita fuelwood consumption and biomass productivity.

As already discussed in Chapter 3, the closely knit inter-relations between crop production, livestock, and forest resources of the average traditional subsistence agriculture of the *pahar* indicate that one or more hectares of forest are required to 'support' one hectare of arable land. To be reviewed now is the widespread claim that growing population pressure leading to increased demands for fuelwood is the primary cause of forest depletion – whether through actual reduction in forest area, which has been contested already, or in reduction in the quality (density, species distribution, and age structure) of the forests within their existing pre-1950 approximate boundaries. One of the especially intellectually appealing elements of the Theory of Himalayan Environmental Degradation is the fuelwood-demand vicious circle. This graphically depicts thousands of mountain villages with rapidly receding forest perimeters; the womenfolk must walk further and further to collect firewood until a critical threshold is reached whereby more than two man-days (actually, the work is most often done by the women!) per week per family are needed to obtain the necessary fuel supply. At this point increasing amounts of dung are used for fuel and the agricultural terraces are deprived of their only fertilizer; this in turn reduces crop yields and weakens soil structure. More landslides occur and the need arises to cut more terraces on steeper slopes in order to grow more food at still lower yields.

This scenario has attracted the input of large sums of money and human resources in an effort to save the remaining forests. The series of measures introduced includes the development of alternative energy sources and more efficient methods of using the wood fuel that is burned for cooking. Passive solar energy devices, metal and ceramic wood-burning stoves with stove pipes which heat more efficiently, 'micro' and 'mini' hydroelectricity projects, and wind generators have been proffered to the village people. In addition, a stated priority for massive reforestation is resulting in the establishment of forest plantations, often of quick-growing pines and exotic species. This complex of issues and responses needs to be examined in some detail. This is because we seek to demonstrate that, not only has cause and effect been confused generally within the context of the Himalayan Problem, but to treat one perceived cause on a priority basis, such as increased demands for fuelwood, is to risk long-term damage if it turns out that the wrong 'cause' was being treated.

In any mountain subsistence-agricultural system, the demands on the forest are numerous and fairly self-evident: fuelwood, undoubtedly, and construction timber, house shingles, timber for house and and farm utensils, and, of no less importance to the villages, fodder, thatch, and animal bedding. To these must be added medicinal herbs, nuts, fruits, mushrooms, and a host of secondary products such as the use of *Daphne* spp. (Nepalese: lokta) and *Quercus semecarpifolia* for the ash and fuel respectively, for paper making in Nepal. These demands on the forests are augmented by the requirements for charcoal, commercial firewood for sale in the cities and towns, and large commercial construction timber. A curious and decidedly non-traditional requirement, almost a fluke of the shotgun approach to development and aid projects, but having considerable impacts in local situations, is the large consumption of fuelwood by cheese factories, in areas, even countries, where cheese is not eaten (in this case tourism supplies the market). But, above all, it is the fuelwood depredations that have caught the attention of the international community.

Bajracharya (1983a and b) believes that this preoccupation with fuelwood is misguided and especially unfortunate. In a detailed study of the fuelwood issue in Pangma Panchayat in eastern Nepal, Bajracharya effectively demonstrates that there was no shortage of fuelwood in 1980. He goes on to show that the reduction in forest cover in that particular panchayat is the result of the needs of an increasing subsistence population to develop more agricultural terraces or grazing land to ensure the production of more food. Nevertheless, he does point out that, given the current rate of forest reduction, a fuelwood shortage in the next two decades must be anticipated. However, his study shows that in this area, and presumably in much of the *ṕahar*, the wrong problem (perceived or real, depending on the definition) has received undue priority.

The same argument can be developed for the considerable efforts expended by USAID, UNICEF, FAO, and other agencies, in conjunction with the Nepalese government, to introduce alternative sources of energy. The construction and deployment of various metal and ceramic wood-burning stoves, for instance, raises many questions. Some of these can best be examined by returning to Khumbu Himal again, one of the notable targets for such development aid. An understanding of the ambiguities and difficulties associated with alternative energy sources, or more efficient methods of burning wood, requires a more thorough knowledge of indigenous ways of living and thinking, and closely related religious beliefs, than is presently available. However, from the pioneering work of Fürer-Haimendorf (1964, 1975, 1984) and others, there has accumulated an impressive understanding of the Sherpa way of life. Traditionally, Sherpa women cook on an open hearth in the combined living room-sleeping room-kitchen. Since there is no stove pipe, the house is filled with wood-smoke. By western standards the method is highly inefficient and injurious to health because of the high concentrations of carcinogenic particulates within the houses. However, the open fire provides a cosy atmosphere and, of special significance, the hearth is where the traditional house gods reside (Bjønness, 1986).

Many ceramic wood-burning stoves, with pipes, have been introduced into Sherpa and other mountain societies. Reid, Smith, and Sherchand (1986) have demonstrated the beneficial impact of the new stoves in terms of health and efficiency. However, among some of the more traditional families there has been a resistance to the prospect of disturbing the household gods. Claims have been made that the reduction in smoke has resulted in an unacceptable increase in infestations of biting insects, and even that the loss of the curing effects of the smoke on the roof timbers and shingles, and an increase in termites and other boring insects, has raised the timber requirements for rebuilding and repairing houses. As electric lighting has also been introduced in some localities, it has been argued that the more efficient stoves are simply kept burning longer while the family delays going to bed by several hours, thus ensuring the consumption of the same amount of fuelwood (or more) over a longer period. The prospects of the eventual introduction of television are alarming indeed! Moreover, there are other more important immediate consequences. First, it is often claimed that wood-burning stoves require split logs whereas the traditional open hearth consumed twigs and small broken pieces. Thus a requirement for split logs could lead to heavier rather than reduced pressures on the local forests.

A more spectacular problem arising from the attempts to introduce alternative energy sources is the recent policy of constructing 'mini' and 'micro' hydroelectricity plants. Again, Khumbu Himal provides the most devastating and dramatic example. An Austrian Aid Project involved the construction of a 'small hydel' plant on the Bhote Kosi to service the villages of Namche, Khunde, Khumjung, and Thamo with electric power. The prospect of towers and overhead transmission lines in this ultimate of the world's mountain sanctuaries notwithstanding, the hydroelectric facility, approximately 90 percent complete at a cost of several million US dollars, was destroyed one afternoon in August of 1985. The cause of the destruction was the catastrophic outburst of a moraine-dammed lake some 12 km upstream of the weir and intake canal that had not been taken into account during the site survey (Ives, 1986; Vuichard and Zimmermann, 1986, 1987). While this disaster cannot be used to criticize the *policy* of seeking to provide alternative energy to fuelwood, it is a criticism of the way the policy is being carried out and it does raise the question of who perceived the need for electricity in Khumbu Himal, the Sherpas or a development agency?

In a similar vein are the widespread anecdotes about the large number of wood-burning stoves to be found dumped as scrap behind the houses of mountain people. Is this a reflection of the government's 'target mentality' in terms of measuring the success of development projects? This is a term used to describe the bureaucratic system, in this case of ascribing success to a particular policy by counting the number of units (i.e. stoves) that have been dispatched, rather than determining whether and how they are used. In this case, the number of wood-burning stoves distributed does not indicate the success level claimed by a bureaucrat's records if a high proportion of them were thrown out upon receipt, or shortly thereafter. A further dimension of this problem is the question of proper maintenance. It has been shown that careful and constant maintenance is essential for efficient operation (Reid *et al.*, 1986); if attention is not paid to this, as is often the case, the advantages of the new wood-burning stove project are severely curtailed.

A personal insight into the topic of smokeless, fuel-efficient stoves is provided by Messerschmidt (personal communication, March 1987). While on a field trip in Gorkha district in 1982 with an expert from a 'women in development/women in forestry' project and a consultant, Messerschmidt recorded some highly relevant observations. Khumal women (that is, exceedingly poor, low-caste women) who had new ceramic stoves (*chula*) installed in their homes did not like them while the nearby tea-shop owners prized them highly. The housewives claimed that they were not hot enough, or fast enough, and that they used as much, or more, fuel than their traditional methods of cooking. The tea-shop owners said exactly the opposite. The difference in reaction was found to be based on the contrasting patterns of use. The women in the home light up their fires twice a day – early morning and late evening – for the two hot meals of the day: they need quick, hot fires. The tea-shop owners fire their stoves early and keep them burning all day until late at night; they need a constant, steady heat, for which a good flue with a steady draw is required. This takes some time to develop as the chimney must become hot. In the case of the housewives, their food is cooked before the chimney is hot enough to provide a steady draw.

Again, we do not wish to imply that efficient stoves, or electricity, or a more pleasant and healthy work space for housewives is undesirable; on the contrary. Nor do we accept the 'anti-development' anecdotes as proven on a general scale. What we believe is necessary, however, is a critical analysis of success and counter-success claims, and an assessment of the so-called 'appropriate technology transfer' within the context of local practices and preferences.

The 'target mentality' problem is also relevant to another area of response to the perceived fuelwood-deforestation crisis – namely, reforestation. This is a complex topic in itself and to illustrate, at least partially, its widespread nature we shall go much further afield than even the Himalayan region *sensu lato* – specifically to southern India.

Reforestation with Eucalyptus (and other fast-growing exotic species) has long been regarded as a panacea for reversing world trends toward deforestation because of their rapid growth capability and ability to flourish in areas of poor soil. Large sections of the mountain landscapes of Ecuador, for instance, would be hard to imagine without the ubiquitous scatter of groves, forests, and hedgerow alignments of *Eucalyptus globulus*. The same is true across vast tracts of China and India; Ethiopia is another dramatic example of a Eucalyptus landscape. Nevertheless, a very complex controversy has arisen over the wisdom of using eucalypts, and other exotics, in this manner. Without going into this much larger problem, we will review here the findings of a study by Shiva, Sharatchandra, and Bandyopadhyay (1981) on the impacts of 'social forestry' in the Kolar district of Karnataka State, southern India.

Social forestry has become both a major concept and leading policy over the past decade or so. It is strongly supported by such powerful agencies as The World Bank and FAO. The primary objective is to save and render more efficient the remaining natural forests, and to develop new forest plantations to improve the living conditions at the village level. This 'social forestry' aims to produce improved fuelwood supply for rural areas and to replace use of animal dung, to augment small timber supply and animal fodder, to provide protection of agricultural fields against the wind, and to serve recreational needs. By involving the rural poor in the development of these objectives, many of the negative aspects of traditional forestry policies will be avoided, since it has long been recognized that successful forest policy will remain an illusion without the active and positive input of the local people. This input, ideally, should involve the local people in all levels of project development and operation, including identification of their own felt needs, selection of land, choice of species (in the case of forestry), control and monitoring, and in the development of a clear perception and guarantee of the envisaged benefits. The enthusiastic support and commitment of the man-on-the-spot is necessary to ensure that seedlings are properly maintained and especially that they are protected from domestic livestock trampling and grazing. The ideal is not universally achieved as one particular example will show.

During the sixth Five Year Plan (1980–85) the Government of India earmarked one thousand million rupees to promote social forestry projects throughout the country. This was intended to raise fuelwood plantations over 260,000 ha and to supply 530 million seedlings to be distributed via the state governments in one hundred selected districts. Kolar district was one of the hundred selected, and it was also to be used to determine the feasibility for an even larger World Bank project.

The Shiva *et al.* (1981) study in Kolar shows that the specific objective of bringing a large area of land under fast-growing eucalypts has caused a deterioration rather than an improvement in the local rural conditions. In practice the farming families with more land than the minimum required to satisfy subsistence needs converted cropland to eucalyptus woodlots and forest plantations. Once planted, the seedlings required little or no labour. This meant that poorer families, who depended upon the proceeds of their labour on these larger farms to supplement their own inadequate subsistence production, were forced off their land to add to the growing hordes of urban slum dwellers. The beneficiaries were the richer farmers, but especially the commercial timber, wood products, and paper and pulp industries, which harvested the eucalyptus plantations. Shiva *et al.* (1981) demonstrate the fallacy that eucalyptus could fill fuelwood needs when most of these needs are satisfied by dead twigs, shrubs, and general litter – products that have no apparent economic value because they do not enter the commercial market sector and therefore are not included in the narrow financial analyses that are frequently the basis of such 'development' projects. Furthermore, the project in Kolar district is resulting in the conversion of 12 percent of land that originally produced food crops into commercial timber plantations, producing shortages in traditional village grains such as *ragi*, the staple millet. Concomitantly there was a much larger proportion of plantation on private land than on communal land. The project also has increased, rather than decreased, pressures on the remaining forests since purchase of eucalyptus for fuel is ridiculously beyond the capacity of most local people.

There are three additional negative aspects of Eucalyptus introduction, at least from the point of view of many local people:

1. the perception that eucalypts take up more water than indigenous species and therefore cause wells and springs to dry up;
2. they have a medicinal odour which is offensive;
3. the 'Vick's Vapour Rub' medicinal flavour is transmitted to food cooked over Eucalyptus fires.

The authors draw us back to our Himalayan Region by the following statement in their concluding section:

> This trend of inappropriate choice of [tree] species seems to be the centre of political controversies around several other World Bank aided major forestry projects.... In the Himalayas the Chir Pine mono-culture, and the Tropical Pine in the Bastar region of Mahya Pradesh, are viewed by the local people as an attack on their local life-support system. In the light of the project [World Bank], the empirical trends in shifts in land use as analysed by the present study are found to be the result of conscious policy orientation ... guaranteed sabotage ... bureaucratic anarchy.
>
> (Shiva *et al.*, 1981: 77–8)

The Chipko Movement has made the challenging accusation that one of the biggest threats to the Himalayan environment and the mountain forest peoples is the World Bank (S. Bahuguna, personal communication, 1986). At issue is not the basic concept of 'social forestry' but rather its implementation. On one important detail Budowski (personal communication, 1986) insists that the growing attacks on the use of Eucalyptus are also often misdirected and there is a pressing need for systematic study of *appropriate* forms of general policy implementation. Hamilton (personal communication, March 1987) goes a long way to clear up much of the misunderstanding that has arisen in the emotionally charged 'Eucalyptus Debate' which is included here verbatim:

> In India (particularly in the state of Karnataka), in Bangladesh, in Portugal, and several other places, there has grown up a substantial public feeling that the establishment of eucalyptus plantations is creating a host of undesirable social, water, and soil consequences. In Portugal, it has been nicknamed the 'fascist' or the 'capitalist' tree because it supposedly impoverishes the countryside and the people. In Sri Lanka, it is accused of causing droughts. In India, it has been claimed that it leads to desertification, and makes the rich richer and the poor poorer. In that country, professional foresters who have taken opposite sides on the question are no longer speaking to each other, and two major research institutes have come down on opposite sides of the question of whether eucalypts are 'good' or 'bad'.
>
> It is my professional judgment that a number of attacks on eucalyptus are misdirected. In the majority of these cases, we are blaming a species for a land-use decision which was made by land owners or land leasers, to change land use into a commercial plantation. This has often caused serious social and economic disruption for rural people who may have had some kind of traditional use of the lands in question. Pointing the finger at a species is targeting the wrong enemy, and one should be questioning the land-use decision with all its complications. Whether eucalyptus was used, or pine, whether an exotic, or a native, species, the social impact may well have been the same.

In cases where there are supposedly undesirable soil nutrient, erosional, and hydrological effects occurring because of eucalyptus planting, there is certainly need for more systematic study, but again, a number of questions can be raised about the generalizations. Many land owners, or land users, want a fast-growing species which will give quick returns. If they are successful in selecting one which matches the available soil and climate, then of course it will use a lot of water because it is performing successfully. The genus eucalyptus happens to have a very large number of species, and it is often possible to find one which is adapted to the site available for planting. When it succeeds, and therefore does the job of producing wood using nutrients and water and sunlight efficiently, it is difficult to see why one should either blame the species, or even more so, the whole genus. If, however, it produces wood for pulpwood, when what was really needed was cheap, low-cost fodder for livestock, or if it reduces streamflow, when what we really want is more water, then we really have made a poor choice of vegetative cover, and we need to specify what it is we want from the land area in question.

At any rate, it is not a simple situation, because there may well be some allelopathic effects between eucalyptus and interplanted or adjacent plants that are important. The litter characteristics of eucalyptus are somewhat different from those of many other species, and it may be that we want to focus on litter and the soil protection it gives as an important product from the area being planted. These questions indicate that fine tuning is needed, rather than blanket denunciation of all eucalyptus. I personally think we have too much exotic eucalyptus dotting the landscapes of the world, and would prefer to see native species, but that is subjective (or a bias, or a value judgment) and needs to be clearly stated before one is making very much sense in this controversy about eucalyptus.

(O'Loughlin and Will, 1981; Poore and Fries, 1985; Shiva and Bandyopadhyay, 1985; Davidson, 1986; Kardell *et al.*, 1986)

This situation, from a positive viewpoint, is brought out clearly by an analysis of the Nepal–Australia Forestry Project, regarded by many as one of the few successful development projects in that environmentally beleaguered country. But, in this instance, the project managers have realized that if the local people are to become their own managers (the essential criterion for success) considerable time is required, as well as a flexible programme that will provide for trial, error, and adjustment. This is politically difficult in a situation where politicians and decision makers are demanding immediate and demonstrable success, such as a completed report that states that so many hundred thousand seedlings have been planted, regardless of whether or not they have survived – this, again, is the so-called 'target mentality'.

The pressures leading to these policy constraints are brought out graphically in the Asian Development Bank report on the Nepal Agriculture Sector (ADB, 1982). The report indicates, for instance, that for the fifth Five Year Plan, of a target of 20,000 ha for reforestation, only 9,860 ha had actually been planted (due to lack of organization and manpower). During the past fifteen years only about 20,000 ha *in toto* had been planted while 2,000,000 ha of forest land are estimated to have been cleared (ADB, 1982 (2):70). Despite the possibly unreliable nature of these data, the broad pattern is likely to be correct. Equally important is the fact that 'planting' often means simply that. Frequently there is no maintenance, and damage by uncontrolled grazing of livestock is widespread. It follows that seedling losses are unacceptably high. It is to the remarkable credit of the Nepal–Australia Forestry Project (Sindhu Palchok and Kabhre Palanchok districts) that successful reforestation has been implemented – and without the need for costly fencing; with local support 'psychological fences' have been sufficient. Because the day-to-day decisions – choice of species for planting on their own land, choice of location, and so on – are taken by the villagers, livestock are effectively restrained. Nevertheless, the Nepal–Australia Forestry Project has not yet been able to solve the problem of producing preferred species from seedlings on degraded land (David A. Griffin, personal communication, May 1987).

A final comment on the issues of social forestry, perceived and actual needs of afforestation, and size and costs of projects is taken from a personal letter by Norman Uphoff addressed to Donald Messerschmidt:

> the problems of changing and improving resource management in the hills [Nepal's Middle Mountains] are *so* immense that size and cost are not relevant quarrels *if* the approach works, and if it gets the change process started on a sound footing. What is it worth to HMG [His Majesty's Government of Nepal] to get these improvements? If HMG spends less and ends up with nothing was this an economical investment? Hardly.
>
> (personal communication, Norman Uphoff to Donald Messerschmidt, 1983)

How Much Fuelwood is Consumed and How Much Biomass do the Forests Produce?

The next step in the Himalayan forest-cover discussion relates to the data that have been published to demonstrate quantitatively that Nepal *must* be incurring massive forest losses when per capita annual fuelwood consumption is balanced against the calculations of average forest biomass production. Thompson and Warburton (1985a) illustrate their 'uncertainty' proposition by use of such data sets. They argue that the available, published data for fuelwood consumption vary by a factor of 67. Even if the extreme figures are eliminated because they may be the result of typographical error, the reduced range still varies by a factor of 26. They point out that the very impossibility of determining whether the highest consumption figure is an error or not adds weight to their insistence that it is the uncertainty that overwhelms problem definition in the Himalaya (Thompson and Warburton, 1985a).

If the sets of data published by the United States Library of Congress (1979) are considered, fuelwood consumption is stipulated to be 546.3 kg per capita/yr against a sustainable forest biomass yield of 77.9 kg per capita/yr. Given that these figures appear to be imposingly accurate, we may be tempted to accept the simple subtraction: total biomass production per annum minus total fuelwood consumption equals loss of forest resources, leading to the conclusion that Nepal is incurring a country-wide loss of 6.5 million tons of wood per annum. In this case the pseudo-scientific approach and the spurious accuracy of the data lead to the conclusion that we have been predisposed to accept from the welter of conservationist and scientific literature. Thus it is the more easy to accept the World Bank prediction that no accessible forest will remain by about AD 2000.

But let us return to Thompson and Warburton's factor of 26 (or 67 – there is no significant difference in the context of this discussion). They maintain that if the highest calculations for fuelwood consumption are matched against the most conservative estimate of biomass production (most pessimistic case), it follows that the Himalaya 'will become as bald as a coot overnight'. If, on the other hand, the opposite sets of data are matched (lowest estimate of fuelwood consumption against highest estimates of biomass production), then the Himalaya, due to the inexorable forces of gravity and isostasy, 'will shortly sink beneath the greatest accumulation of biomass the world has ever seen.'

It is emphasized that this seemingly facetious discourse is very serious indeed. One of its important ramifications is brought out by the question: since the data used are the result of taking presumably accurate measurements of things that can be measured, unlike the data base for the much more acceptable calculations of the world's unproven natural gas and petroleum reserves, why do the results become such a travesty of scientific rigour? We believe that an attempt to answer this simple question is vital as an insight into the Himalayan Problem. The answer is based upon recent work by Mahat, Griffin, and Shepherd (1987a).

Donovan (1981), a source of much of the discussion developed by Thompson and Warburton (1985a), has emphasized the great variability in the estimates of fuelwood consumption, which she indicates as ranging from 0.1 to 2.57 (or even 6.67 – the source of Thompson and Warburton's magic factor of 67) m^3 per capita/yr. Many of the possible sources of error relate to conversion factors. What is the weight of a headload (*bhari*)? If volumetric measures are used, what is the density? Then there are the questions relating to the condition of the material – is it green, air-dry, or something in between? Do calculations based on 'dry weight' refer to 'air-dry' or 'oven-dry'? What is the energy value of the material? Accurate comparisons, and especially reliable extrapolations, demand figures based upon oven-dry weights of biomass and determination of the energy equivalent of the material which has been weighed. For any given study, few of these conversion factors are known with certainty and, consequently, different investigators have adopted widely differing values, often for reasons of doubtful validity.

There are additional problems, however, as indicated by Mahat *et al.* (1987b), that are often embedded in the methods of a survey from which prime data are obtained. Standard methods involve the interview technique with estimates being dependent upon the respondents' recollection. Not only is memory fallible, another source of error, but it is well known that responses of interviewees often reflect what they perceive to be the wishes of the interviewing investigator. Direct measurement, as an alternative approach, is time-consuming and requires an even greater willingness on the part of the local people to participate in the investigation. If an investigator chooses to take the approach of directly weighing headloads carried into the village the sample size is usually much smaller than that of the interview method.

Another problem, to which Mahat *et al.* (1987a) attach great importance, relates to the actual words used during the interview. Thus *balne* or *balnekura* (fuel, something to burn) are the most general words encompassing the notion of burnable biomass in Nepal. Fuelwood, however, is *daura*. And where deadwood and fallen branches are referred to, *suka daura* is used to distinguish it from green wood, *kacho daura* (that is, the trunks or major branches of living trees) which, after cutting or felling, are dried for use as fuel, especially during the summer monsoon season. A further complication is that, within Sindhu Palchok and Kabhre Palanchok districts at least (see Figure 3.1), the terms *sitapita* and *jhikra* are used to refer to any combustible plant material of small dimensions, such as woody residues from fodder and bedding, forest weeds (for example, *Eupatorium adenophorum*), and old fences and bamboo. Mahat *et al.* (1987a) explain that crop residues often form a significant component of fuel, yet there is no single word or phrase encompassing this group of materials. Each would need to be identified individually; for instance, maize cobs, maize or sugar stalks, and so on. Finally, this discussion, based upon experience in the districts of Sindhu Palchok and Kabhre Palanchok, does not take into account the large number of ethnic groups and languages that characterize Nepal.

Bajracharya (1983a), Fox (1983), Wiart (1983), and Mahat *et al.* (1986, 1987) have examined the implications of the nomenclature problem in more detail. They also show that wide variations occur in seasonal use of different types of fuel and between villages at different altitudes. Generally fuelwood (that is *kacho daura*) that is cut green from the forest and dried, is the primary fuel source during the wet season (summer monsoon), but the proportion of forest to non-forest combustible material will vary from season to season and from one altitudinal belt to another (Mahat *et al.*, 1987b). Thus it becomes clear that the phrasing of the question about fuel use by the interviewer will significantly affect the answer obtained. Only if *balne* is used will the answer be comprehensive, encompassing all fuels. If the emphasis is placed on *daura*, it is most likely that only a small proportion of the total combustible material consumed will be reported, yet *daura* is the closest translation to *fuelwood*, the English word most commonly employed in this context.

From the foregoing discussion, Donovan's (1981) efforts to estimate average fuelwood usage are seen to be fraught with difficulties greater than those of which even she appears to have been aware. In addition, actual interview methods and data conversion methods are not generally provided in published reports, so that the margins of error cannot be accurately assessed. The effective dramatization of the factor 67 by Thompson and Warburton (1985a), therefore, becomes completely understandable. Donovan concluded that a reasonable figure for average fuelwood consumption (assuming 5.5 persons per family and a weight of 600 kg for one cubic metre of fuelwood) is 1.4 m^3 per capita/yr. She also concluded that lower values than this probably characterized exceptionally fuel-deficient areas where demand actually exceeded consumption. Mahat *et al.* (1987b) question the validity of this assumption because of the uncertainty of differentiation between the various components of the total fuel materials in Donovan's original surveys and in her synthesis of them. There is also the problem of fuel substitution; for example, maize roots, corn cobs, and other agricultural 'residues' are used instead of wood as the latter becomes scarcer. This renders extrapolation well nigh impossible (Griffin, personal communication, April 1987). See Stone and Campbell (1986) for a discussion on the use and misuse of surveys in international development.

From the discussion so far we can conclude only that attempts to determine fuelwood consumption are so clouded with uncertainty that any figures used as the basis for extrapolation into total national annual consumption are meaningless, as Thompson and Warburton's (1985a) approach so effectively demonstrates. To stengthen this conclusion further it is emphasized that, if attempts to quantify fuelwood consumption produce preposterous results, estimates of biomass production are even less reliable. We see no alternative but to discount totally this line of reasoning in support of what appears as a preconceived conclusion, that Nepal is losing large areas of forest cover because fuelwood consumption exceeds forest biomass productivity.

Perceptions of the Forests

The discussion on the fuelwood consumption–loss of forest cover linkage should not be taken to support a counter argument to the effect that the mountain communities of the Himalaya have no problem in terms of access to forest resources – they do. The discussion is rather an attempt to demonstrate that, in yet another area, the bureaucratic and conservationist approaches to demonstrating the extent of a crisis facing the Himalaya are frequently not based upon facts. The claim that the Himalaya are in a state of crisis is thus in danger of being countered as spurious. The discussion also subsumes the critical question of how the forests are perceived by the different sectors of society that are concerned about, or involved in, their use. It is apparent that government agencies, development and aid organizations, conservationists, and western-trained elites regard the forests as a *renewable* resource. Hence their efforts are directed toward preserving the resource by trying to ensure that a balance is achieved between use and biomass reproduction so that the use becomes sustainable. However, many of the subsistence farming communities perceive them as *convertible* resource, that is, as a source for new agricultural land. Thompson and Warburton (1985a) pursue this line of thought further by claiming that such contrasting perceptions of the value of the forests will remain a major obstacle in the path of the possibility of success of the perceived solutions (reforestation) that have tended to have been conceived and imposed from the top down.

The Chipko Movement slogan:

> What do the forests bear?
> Soil, Water, and Pure Air.
> (Sunderlal Bahuguna, personal communication, April 1986)

would indicate, however, that a large component of subsistence forest dwellers do perceive the forests as a renewable and renewing resource, hence the 'hug-the-trees' non-violent protest throughout the Garhwal and Kumaun Himalaya. And this is opposed to the perceptions of some agencies in India whose view of the forests as a source of sustainable monetary revenue largely set off the Chipko Movement (Shiva and Bandyopadhyay, 1986).

From a broad historical view, nevertheless, forests have been perceived and used as a convertible resource. Tucker (1986, 1987) indicates that the present assault on the forests of the Himalaya and the Terai is merely the last stage in the historic rolling-back of the vast former forest cover of the subcontinent. Regardless of the difficulties of determining causes and effects and the differing points of view about rates and localities of forest cover, Moench and Bandyopadhyay (1986) have effectively demonstrated that, even where biomass production comfortably exceeds fuel and fodder consumption, progressive forest recession can occur. This conclusion derives from a detailed study of village subsistence activities in a heavily forested region north of Mussoorie in Uttar Pradesh State. The pattern of fodder and fuelwood collection, concentrating on the proximate forest perimeter, is producing what they refer to as a 'nibble effect' or 'peripheral degradation.' If the patterns of forest usage were to change in the direction of spreading the impacts of human usage throughout a wider area, then, lack of precise figures notwithstanding, annual forest biomass production would easily encompass local subsistence demands, at least in this study area.

Fuelwood Collection and the Villagers' Use of Animal Dung

Another small link in the population growth–fuelwood demand–deforestation chain of the Theory of Himalayan Environmental Degradation is the damage stipulated as a result of the assumed increased use of dung for fuel. Obviously, if this claim is correct, then the agricultural terraces in many places are being deprived of their primary source of nutrient replacement in the form of animal manure. It follows from this that crop yields would fall and the weakened soil structure would lead to increased susceptibility to landsliding, which would set up another round of deforestation to make room for more agricultural terraces on which to grow more food. We have tried to show that fuelwood demands, at least in some areas, are not a primary cause of deforestation. Thus it would appear that the claim that dung is being increasingly used for fuel must also be challenged. It would also appear that in many areas of the Himalaya, and especially in Nepal, the use of dung for fuel is more closely related to the ethnic origin of the community rather than shortage of fuelwood (Panday, 1984, personal communication). In the Kathmandu Valley, for instance (worthy of comment despite its uniquely urbanized character), dung is used for fuel during the winters when it would otherwise go to waste. Its primary use during the summer wet season is for fertilizer.

Fox (1983) and Bajracharya (1983a), as well as Messerschmidt (personal communication, March 1987) have also noted that before any switch to using dung as fuel occurs, hill villagers will use increasing amounts of waste stubble, corn cobs and stalks, and old fencing, and that the use of dung *per se* is a last resort not yet widely taken up in the Middle Mountains. Messerschmidt stipulates that, in his extensive travels over a twenty-year period, he has never seen dung fires, except in the Terai, and in the highest alpine pastures, where yak dung is the traditional fuel and often the only fuel source available.

Thus the assumed linkage between the burden of carrying fuelwood over increasingly greater distances and the progressively greater dependency on dung as fuel (Eckholm, 1975, 1976) instead of fertilizer, can be challenged on two counts.

What Are the Physical Effects of Deforestation?

Let us proceed from the labyrinth of uncertainty about the causes and effects of the pressures on the mountain forests to a discussion of the hypothetical impacts of deforestation, assuming that the World Bank assessment is correct and the mountains of Nepal will indeed be laid bare by AD 2000. The view that deforestation automatically and most certainly will produce devastating soil erosion, overland flow of water, rill and gully development, rainy-season flooding and dry-season water shortage is so widespread that it is usually taken as one of the fundamental truisms within the concept of 'one small earth'. Where deforestation occurs on mountain slopes, the environmental degradation is assumed to be all the greater than deforestation on the plains, with local damage being proportionate to angle of slope. Linked to this is the assumption that the greater the slope angle the greater will be the impacts on the plains downstream – more destructive floods, siltation, extension (in our case) of the Brahmaputra–Ganges delta, and so on. Hamilton (1987), under the challenging heading 'Myth and Misunderstanding' develops a reasoned attack on these important conventional assumptions.

Hamilton (1987) first argues that the very term *deforestation* is used so ambiguously that it is virtually meaningless as a description of land-use change: it should be replaced with more precise terms that would be less charged with emotion and better suited to specific areas. It has been used to refer to any or all of the following activities with respect to existing forest: fuelwood cutting; commercial logging; shifting cultivation; forest clearing for continuous annual cropping, for grazing, for food, beverage, or industrial tree-crops, or forest plantations. It has also been used to describe the process of burning, for various reasons, with a degraded forest as the end product. Other activities could be added. For our present purpose – discussion of the hypothetical effects of deforestation – since these various activities produce different hydrological and erosional results, it is emphasized that lack of differentiation only confuses attempts to examine the effects.

Fuelwood cutting, where the products are carried out of the forest manually, will have very different impacts to those resulting, for instance, from commercial logging with heavy equipment and where skidding trails and logging roads may occupy as much as 16–30 percent of the logged area. Similarly, the conversion of forest to sustained annual cropping will have very different impacts depending on the type of crops, and the development (or lack of development) of properly constructed and maintained agricultural terraces, and the application of other conservation practices. 'The aesthetically pleasing, hydrologically and erosionally benign, productive paddy terraces that characterize many parts of the lower slopes of the Himalaya are *deforested.* Yet generally deforestation is perceived to be always destructive, if not evil' (Hamilton, 1987:257).

From this we must conclude that it is not the cutting of trees *per se* that causes large increases in on-site erosion, but the methods of cutting and of transporting the products out of the area, and finally the subsequent form of use of the cut-over land (Hamilton, 1983). In most of the humid Himalayan area, harvesting activity should not be described as deforestation unless it is labelled as *temporary* deforestation. If the affected area is left, and not burned, grazed, or cultivated, a rapid establishment of new reproduction will occur and a spurt in understorey growth will ensure that little bare ground is exposed to high intensity rainfalls. Realization of the confusion caused by semantics led Hamilton (1983) to develop the following headings for chapters of a book on the hydrologic and pedologic aspects of tropical forested watersheds:

USES OR ALTERNATIVES	1. Harvesting minor forest products
	2. Shifting agriculture
	3. Harvesting fuelwood and lopping fodder
	4. Harvesting commercial wood
	5. Grazing on forest land
	6. Burning forest land
CONVERSIONS OF FORESTS	7. Conversion to forest tree plantations
	8. Conversion to grassland or savanna for grazing
	9. Conversion to food or extractive food crops
	10. Conversion to annual cropping
	11. Conversion to agroforestry
RESTORATION OF FORESTS	12. Reforestation or afforestation

Hamilton argues that even with this degree of specificity, the biophysical consequences of these activities are complex, unknown, or ambiguous. He believes that the emotive term *deforestation* only compounds the problem and should be eliminated from both technical and popular thinking unless used with qualifications to indicate the actual pattern of forest land-use change.

Having quickly reviewed the semantics let us now go on to address some of Hamilton's stipulated myths and misunderstandings. One example of a widespread misunderstanding is the oft-repeated assertion that 'tree roots are a sponge' which soak up rainwater during wet seasons and release it slowly during dry seasons – that is, that the forests act as hydrological regulators (Spears, 1982; Myers, 1983). The implications are that cutting down forests causes floods during the summer monsoon season or during major storms, and results in lower water levels, or in rivers totally drying up during the dry seasons (Eckholm, 1976; Sharp and Sharp, 1982; and many others). Watershed research, however, suggests that tree roots behave more like a *pump* than a *sponge*. Forest cutting on 94 small catchments world wide, compared with uncut control catchments, resulted in more total water yield from the cut-over areas, with the increase being proportional to the intensity of the cutting (Bosch and Hewlett, 1982). This occurs because tall forests evaporate and transpire more water than other types of vegetation. Furthermore, Hamilton's (1983) review of catchment research showed that in most cases following cutting, increased flows occurred throughout the year, including the dry season. Gilmour (1977), for instance, demonstrated in the humid Australian tropics that a stream that previously had ceased to flow during many dry periods flowed perennially after logging. Hamilton (1987) points out that a notable exception to this effect is the case of cloud forests that intercept precipitation from wind-driven clouds.

The line of reasoning introduced above can be applied to a variety of other myths concerning deforestation. These include: the alleged role of the upper forest canopy in protecting the soil from raindrop impact; lowering of the groundwater table; water yield; and timing and distribution of stream flow. To provide one more example, we will examine the contention that removal of a high-forest canopy by exposing the soil surface to direct impact of raindrops will result in increased soil erosion. It is necessary to take into account the fact that, with an intact high-forest canopy rainwater will accumulate on the leaves and fall to earth as much bigger drops than raindrops. Also, given a height in excess of 20 m the terminal velocity of the larger water drops will not be significantly different from that of raindrops falling on the ground unimpeded. Furthermore, it is the litter and the low vegetation that protects the soil from particle detachment by raindrop impact. Fire or grazing, which remove low vegetation, and litter gathering that exposes bare soil should be reduced, rather than worrying about reduction in the tree canopy if the principal concern is sheet erosion. With a surge in productivity of a shrub and herb layer following forest clearing with more solar radiation reaching ground level, the soil surface may be actually better protected against raindrop impact. Once more, therefore, we are prompted to caution against the dangers of generalization. Hamilton (1987) concludes with the following summary:

1. Cutting of forests, and also their replacement with a different land use (unless the soil surface is sealed), have usually resulted in the free water-table moving closer to the surface, because less water is lost to evapotranspiration.
2. Most forest-cutting catchment research has shown greater streamflow *throughout the year*, and many studies have shown proportionately greater increases in dry-season flow than in wet-season flow.
3. If cutting of tall forests is followed by conversion to other conservationist land uses then gains in streamflow yield persist.
4. Tree cutting alone does not lead to increases in surface erosion – what is important is the mode of forest harvesting (dragging, skid trails, roads, and especially large-scale mechanization).
5. Much of the sediment in a given river basin is in various temporary storages; thus claims of immediate sediment reduction on a large scale and with far-distant effects following any reforestation project should be critically re-examined.

> Some people have blamed 'deforestation' for the catastrophic flooding in June 1985, that killed 237 persons in six Indian states (102 in Kerala alone and 400,000 homes destroyed or damaged) (Reuters, 1985). The real cause was too much rain occurring in too short a time. Some areas received 414 mm in 24 hours, in the monsoon season when soils were already saturated. Even if the entire basin had been forested there would have been disastrous floods. Forests are beautifully functioning ecosystems, capable of producing many products and services on a sustained basis. They must in places be totally protected, and in other places wisely managed. But they will not *prevent* floods or sedimentation in the lower reaches of major rivers, nor significantly *reduce* flooding in major storm events.
>
> (Hamilton, 1987:262)

The deforestation → changes in hydrologic conditions → soil erosion → sediment-transfer linkages make up yet another instance where assumptions, emotions, and widespread generalizations have been pulled into the Theory of Himalayan Environmental Degradation regardless of the available facts. The difficulty of countering these assumptions, or attempting to expose the myths surrounding them, is exacerbated by the fact that reliable data are available only from a very few sites, most of them located in other parts of the world. A compounding problem is that deforestation, in its many forms, and forest degradation are occurring, and these processes are inextricably linked with the problem of sustaining a viable subsistence agriculture for the great majority of the population that inhabits the lower altitudinal belts of the Himalayan region. Nevertheless, we will conclude this section with the presentation of the results of a detailed and narrowly site-specific study that we believe throw valuable light on several of the critical points discussed above.

Gilmour (1986) and Gilmour *et al.* (1987), within the overall scope of the Nepal–Australia Forestry Project, set out to examine the rather important relationships between rainfall amounts and short-period intensities and soil erosion in the context of type of surface vegetation cover and land use. Gilmour *et al.* (1987) instrumented five study plots close to the Sindhu Palchok district headquarters of Chautara, about 40 km northeast of Kathmandu. The study area lies at about 1,660 m above sea level and experiences a typical monsoon climate: about 85 percent of the total annual precipitation occurs between June and October.

The five study sites were chosen to represent various stages in the process of reforestation, ranging from very heavily grazed and trampled grassed land (typical of much of the degraded grassland that is available for reforestation) to a relatively undisturbed religious[1] forest which is as close to a natural forest site as could be found. Intermediate sites included a formerly heavily grazed and trampled grassland that had been planted with *Pinus patula* five years earlier and from which grazing animals are excluded, a site similar to the previous one but which had been planted with *Pinus roxburghii* twelve years earlier, and a site, also planted with *P. roxburghii* twelve years previously but which carried a cover of low shrubs and herbs about 40 cm high plus a few scattered remnants of the original forest. Here too grazing animals are excluded. The field programme consisted of determining values of near-saturated hydraulic conductivity for various layers of the soil profile of each of the five sites at increments of 0–0.1, 0.1–0.2, 0.2–0.5, and 0.5–1.0 m depth from the soil surface; in other words, an experiment was set up to determine whether or not soils under different surface conditions, including different degrees of compaction, could absorb the heaviest rainstorms to be expected in their vicinity. It was considered that such measurements would facilitate characterization of the nature of water movement through the soil profile.

At each site twenty-five measurements were taken for each of the four soil profile levels. For all sites the hydraulic conductivity values showed little variation within the deepest layer (0.5–1.0 m). Variability increased with closer proximity to the surface, with maximum differences occurring in the 0–0.1 m layer which showed a range of 39–524 mm/hour. The site with the lowest value represented the totally deforested area that had been heavily grazed and trampled for many decades. Site 2, reforested five years earlier showed a slight improvement, while Site 3, reforested twelve years earlier showed further improvement with a value of 183 mm/hour. As near as can be determined from examination of old photographs and discussions with the local villagers, these three sites had been used in a similar manner for several decades prior to the reforestation of Sites 2 and 3. Gilmour *et al.* (1987) concluded that the measured differences are due to changes in soil surface conditions brought about by the change in land use (reforestation). Of particular importance was the reduction in compaction by the exclusion of grazing animals and an improvement in soil aeration through greater activity of soil micro-flora and fauna associated with the higher levels of soil organic matter.

From these results, however, it still could not be determined how much time is required to return heavily grazed and trampled sites to near natural conditions, but it is assumed that it would be many decades. However, the data obtained were adequate for comparison with the actual characteristics of rainfall events. Thus the authors postulate that if, for instance, the hydraulic conductivity of the 0–0.1 m soil layer at Site 1 is only 39 mm/hour yet the rainfall intensity never exceeds this value, then all the incident rain would infiltrate the soil profile and no surface runoff would occur. This would apply provided the soil was not already saturated before the onset of the specific rainstorm. The fact that the surface layer value of Site 4 is thirteen times higher than that of Site 1 would have no practical significance. Thus the next step was to consider rainfall characteristics.

The annual precipitation for the experiment site is 2,008 mm, with 85 percent occurring between June and October inclusive. It is necessary, therefore, to isolate the summer high-intensity events. Since rainfall short-term-intensity data are not available for the actual field area, data from Kathmandu were used. This data set was considered by the authors to give an adequately close approximation to actual field conditions. This is based upon the many years of project personnel experience in the two areas.

Examination of the Kathmandu climatological record (1971–79) allowed the compilation of 5-minute rainfall intensities expressed as equivalent hourly intensities in mm/hour to enable a direct comparison with the measured soil hydraulic conductivity values (also in mm/hour). The highest 5-minute intensity value during the period 1971–79 was 120 mm/hour and was recorded on four occasions. Maximum 24-hour rainfall totals during the period 1972–82 varied from 59 mm to 117 mm. A summary of all the data on 5-minute rainfall intensity peaks shows that incident rainfall intensities exceed the soil absorption capacities for Site 1 on an average of 6.7 days per year and for Site 2 on an average of 3.7 days per year. The remaining three sites, according to these data, have hydraulic conductivities that exceed rainfall intensity peaks on all occasions. Thus, in the worst possible situation (that is, heavily grazed grassland as represented by Site 1), overland flow is likely to occur on 6.7 occasions for an average summer monsoon period.

Gilmour *et al.* (1987) next use Caine and Mool's (1982) data to show the effects of a 1-hour duration storm of 40 mm/hour. A storm of this magnitude would have a recurrence interval of only once in twenty-five years and is the type of event likely to cause a degree of local flooding. At the most degraded site (Site 1) such a severe storm would generate only 1 mm/hour of surface runoff. The inference is that deforestation under these conditions would have little effect on the generation of any major flooding. Conversely, the improvement in soil surface conditions caused by reforestation would result in the absorption of an additional 1 mm/hour of rainfall – hardly the kind of result likely to have any noticeable beneficial effect on local downstream flooding and certainly not on distant downstream flooding, such as on the Ganges Plain. The authors conclude that their findings from a select research site in the Nepal Middle Mountains support the results of the majority of researchers (see reviews by Boughton, 1970; Hewlett, 1982; Hamilton, 1983; Gilmour, 1986) that deforestation and reforestation would not, by themselves, influence major flooding. 'Major floods are primarily influenced by precipitation factors in association with the geomorphology of the catchment' (Gilmour *et al.*, 1987:247).

The other factor to be considered is the extent of poor-quality catchment cover and the degree to which this can be improved. Data collected by Nield (1985) indicate that about 7 percent of Nepal's Middle Mountains had a cover of degraded grassland. Even if all of this could be reforested (a mammoth task in itself) the likely effect on flood reduction, partly on account of the broken and scattered nature of the grassland, would be insignificant.

The remaining point to be considered is the degree to which a well-vegetated and protected soil surface would decrease the incidence of overland flow and thereby reduce local soil erosion. Planting trees on mountain sides can decrease the number of shallow landslides, or conversely, tree removal can increase the likelihood of shallow landslides (Hamilton, 1987). Surface soil erosion is widespread both in the local field area of Gilmour *et al.* and more widely throughout Sindhu Palchok and Kabhre Palanchok districts, especially on the over-grazed and trampled grassland areas characterized by Site 1 of the study. Reduction in this process of local environmental degradation would undoubtedly be achieved as reforestation is extended. Moreover, establishment of trees can impart a greater margin of safety against shallow landslides (Hamilton, 1987). This would be of considerable importance in terms of improved maintenance and site productivity and is probably one of the major benefits to accrue from regional and local reforestation programmes. A more general consideration of reforestation, in terms of species selection and methodologies is beyond the scope of the present discussion.

A number of general conclusions can be drawn from this discussion on the hypothetical and perceived impacts of deforestation and reforestation in the context of the wider Theory of Himalayan Environmental Degradation. First, as Gilmour *et al.* (1987) infer, rainfall total amounts are lower and the incidence of storm periods with high-intensity downpours is much less frequent than has been widely assumed for a monsoon climate. While the data set upon which this conclusion is based, involving extrapolation from Kathmandu to Chautara, is not completely satisfactory, it does contrast with the earlier assumptions that deforestation *must* be disastrous in terms of increased soil erosion due to high-intensity rainfall on an unprotected or compacted soil surface. This is perhaps another assumption that has contributed to much misunderstanding, although certainly much more data on rainfall amounts and short-period intensities and a much longer period of observation are needed. The short period of observation and the single study site should be carefully borne in mind; extrapolation to other areas is not yet warranted. We should also exercise extreme caution in this regard because, as indicated in Chapter 2, as more precipitation-recording stations yield new data, examples of extreme annual totals (that is, in excess of 5,000 mm) increase in number. Nevertheless, the combined results and reasoning of Gilmour *et al.* and Hamilton must be considered as cautionary tales leading to the recommendation that even on a local scale the widely cherished myths concerning the evils of deforestation, and conversely, the beneficial effects of reforestation are likely to lead government and development agencies into policy decisions with unattainable objectives. There are enough real benefits from maintaining natural forest, and from reforestation of degraded lands, without having to rely upon questionable values that simply sound attractive – specifically, the dubious promise of reducing the annual floods and siltation far downstream on the floodplains of the Ganges and Brahmaputra.

NOTE

[1]Many small forest areas throughout the Himalaya are protected for religious observances or because they harbour shrines or temples. They may be occasionally disturbed by domestic grazing and collection of litter, but are as close to primeval forest as can be found over wide areas; a wide gap, nevertheless, remains, so that they are best classified as 'relatively' undisturbed.

5 MOUNTAIN SLOPE INSTABILITY: NATURAL PROCESSES OR HUMAN INTERVENTION?

INTRODUCTION

The previous two chapters dealt with questions and assumptions about processes of change in the area of forest cover, the causes and effects of deforestation, and the relationship between soil erosion and the transformation of forested hillslopes to other forms of land use. The conventional assumptions (myths) concerning the linkages between these dynamic mountain processes were heavily challenged and the argument was made that a much longer-term perspective and much more reliable data are needed.

Another group of linkages that hold together several components of the Theory of Himalayan Environmental Degradation relate deforestation on steep slopes and construction of agricultural terraces to a rapid acceleration in gullying and landslide incidence and increased soil erosion. It is claimed that these in turn produce serious deleterious downstream impacts. We ourselves (Ives, 1981; Ives and Messerli, 1981) held this assumption on the initiation of the Nepal Mountain Hazards Mapping Project in 1978; we have subsequently reversed our position.

As somewhat representative of 'expatriate experts' making short visits, we were able to observe the Middle Mountains of Nepal from brief road traverses out from Kathmandu in 1978 and 1979. Like many other visitors, we timed our presence to coincide with good weather – March/April and October – and were duly impressed with the large number of landslide scars (strictly – debris flows) and gullies that obviously had engulfed significant amounts of agricultural terrace land during the preceding summer monsoon periods. Within this context the 'Kakani Phase' of the Nepal Mountain Hazards Mapping Project was set in motion. This included studies of stream channel morphology and landslide and gully dynamics (Caine and Mool, 1981, 1982), investigations concerning hazard perception of indigenous subsistence farmers and their coping strategies (Johnson *et al.*, 1982; Gurung, 1988), and systematic mapping of land use, geomorphic features, and mountain hazards on a scale of 1:10,000 (Kienholz *et al.*, 1983, 1984). Since the project necessitated repeated visits of considerable duration to the field area and fieldwork throughout the agricultural cycle over a five-year period, we were afforded the hitherto uncommon perspective of time. We were also

Figure 5.1 This magnificent example of terrace architecture has been emplaced on the deposition area of a former landslide. Note the precise, narrow bunds of the *khet* terraces and their extremely narrow treads. Rice has just been transplanted. Nepal-MAB/United Nations University Kakani field area, Nepal.

able to develop communication with the subsistence farmers. This gave us a greater understanding of the landscape changes that have been occurring over several generations and an appreciation of local attitudes to dynamic slope processes and local responses to them.

It became apparent that many landslide scars are eventually re-terraced and stabilized and that irrigation systems are repaired: this is the most important stabilizing process (Figure 5.1). In certain instances the local people perceive a landslide to be a beneficial occurrence because the more easily worked earth of the landslide scar actually facilitates terrace construction. In other instances landslides are deliberately triggered by water diversions in order to

Figure 5.2 Photograph from the Trisuli Road below Kakani in the Nepal Middle Mountains. The landslide occurred during the preceding monsoon season (1978) and was regarded as exceptionally unstable and likely to enlarge.

facilitate new terrace construction (Kienholz *et al.*, 1984; Sumitra M. Gurung, personal communiction, May 1985). We began to appreciate the complicated balance between slope stability, the particular stage of the agricultural cycle, rainfall incidence, availability of emergency labour, and type of terrace that is threatened. Thus our early predictions on rates of progressive land loss and estimated population growth (Caine and Mool, 1981, 1982; Ives and Messerli, 1981) had to be adjusted. We also observed that the indigenous farmers had evolved an intricate set of coping strategies that, in addition to subsequent re-terracing of collapsed slopes, included changes in land use to match changes in slope stability; Johnson *et al.* (1982) introduced the concept of agricultural de-intensification, as an adjustment to the threat of slope instability. Moreover, in strictly physical terms, it has been demonstrated that

Figure 5.3 The same view as shown in Figure 5.2 but taken in October 1987. It is almost impossible to determine the location of the 1978 landslide and it is unlikely that, without access to Figure 5.2, a visiting "expert" would ever understand the land-use history of this site. A very small central tract has not yet been re-terraced but the land is basically stable and productive. Several houses visible in Figures 5.2 have been moved or pulled down and the forms of many of the undamaged terraces have changed markedly. This is a very dynamic landscape.

many of the bedrock types in the Middle Mountains, and specifically in the Kakani test area, undergo very rapid weathering and a high incidence of soil formation; thus they can withstand a relatively high rate of soil loss (Peters and Mool, 1983).

Over the slightly longer period (1978–87) frequent visits to Kathmandu facilitated repeat photography of original landslide scars and slope segments at different times of the year. Figures 5.2 and 5.3 illustrate only a single example. Nevertheless, it is thought-provoking to see that an inherently unstable landslide scar photographed in 1978 is almost invisible in 1987.

Furthermore, the new terraces that have been cut into the original landslide scar were supporting vigorous crops of maize and rice in August 1986 (Ives, 1987). The kind of data contained in Figures 5.2 and 5.3 are not adequate for regional extrapolation. They are introduced here to explain how experts, who depend on short-term visits, can be led to believe that the Middle Mountains are on the point of collapsing into the Ganges River when their observations are confined to a single tourist season (which is frequently the case) and lack a longer-term perspective and a close communication with the local people, particularly in the local language.

The experiences illustrated above are not considered as proof that there are *no* problems of landsliding, gullying, and soil erosion in the Nepal Middle Mountains; most emphatically, there are. Rather, our intent is to argue that the worst scenarios that have been used to permeate the conservationist, development/aid, and scientific literature may be exaggerations and, possibly, gross exaggerations that are emotionally or intuitively based. As human population numbers have increased over the past hundred years or so, leading to conversion of forested land to agricultural terraces on steeper and more marginal slopes, presumably more energy per unit of land is required to maintain a balance between stability and instability. The indigenous population may be losing ground, but not nearly so rapidly as has been assumed (Eckholm, 1976). Furthermore, it would be inappropriate to create the impression that subsistence farmers enjoy the landslide activity. Houses, human lives, and livestock are lost to landslides, and the terror of sleepless nights in small houses on steep slopes during periods of torrential rain is not to be dismissed lightly. Our aim here is to establish a better sense of proportion. It is proposed, for instance, that some of the most densely populated and extensively terraced land in the Middle Mountains probably experiences some of the lowest rates of soil erosion and land loss. A very real danger of soil erosion and slope collapse would arise if such areas were abandoned, a situation well known in the European Alps. Moreover, poor location, design, and maintenance of roads in this type of terrain has led to landslide regimes that are an intermittent or continual problem and source of stream sediment. If one looks closely at most photographs that show landslides, purportedly the result of deforestation and poor farming practices, one can almost invariably see that they have been initiated by a road or heavily used trail (Hamilton, personal communication, April 1987).

These more recent observations prompted a modified approach to geomorphic process studies during the 'Khumbu Phase' of the Mountain Hazards Mapping Project. Thus, in association with the production of hazards maps at a scale of 1:50,000 (Zimmermann *et al.*, 1986), Byers, Thorn, and Ives (1985), and Byers (1986, 1987c) planned for detailed soil erosion process studies at more than thirty plots through a range of altitude of over 1,000 m during the entire 1984 summer monsoon. The field experience in the Kakani area also prompted our close questioning of all the linkages relating deforestation and increased agricultural terracing to soil erosion, gullying, and landslide activity.

THE NATURE OF MOUNTAIN GEOMORPHOLOGY: WHAT IS KNOWN ABOUT SLOPE PROCESS IN DENSELY POPULATED MOUNTAIN TERRAIN?

Before a reasonable perspective on erosion and sediment transfer in the Himalaya can be developed, we should consider the broader issue of mountain geomorphology itself and its evolution over the past thirty or forty years. This will also facilitate the introduction and definition of a few key concepts and terms, the misuse of which has exacerbated the confused interpretation of processes affecting the Himalaya (and other mountain areas) today.

Geomorphology evolved steadily as a semi-independent discipline on the boundary between geology and geography from about 1850 to 1950. The broad concepts of landscape evolution were developed during this period. These included the concepts of base level, the geographical (or Davisian) cycle, and interruptions in the cycle, either by changes in climate or changes in base level brought about by tectonic adjustments, but progress was constrained by an overall reliance on a descriptive approach. Precise measurements of actual geomorphic processes remained a rarity. There was also an intellectual barrier in that the earth's geological history was perceived as being characterized by relatively short periods of instability, the periods of mountain building, separated by relatively long periods of quiescence during which the mountains were eroded until they were almost reduced to plains (i.e., peneplains = almost plains). This basic concept of descriptive geomorphology argued that, following a major mountain-building episode (orogenesis) sufficient time usually elapsed for streams and rivers, and the many other agents of erosion, to wear down the higher elevations until they were graded to a relatively stable sea level (base level). Thus most of the steeper slopes were eliminated and a large proportion of the terrain was reduced almost to a plain. The next cycle of orogenesis then uplifted the landmass and reactivated (or rejuvenated) the processes of erosion as the increased vertical difference between the height-of-land and sea level made available so much more additional energy. The peneplains were recognized in the accordance of summit levels in many of the most recent mountain ranges stemming from the last or Tertiary (Alpine) episode of mountain building, of which the Himalaya are amongst the most recent.

Dissatisfaction with the descriptive approach led to the 'quantitative revolution' heralded by the seminal study of Anders Rapp in a glaciated valley (Karkevagge) in northern Sweden (Rapp, 1960). The objective of this study was to rank, both relatively and absolutely, the slope-modifying processes responsible for landscape evolution in this maritime arctic area. Rapp was able to rank the dominant geomorphic processes as follows:

1. transport of salts in running water
2. earthslides and mudflows
3. dirty avalanches

4. rockfalls
5. solifluction
6. talus creep.

These are all processes characteristic of high mountain regions, regardless of latitude, including the Himalaya. Of course, any detailed consideration would emphasize an array of differences between the mountains of northern Sweden and the Himalaya, the major ones being climatic regime, vegetation cover, altitude, and rock type and structure. Human activities are also vastly more important in the Himalaya. It must also be borne in mind that there are considerable differences from one part of the Himalaya to another. A major question concerns the extent to which these processes and their ranking vary from one mountain area to another. Moreover, Rapp, for his analysis of the Karkevagge area, excluded consideration of glacial erosion (probably the most effective erosion process where glacier systems occur) and it was not possible to rank slope wash because of practical difficulties in collecting adequate data. Frost-bursting, the prying-off of rock particles from the valley side cliffs by freezing and thawing of water in joint planes and crevices, was not included, partly because it is not considered as a transporting process and partly because it is difficult to measure directly. The annual production of rock-waste by frost-bursting on rock walls, however, was calculated at 100–400 tons/km^2 of wall surface, which would probably give it top ranking over the processes listed above. Rapp also concluded that, without many more data from other valleys, with other types of slope, these results could not be extrapolated to the immediate local region; the ranking of the six processes could be significantly different in adjacent valleys with somewhat different slope combinations even though they were subject to the same type of climate.

Rapp's work set the stage for process studies in many parts of the world. These ranged from Alaska to Tasmania, from New Zealand to the Canadian and Colorado Rockies, from the Alps and the Tatra to the Khumbu Himal. But already in Rapp's conclusions the problems were recognized that still beset any full understanding of what is often referred to as climatic geomorphology. To complete his treatise, Rapp and assistants had collected data during eight years, yet the results, as indicated above, cannot be extrapolated with confidence to Karkevagge's neighbouring valleys, let alone to the Dudh Kosi of Nepal; nor can they be extrapolated backward in time, nor used as a basis for prediction. In addition, despite Rapp's top ranking of transport by salts in running water, most subsequent research has concentrated on mechanical weathering and transport of coarse debris, with a very heavy emphasis on talus slopes and solifluction, the two lowest-ranked processes. Moreover, heavy rains in October 1959, Rapp's final year of data collection, set in motion processes, principally mudflows and debris flows, representing by far the largest geomorphic event to occur during the entire period of fieldwork. Thus the concept of the large event with a long recurrence interval was introduced very early in the 'quantitative revolution'; it remains a major dilemma for any assessment of the development of mountain slopes in the Himalaya and raises the problem of representativeness in *time* as well as in *space*.

During the period 5–7 October 1959 the Riksgransen weather station on the Narvik–Kiruna railway recorded 107 mm of rainfall in 24 hours and 175 mm in 72 hours. Total precipitation for the four months, July to October 1959, was 794 mm compared with the 1901–30 annual average of 308 mm. The October rainstorm was the heaviest since the Riksgransen station was established in 1904; the recurrence interval of such a downpour may exceed a hundred years. (Rainfall intensities such as this are not uncommon in the Himalaya.)

The effects of the hundred-year climatic event, setting in motion catastrophic erosive activity, also may be dwarfed by even more spectacular occurrences. For example, Heuberger *et al.* (1984) have documented a giant landslide in the Langtang Himal, Nepal, which occurred about 25,000 years ago. This landslide, which generated fused rock (frictionite) along its sliding surface, displaced approximately 10 km^3 of debris through a vertical distance of up to 2,000 m.

One of the most spectacular geomorphic events to have occurred in historic time in the Nepal Himalaya was the outburst of a moraine-dammed lake behind the mountain Machapuchare between 600 and 800 years ago. This caused a flood surge down the Seti Khola which deposited 5.5 km^3 of debris in the Pokhara Valley, damming Lake Phewa (Fort and Freytet, 1982). On a somewhat smaller but still catastrophic scale, in October 1968, rainfall, varying in amount between 600 and 1,200 mm, fell on the Darjeeling area, West Bengal Himalaya, during a three-day period at the end of the summer monsoon when the ground was already saturated. It is estimated that some 20,000 debris flows were released; the 50-km road between Siliguri on the plains and Darjeeling at 2,200 m was cut in 92 places and approximately 20,000 were killed, injured, or displaced (Ives, 1970). While there is some disagreement concerning the estimate of the recurrence interval of this event, Starkel (1972a and b) concluded that occurrences of this magnitude are the primary slope-forming processes and calculated an average denudation rate for the Darjeeling area of the order of 0.5–5 mm/yr and up to 20 mm/yr for the individual years when such catastrophes occur. These are amongst the highest denudation rates ever proposed and the implications are disussed in more detail below.

The regularity, or irregularity, of occurrence of the extraordinarily large events, such as the Seti Khola torrent, the Langtang landslide, or even the Riksgransen–Karkevagge rainstorm, poses a serious problem for any attempt to rank geomorphic processes and to deduce long-term denudation rates. A primary difficulty is the problem of estimating the magnitude of very long recurrence intervals when there is no historical record to give adequate control; another is the actual identification of the enormous deposits that result from such large-scale events. From all of these considerations it will be appreciated that the task of determining the relative importance of catastrophic events to total landscape evolution in space and time faces severe difficulty.

Within the same context as variations in large-scale geomorphic activity through long periods of time, rates of slow mass-wasting that operate continuously in millimetres per year also appear to vary through time. For instance, Benedict (1970) demonstrated for the Colorado Front Range that rates of mass-wasting are lower now than at any other time during the Holocene (the last 10,000 years or so) and, within this period, have varied by an order of magnitude. Current rates of movement range from 4 to 43 mm/yr in the uppermost 50 cm of surface weathering mantle, and up to an order of magnitude higher during the close of glacial episodes when higher soil moistures can be assumed (Benedict, 1970).

The foregoing discussion amply illustrates the difficulty of ranking slow mass-wasting processes, such as solifluction and soil and frost creep, with medium-scale events, such as the 1959 Riksgransen–Karkevagge and the 1968 Darjeeling rainstorms, and with the giant events such as the Langtang landslide and the Seti Khola torrent. These are problems that face geomorphologists in areas such as the Colorado Front Range, the Alps, and Karkevagge that are blessed with ease of access, a decade or more of accumulated data, excellent topographic maps and air photograph coverage, and even permanent mountain research laboratories (Ives, 1980). How much more difficult, therefore, is such ranking of geomorphic processes in the relatively inaccessible Himalaya?

Much of the present-day knowledge concerning the effectiveness of different geomorphic processes, especially mass-movement processes, has been gleaned from field observations in high mountains. By this we mean from areas above the upper timberline and more especially from sites picked because they are largely unvegetated and thereby characteristic of the mountain landscape where processes are operating most rapidly, where there are talus slopes, rock glaciers, block fields, and solifluction terraces on debris slopes with a broken vegetation cover. The reasons for this are pragmatic. If the field scientist is to obtain observations on the movement of rock particles that are significantly larger than the magnitude of error inherent in his instrumentation, and to complete this in a reasonable length of time – we should remember that Rapp needed eight years, and data sets collected over a decade are not unusual – then perforce his field site must experience high rates of mobility. While this over-emphasis on high mountain, largely unvegetated sites can be offset by estimation of regional denudation rates from river sediment load at lower elevations, as Caine has demonstrated in his concept of the alpine cascade of sediment fluxes (Caine, 1974; Barsch and Caine, 1984), there is frequently a lack of coupling between alpine hillslope systems and the fluvial system. In other words, much of the geomorphic 'work' measured relates to mass movement of debris that remains within small alpine watersheds and does not pass through the system as fluvial sediments. It is perhaps understandable, therefore, that data sets and hypotheses acquired from field studies in high mountain environments must be applied to middle and low mountain belts with great care. This is especially important in an area such as the Himalayan Middle Mountains where frequently landscapes are

Figure 5.4 A totally transformed landscape. In this Nepal Middle Mountains scene *bari* terrace construction has virtually changed every natural slope facet and no original vegetation remains. Yet these areas are extremely stable as long as the subsistence farmers continue to grow crops and maintain the terraces.

totally transformed by subsistence agriculture (Figure 5.4).

Despite this qualification all available geomorphic studies have demonstrated a difference in degree, rather than in kind (the glacial system excepted) of erosion between high mountains and regions of less pronounced relief. In very simplified terms, this means that the same basic processes operate more rapidly (= effectively) on steep slopes than on less steep slopes. This difference in degree appears to be consistent and measurements, however sparse and unrepresentative they may be, range from five times to one or two orders of magnitude, based upon river sediment load estimates (that is, the denudation rate, or overall rate of landscape lowering), reservoir sedimentation, and geological data (the sediment record in the plains representing the long-term accumulation of material eroded from the neighbouring mountains).

Another perspective necessary for this review is that of the rates of mountain building and regional denudation within the context of geologic time. With the widespread acceptance of the theory of plate tectonics over the past twenty years our appreciation of the dynamism of continued mountain building has advanced considerably. Several estimates of present-day rates of Himalayan uplift have been published. Zeitler *et al.* (1982) indicate a rate of uplift for the Greater Himalaya of about 1 mm/yr; Low (1968) estimates 1–4 mm/yr since the close of the Lower Pleistocene; Iwata *et al.* (1984) about 1 mm/yr for the Nepalese Himalaya; and Zeitler *et al.* (1982) about 9 mm/yr for the Nanga Parbat region. Extensive geophysical work by the Chinese Academy of Sciences (Liu and Sun, 1981) on the Tibetan Plateau and the northern slope of the Himalaya has resulted in figures of 4–5 mm/yr over the past 10,000 years, continuing today. Precise measurements in the vicinity of Garm, Tadjik SSR, over the past thirty years indicate that Peter the First Range, an outer range of the Pamir Mountains, is rising at a rate of 15 mm/yr. The very crude estimates of regional denudation rates, discussed below, barely match those of the uplift estimates. It is argued, therefore, that at present, uplift equals or even exceeds denudation in some areas, thus implying that the Himalaya–Ganges system over the past 10,000 years has continued to be an extremely dynamic section of the earth's crust.

Because the Himalaya and the Tibetan Plateau are being uplifted as the Indian plate thrusts beneath Central Asia, the enormous masses of eroded sediment are deposited in the foredeep to the south which, over the past several million years, has become, in effect, the great plains of the Indus, Ganges, and Brahmaputra. This point is nicely emphasized if we consider that drill holes have penetrated more than 5,000 m of alluvial sediments beneath the Ganges Plain (C. K. Sharma, 1983). On a more recent time frame, the (Sapta) Kosi River has shifted its channel across its great alluvial fan, which forms much of Bihar State, through a distance of more than 100 km in the past 250 years. There is, therefore, abundant evidence of massive erosion and regional denudation and equally massive sediment transfer and deposition that has occurred over the past million or more years. Present-day evidence and geophysical hypothesis would indicate that the height of the Himalaya is equal to, if not higher than, that of a million years ago, and that the relief energy between the crestline and the Ganges Plain remains undiminished over recent geological, as well as historical, time. Thus, without very convincing evidence to the contrary, it would seem reasonable to argue that the contribution of human interventions over the past three or four decades, or even centuries, has been insignificant when balanced against the natural processes at work. Before this point is examined further it will be helpful to bear in mind a few broad concepts and to reiterate some of the dilemmas facing geomorphic and hydrological research.

1. The Himalaya–Brahmaputra–Ganges–Indus system is one of the world's most dynamic mountain-building and sediment-transfer systems, processes that have continued unabated over recent geological time and

will likely continue into the future.

2. These processes, the endogenous, tectonic/isostatic activity and the exogenous, climatic/weathering/hydrological ones, have created an unstable landscape of the utmost complexity.
3. Given the massive scale of relief, from more than 5,000 m below sea level to nearly 9,000 m above sea level, and the enormous variations in climate, vegetation, and topography, and the variability of major geomorphic events in time and space, the present data base is completely inadequate for determination of actual rates of activity of the various processes affecting the land surface. Thus, determination of the impacts of human intervention, including deforestation, land-use changes, and manipulation of water flow, and their differentiation from the natural processes as a proportion of the total rate of change, is not possible.

Regardless of this apparent total obstacle to geomorphic evaluation, some important contributions are feasible provided they are set in the context of recent geological time. It is equally important to question the widespread tendency over the past thirty years or so to assume that problems of erosion affecting either the plains or the mountain slopes themselves are entirely, or largely, the result of human intervention, in terms of misuse of the Himalayan environment.

Some critical definitions are needed. First, soil erosion (or surface erosion), a widely used term, should be restricted to the secular loss of soil, especially the A-horizon in which most organic matter is concentrated. Soil erosion occurs in entirely natural environments as well as in environments transformed by human intervention; for the latter it is convenient to use the term *accelerated erosion*, which implies a combination of a natural process and a man-induced process. Soil erosion should be distinguished from mass-wasting, which is the down-slope movement of the mass of fractured and weathered bedrock, the weathering mantle, on which the topsoil forms as the end-product of that weathering process. Mass movement includes such almost imperceptible, continuously operating processes as soil creep, frost creep, and solifluction by which the weathering mantle moves downhill under the influence of gravity at rates of a few millimetres per annum. Mass movement also includes more dramatic, intermittent processes such as landslides, mudflows, rock falls, and rockslides with short, or long, indeterminate, recurrence intervals.

Soil erosion, whether natural or accelerated, and mass movement, in practice, grade into each other, but to facilitate a clearer understanding of this section of the Theory of Himalayan Environmental Degradation, they should be retained as conceptually separate processes in landscape change. It should also be borne in mind that the weathering mantle and the topsoil are continually forming as the bedrock is broken down by a combination of mechanical and chemical processes. In certain instances in nature the topsoil and weathering mantle may be shed from a slope (for instance, during a cycle of landsliding) and the partially weathered bedrock exposed. This type of rapid mass movement will usually be followed by a long period during which the weathered mantle and its vegetation cover will be replenished. On steep mountain slopes climax, or zonal, soils may never develop because the slopes are too unstable to allow a mature soil cover to evolve. In these circumstances agricultural terraces actually reduce slope instability and the soils developed on them are at least partially, and in some cases largely, man-made. In other words, the immature azonal mountain soils receive much of their organic matter from the addition of crop residues and domestic animal fertilizer. Soil loss is characteristic of all natural and man-modified slopes; it becomes a problem for subsistence farmers only when the decline in soil productivity due to topsoil losses cannot be compensated for by addition of nutrients from organic matter, and continued accumulation of inorganic matter.

Effective erosion, if the soil or weathered material is to be moved out of the immediate field area (small watershed, or hillslope segment), requires the assistance of a transporting agent. The most effective agent of transport is running water (hence the formerly popular concept of the 'normal' cycle of erosion, that in which transport by running water predominates: the fluvial cycle of erosion). Thus information on the relationship between mass movement on slopes and water transport is important. The river itself is an agent of erosion as well as an agent of transport and, by cutting its channel and undermining its banks, the river is the principal force in maintaining local relief energy in an orogenically active mountain range. The river is also responsible for deposition of its transported load, and hence for the formation of the plains. Glaciers and wind are effective transporting agents but will receive little attention here because of their minimal spatial significance in the intensely used Himalayan belts – the Middle Mountains and the Siwaliks.

Denudation is a term used to describe the overall lowering of the landscape resulting from the erosive and transporting activities of all operating processes. In practical terms this is usually calculated as mm/yr in surface lowering averaged over entire regions, usually watersheds (drainage basins), despite the fact that actual surface lowering will be extremely variable in space, as it depends upon many factors, including the underlying bedrock lithologies and structures, and slope angle. Regional denudation estimates are derived in two ways: they are obtained from numerous observations on the principal erosion processes characteristic of a watershed multiplied by the total area (as exemplified by Rapp's (1960) study of Karkevagge), or they are extrapolated from measurements of sediment being moved out of the watershed through the main stream channel. In either approach the *sediment delivery ratio* must be taken into account. This is the ratio of sediment yield in a river – the actual volume of material transported out of the watershed – to the gross sediment production upstream, much of which goes into temporary storage within the watershed. Mass-movement data cannot be directly translated into denudation because much of the material moved remains within the watershed in storage, for instance in the form of lake sediments or as accumulations (talus, glacier moraines, landslide deposits, footslope colluvium) lower on the slopes or on the valley floors.

River transport is conveniently divided into the suspended load, the load carried as dissolved salts, and bedload. The sediment yield that is actually measured is often only the suspended sediment. Even this is difficult to measure accurately for rivers that experience enormous variations in volume over the course of the year and from year to year. This is especially characteristic of rivers of monsoonal climates where low flow in late winter and spring may be several orders of magnitude below peak rainy-season discharge. This problem of measurement is exacerbated when we consider that Himalayan river channels are frequently dammed by landslides; the ensuing ephemeral lake, when it breaks the dam, will produce a peak discharge and be capable of carrying much larger sediment loads, sometimes an order of magnitude or more, than that of normal summer monsoon peaks. Even these periods of high sediment yield may be totally eclipsed by peak surges resulting from the outbreak of ice-dammed and moraine-dammed lakes. The critical importance of such catastrophic floods to Himalayan water resource development have only been recognized in recent years (Hewitt, 1982; Xu, 1985; Galay, 1986; Ives, 1986; Vuichard and Zimmermann, 1986, 1987).

Measurements of bedload have not been recorded on any Himalayan river, even under conditions of 'normal' summer flow. An estimate is usually made for this component of the total sediment transfer in calculating the design life of reservoirs. Bedload is now regarded as having been grossly underestimated systematically throughout the Indian and Nepal Himalayan foreland (this is a particularly critical observation since hundreds of millions of dollars have been expended on dam and reservoir construction with the design life over-estimated two-, three- and fourfold). Finally the dissolved load of rivers in the Himalayan region is completely unknown. The importance of this lack of knowledge can be understood if we refer back to Rapp's (1960) ranking of the transport of dissolved salts in running water in northern Sweden. In the Himalayan region, which at lower elevation is sub-tropical, this component of a river's sediment load will not be of less significance.

It is now appropriate to consider the data that are available on sediment yield, erosion, mass movement, and denudation.

ESTIMATION OF DENUDATION RATES IN THE HIMALAYA

Any attempt to understand the significance of rates of denudation that have been determined for the Himalaya not only requires an appreciation of the status of mountain geomorphology, as has been outlined in the preceding section, but also must be founded on at least a general overview of the region's physiographic divisions, bedrock geology, and tectonic structure. To keep this section within reasonable limits the discussion is restricted to the Himalaya *sensu stricto*. Reference also should be made to Chapter 2, and particularly its treatment of climate and vegetation.

The entire central part of the mountain system can be divided into nine strike-oriented physiographic units that trend approximately from west-northwest to east-southeast (cf. Figure 2.3 page 20). The four northern units include the southern portion of the Tibetan Plateau, the Tibetan Marginal Range (6,000–7,000 m), the Greater, or High, Himalaya (with numerous summits exceeding 7,500–8,500m – maximum 8,848 m), and the Inner Himalaya (sometimes referred to as Trans-Himalaya), an intervening system of high plateaus and valleys lying between the two great mountain ranges. These four units contain the 'high mountain' belt and, except for very small areas in the more deeply cut valleys, extend above the montane forest belts and timberline at approximately 4,000 m. This is an extremely complex mountain landscape, heavily sculptured by glacial erosion during the Late Cainozoic and carrying a considerable cover of snow and ice today. Lithologies range from sedimentary and metamorphic rocks to granite intrusives: the whole has been subject to extremely complex faulting, folding, and overthrusting. The higher, north-facing slopes tend to be semi-arid to extremely arid because of high altitude and rain-shadow effects. The south-facing slopes intercept the summer monsoon although, as discussed in Chapter 2, precipitation decreases with increasing altitudes, and the very large precipitation totals are generally confined to the outer ranges and plains. Until recently Cherrapungi in the plains of Assam claimed the world's record annual rainfall total (11,615 mm). However, long-term records from high-mountain stations, or any precipitation data at all, are rather scarce. Several more recently established stations are showing totals in excess of 5,000 mm/yr; nevertheless, these are all located on south-facing slopes highly exposed to the monsoon.

The southern flanks of the Greater Himalaya merge via a transition belt with the so-called Middle Mountains. The latter have been the traditional centre of high population densities for much of the Himalaya. The transition belt, the Middle Mountains, and the northern flank of the next, outer unit, the Mahabharat Lekh in Nepal, or the Lesser Himalaya in Uttar Pradesh, are collectively referred to as the *pahar* in Nepal (*pahad* in Mahat *et al.*, 1986a). The upper northern sections of this tripartite division remain largely under upper montane forest (2,900–4,000 m) below which is the belt of intensive agriculture. Lithologies are extremely varied, including sedimentaries, metamorphics, and granites. However, there are extensive areas of phyllites and schists; these are deeply weathered and this, together with the prevailing steep slopes, renders them highly susceptible to erosion.

The Mahabharat Lekh, together with the Siwaliks (or Churia Hills), constitute the Lesser Himalaya, or outer ranges, bounded against the Ganges Plain (Terai) by the Main Frontal Thrust (a major tectonic translocation). Highest summits of the Mahabharat Lekh approach 3,000 m; the Siwaliks are much lower. Between and within them occur the famous 'dun' valleys, tectonic depressions which support some of the richest agricultural land, ranging from Dehra Dun in the west and including the Rapti Valley at Chitwan in Nepal. The lower, outer duns are locally referred to as the Inner Terai. The Mahabharat Lekh and Siwaliks are composed of the youngest,

Tertiary strata and contain some of the most easily erodible lithologies (including unconsolidated sands and gravels) of the entire Himalayan Region. Where these ranges have been extensively deforested, as in parts of Uttar Pradesh and Himachal Pradesh, there has been a catastrophic production of sediments and development of badland topography.

The Siwaliks abut the Ganges Plain abruptly at an altitude ranging from about 200 m in the east to 500 m in the vicinity of Dehra Dun. The alluvial sediments of the Ganges Plain can be divided into the high Terai (Nepali, *barbar*, or porous place) composed of massive coarse-grained alluvial fans and torrent fans, and the low Terai, underlain by finer sediments of the Ganges flood plain proper. As mentioned above, these sediments are up to 5,000 m thick in places and represent the vast accumulations of Himalayan weathering products over the past several million years.

Central to an understanding of the efficiency of erosional processes in the Himalaya, as well as the significance of the available estimates of denudation rates, is the concept of *slope-channel coupling* (Brunsden and Thornes, 1979). This concept was introduced in the previous section (pp. 96–97) in the discussion of the lack of complete transfer of the products of alpine (that is, the altitudinal belt above timberline) mass movement from the mountain slopes and out through the fluvial system (Barsch and Caine, 1984). Given these qualifications, and the problems of representativeness of the study sites in both time and space, we can proceed to a review of the denudation rates that have been derived for the Himalaya region.

Table 5.1 provides selected denudation rates for the Himalaya as a whole, or for various watersheds and areas of differing size. Despite the wide range (from 0.5–20 mm/yr), and the inherent errors discussed above, these figures are very high when compared with rates from other parts of the world. The Alps, for instance, have provided an overall rate of 1 mm/yr; figures above about 6–7 mm/yr are amongst the highest ever recorded or estimated. At the outset, therefore, they are suggestive of a very dynamic environment, especially when coupled with the high rate of tectonic uplift and the intensity of seismic activity.

The work of Brunsden *et al.* (1981), Starkel (1972a and b), and Ramsay (1985, 1986) represent almost the only systematic attempts in recent years to derive denudation rates from actual field measurements. The study by Brunsden *et al.* provides especially valuable insights for a section of the Lesser Himalaya in eastern Nepal, including observations from two field surveys for a road alignment between Dharan on the Terai to Dhankuta at 2,200 m. Theirs is the first comprehensive description of current hillslope and fluvial processes. They concluded that the relative relief of their study area has been increasing throughout the past million years or so because stream downcutting has exceeded lowering of the local interfluves. Valley side slopes, therefore, have progressively lengthened and their parallel retreat is being effected by stream undercutting and landsliding. Thus, according to Brunsden *et al.* (1981), debris mobilized on the steep slopes flows directly into the rivers giving a very high sediment-delivery ratio, and is subsequently moved out

Table 5.1 *Selected denudation rates for the Himalayan Region (from various sources, after Ramsay (1985:9).*

Location	*Denudation rate (mm/yr)*	*Comments*	*Author*
Himalaya	1.00	Regional	Menard, 1961
Ganges/Brahmaputra watershed	0.70	From present rate of influx to Bay of Bengal Fan	Curray and Moore, 1971
R. Hunza watershed	1.80	From sediment yield	Ferguson, 1984
R. Tamur watershed	5.14	From sediment yield	Seshadri, 1960[1]
R. Tamur watershed	4.70	From sediment yield 1948–50	Ahuja and Rao, 1958[1]
R. Tamur watershed	2.56		Williams, 1977
R. Arun watershed	1.90	1947–60	Pal and Bagchi, 1974[1]
R. Arun watershed	0.51		Williams, 1977, after Das, 1968
R. Sun Kosi watershed	2.50		Pal and Bagchi, 1974[1]
R. Sun Kosi watershed	1.43		Williams, 1977, after Das, 1968
R. (Sapta) Kosi watershed	0.98	From suspended sediment	Schumm, 1963, based on Khosla, 1953
R. (Sapta) Kosi watershed	1.00		Williams, 1977, after Das, 1968
R. Karnali watershed	1.50		UNDP, 1966
Darjeeling area[2]	0.50–5.00	Forested/deforested	Starkel, 1972a
Darjeeling area[2]	10.0–20.00	In catastrophic storms	Starkel, 1972a

Notes: [1]Brunsden *et al.* (1981).
[2]These figures are estimates and are not based on precise measurements.

onto the Ganges Plain by fluvial transport.

The two-period fieldwork of Brunsden and his colleagues was serendipitously timed to bracket a summer monsoon period with high rainfall amounts and a flood peak with an estimated ten-year recurrence interval. Thus they were able to appreciate the impact of such a condition on the steep slopes of their field area. They concluded that the Lesser Himalaya of eastern Nepal have one of the highest rates of denudation in the world. They regard the present landforms as characteristic forms in equilibrium with current processes and controlling tectonic, climatic, and base-level conditions. Storage and transport processes and the linkages between them were identified. In particular they conclude that one of the consequences of the very efficient slope-to-stream channel sediment-transfer system is that delivery of weathering mantle to the valley floor is not a continuous process. Debris tends to move in waves during storm events, causing pulses of heavily silted water to move downstream. This conclusion indicates that the system of erosion in the Lesser Himalaya of eastern Nepal contrasts with Barsch and Caine's alpine (mid-latitude) situation in terms of high sediment-delivery ratio for the former and low for the latter.

However, quantification of the various elements of the Brunsden *et al.* (1981) system, and especially the development of even a preliminary sediment budget for a single small watershed is not yet attainable.

Ramsay (1985, 1986) undertook the most exacting and effective study yet available of a specific watershed in the Himalaya. His assessment of erosion processes in the Phewa Valley, near Pokhara, Nepal, is discussed in the following section. Within the context of his attempt to convert erosion process data, together with an extensive review of the geomorphological literature, into denudation rates, however, his conclusion is an appropriate summing-up of the current status:

> The author . . . would like to emphasize that his own figures for failure age, volume, and frequency (and hence the estimate of surface lowering by landsliding [2.5 mm/yr]), should never be used without the prefix 'based on a small sample.' They are probably as valid, or as invalid, as the estimates made by Caine and Mool (1982), and Starkel (1972a and b).
>
> (Ramsay, 1985:130)

Mass-wasting

As emphasized above, the ubiquity of landslides and other catastrophic slope failures throughout the Himalayan system has been not only widely reported but viewed with alarm because it is assumed to be the direct consequence of human land-use changes over recent decades. This section will review some of the more recent work on mass-movement processes and attempt a preliminary conclusion concerning their causes, natural (geological) or accelerated (man-induced).

Kienholz *et al.* (1983, 1984) produced detailed maps of land use, geomorphic processes, and mountain hazard assessment on a scale of 1:10,000 for a small section of the Middle Mountains (Kakani–Kathmandu). Caine and Mool (1981, 1982) examined the channel geometry and stream-flow estimates and the landslide activity of the same area, as part of the United Nations University Mountain Hazards Mapping Project.

Caine and Mool (1981) concluded that the two mountain streams that they studied in detail were similar in their dynamics and form to streams in other parts of the world. This implies, they believe, that hydrological concepts and models derived from long records and detailed studies in mid-latitude regions can be applied to problem resolution in tropical and sub-tropical mountains, where few direct observations are available. They also demonstrated that the direct impact of man on the two fluvial channels is limited to a constraint on channel width in the uppermost 8 km. This results from the construction of agricultural terraces, especially masonry walls, which cause the stream to adjust by cutting a deeper channel. It would appear that as the stream size increases beyond about 8 km from the source, local landscape modification by the subsistence farming community is limited by the available technology and is not adequate to influence the lower reaches.

Caine and Mool (1982) also conclude that landslides presently occupy about 1 percent of the land surface of the map area. They have a total volume of more than 2.2×10^6 m^3 and a mean age of about 6.5 years, that is, number of years since initial failure. This suggests a current rate of lowering of the entire surface (denudation rate when data from other erosive processes are added) of about 12 mm/yr (recalculated by Ramsay (1985) as 11.3 mm/yr).

For the Phewa Valley, Ramsay (1985, 1986) calculated that landslide density was 1.6 per km^2 with 95 percent of the landslides being small, shallow failures. The total area affected is 0.7 km^2, or 0.5 percent of the area. If the area of the landslide deposits is added (as in the case of the Kakani fieldwork), this gives a figure of 3.25 km^2, or 2.7 percent of the catchment area. Ramsay challenges Caine and Mool's estimate of the rate of landslide expansion per annum and notes that many landslide scars in the Phewa Valley appear to be healing. This view is supported by Ramsay's observation that the age of the small, shallow failures tends to cluster at slightly less than eight years, suggesting that they may heal within a decade. Caine and Mool's estimated mean age of 6.5 years is close to Ramsay's estimate. When this is taken in conjunction with Ives's (1987) repeat photography (1978–87), it would appear that the contribution of landsliding to overall denudation, *regardless of the actual cause of the landsliding*, has been overestimated. Nevertheless, at present it is not possible to take into consideration the positive impacts of the farmers (cf. also Kienholz *et al.*, 1984).

Ramsay's larger areas of failure, which he terms *mass-movement catchments* after Brunsden *et al.* (1981), contrast markedly with the larger number of small failures. Their average age is about twenty-four years and they appear to be enlarging rapidly due to feedback mechanisms, such as the extension of bare ground and overgrazed areas that produce a much higher proportion of immediate runoff during heavy rainstorms. These are certainly influenced to some degree by human intervention in the form of mis-management of land for agricultural purposes.

Ramsay (1985, 1986) calculates an annual denudation rate of 2.5 mm/yr from landslide activity alone. This does not include the effects of soil creep, sheetwash, rilling, gullying, and solution. However, neither does it take into account the sediment-delivery ratio and, as Caine and Mool (1982) and Kienholz *et al.* (1984) have noted, a large proportion of the material transported by mass movements remains within the small watersheds and is not transferred to the larger rivers. Thus it appears that there is either a difference in interpretation by Caine and Mool (1982) on the one hand, and Brunsden *et al.* (1981) on the other, or else a difference in actual field conditions between their two study areas.

Several other studies have been undertaken with a view to determining the regional importance of landsliding, and most of these have been reviewed by Ramsay (1985, 1986) and Carson (1985).

Laban (1979) carried out an airborne reconnaissance of most of Nepal. Using a light aircraft he made observations on the number of slope failures per linear kilometre of flight as viewed from one side of the aircraft. These were categorized according to ecological province. Despite these limitations, he was able to conclude that road and trail construction was associated with 5 percent of all landslides observed; that some of the most densely populated and extensively terraced areas contained the smallest frequency of landslides; and that specific regions and lithologies displayed the highest frequences – the Siwaliks and Mahabharat Lekh, and the deeply weathered phyllites. He concluded that geological structure and lithology were far more important (accounting for more than 75 percent of all landslides in Nepal) in terms of landslide occurrence than were human land-use changes. A similar conclusion was reached following an extensive field and remote-sensing survey in a section of the Garhwal Himalaya (Joshi, 1987).

Wagner (1981, 1983) used a statistical analysis and the general characteristics of a hundred landslides, mainly along roads in the Middle Mountains. He sought to develop a site-specific landslide hazard assessment and mapping approach based upon use of 'equatorial Schmidt projections,' a geometric tool for studying the intersection of geological planes. Wagner concluded that the over-riding cause of landslides and rock slides was geological (natural) and not human. These conclusions, derived from the work of Laban (1979), Wagner (1981, 1983), Joshi (1987), and the synthesis of Carson (1985), need to be carefully qualified. It could be misleading to refer simply to the stipulation that landslides are caused primarily by the impact of rainstorms on certain lithologies (i.e. that the cause is overwhelmingly natural); frequently, at issue will be the relationship between land use and lithology.

Brunsden *et al.* (1981) and Caine and Mool (1982) both attributed human activities as an important cause of gullying. However, they also emphasized the importance of the material, particularly the brittle behaviour of the weathered, untransported bedrock. Caine and Mool (1982) did not find a direct relationship between heavy rainstorms and landslide occurrence, noting that a high water-table (within less than 2 m of the surface) was a primary controlling factor. Thus heavy rainfalls, early in the summer monsoon season, did not produce many landslides. As ground-water recharge progressed and the water-table approached the surface, landslide intensity accelerated. An opposing view is that landslides are most common early in the summer monsoon season because that is the time of minimum surface vegetation cover. This serves to emphasize not necessarily conflicting viewpoints, but possibly differences between specific areas, a common obstacle to any balanced overview. For the Kakani area at least, the conclusions of Caine and Mool (1981) have been verified by several seasons of observation and supported by many casual expressions of opinion by informants in the Kathmandu area.

Carson (1985) sums up by noting that many Himalayan landscapes appear to experience a period of relative slope stability when mass-wasting is

Figure 5.5 This hillside was photographed from the air in 1981 by Brian Carson (1985). His picture showed more than a hundred landslides at Lalitpur caused by torrential rain. Many of the landslide scars can be seen in this photograph but they are largely revegetated and careful inspection is needed. A casual visitor to the area would probably notice nothing unusual.

reduced. This is followed by a brief period of excessive instability when large numbers of landslides occur almost simultaneously on otherwise long-undisturbed slopes. Frequently the triggering mechanism is a catastrophic rainstorm with a long recurrence interval, such as the Darjeeling 1968 disaster (Ives, 1970; Starkel, 1972a and b) or a somewhat smaller event, such as the 10-year rainstorms of the Dharan–Dhankuta area of eastern Nepal (Brunsden *et al.*, 1981). It has also been assumed that major landslide cycles may be triggered by large-scale seismic shocks, with or without rainfall[1]. Carson (1985:11) observes that villagers often tell stories about extreme precipitation events, occasionally coincident with earthquakes, that trigger large-scale slope failures. He cites the area southwest of Banepa in the Nepal Middle Mountains where the villagers maintain that all the landslides still visible had occurred during two heavy rainfall events, one in 1934 and the other in 1971. They maintained that, in spite of more recent heavy rains, no major landsliding has occurred since the 1971 summer monsoon. Another case is the heavy rainstorm of September 1981 where mountain slopes near Lele, Lalitpur district, Nepal, failed *en masse* in spite of a thick natural regrowth of shrub. Our photograph (Figure 5.5) taken in October 1987 shows that already, within six years, natural revegetation has all but concealed the landslide scars from the unpractised eye.

Certainly, the above commentary provides the impression that much of the large-scale landsliding not only requires a major triggering event, but that it is followed by a long period of inactivity. This is presumably required for the accumulation of a new weathered slope mantle as a necessary condition for the next major failure. Seismicity is undoubtedly important, but lack of adequate earthquake records precludes any quantification of its influence. There have also been several recent statements to the effect that presence or absence of a forest cover will have little influence on the scale of a major landslide cycle (Carson, 1985). It may even be conjectured that forested slopes merely postpone the occurrence of a major cycle and, by so facilitating the production of a deeper debris mantle, ensure larger-scale events with longer recurrence intervals. In contrast, deforested areas given over to rain-fed agriculture, and especially grazing and overgrazing, are destabilized more frequently but by small, shallow failures. Thus the integrated effect over a longer time-scale may be comparable. Carson (1985:35–6) concludes that 'Mass wasting processes are not usually directly related to man's activities. Consequently, intervention by man to reduce mass wasting can be very expensive with less clear cut results.' He also concludes that 'Flooding and sedimentation problems in India and Bangladesh are a result of the geomorphic character of the rivers and man's attempts to control the rivers. Deforestation likely plays a minor, if any, role in the major monsoon flood events on the lower Ganges.' The second point, however, will be taken up more fully (see pp. 135–144) after we have considered some other specific and important questions related to the mountains.

Erosion Rates in the Greater Himalaya

There is a dearth of precise information on rates of erosion and mass-movement processes in the Greater Himalaya. The prevailing assumption is that, because the high mountains display some of the greatest relief on earth, including very high-angle glaciated slopes, as well as being influenced by tectonic instability, erosion, and hence denudation rates, must be high.

The mapping of mountain hazards in Khumbu Himal (Zimmermann *et al.*, 1986) led to the conclusion that, with the exception of limited areas characterized by severe tectonic disturbance together with susceptible lithologies, the high-angle slopes were remarkably stable. Zimmermann *et al.* (1986) noted that the single most hazardous process (for humans) was sudden peak river discharges resulting from the infrequent but precipitous break-out of ice-dammed and moraine-dammed lakes. This topic is discussed separately (p. 111).

In part to offset this paucity of information, and in part to test the assumptions that high rates of erosion and mass wasting could be substantiated, Byers established thirty-five soil erosion study plots in Khumbu Himal and made precise observations at each of them at weekly intervals throughout the entire 1984 summer monsoon (Byers *et al.*, 1985; Byers, 1986, 1987c). Total precipitation recorded for 1984 was slightly above the 28-year mean of 1,048 mm for Namche Bazar, so that the data obtained are reasonably representative.

Byers employed a standard geomorphic approach to the problem, using unsophisticated instrumentation that could be maintained easily in an isolated high mountain area. The thirty-five study plots of 5 × 5 m extended through a 1,000 m altitudinal interval (3,440–4,412 m) above and below upper timberline. They were arranged in a stratified, replicated design to permit comparison between (1) north-facing forest (*Abies spectabilis*/*Betula utilis*/*Rhododendron* spp.), (2) south-facing scrub grassland (*Cotoneaster microphyllus*/*R. lepidotum*/grass forb), and (3) variations in elevation. The fixed instrumentation consisted of a rain gauge, plastic sediment trough, erosion pins, and two sets of painted marker pebbles. Ground cover was determined for each plot by direct measurement at the time of instrumentation, and seasonal vegetation changes were assessed on adjacent 400 m^2 quadrats using a point-intercept method (800 points). Weekly observations also included air and ground temperature, soil capillary pressure, notes on plot changes and disturbance caused by local people, and weather.

Preliminary analysis of the data leads to the following findings. Summer monsoon precipitation was modest (Khumjung: 3,790 m, 725 mm compared with a 15-year average of 773 mm) and decreased with increasing altitude. However, local rain-shadow effects are marked, so that the simple relationship of diminishing total precipitation with altitude is frequently masked (compare the Khumjung total with that of 1,002 mm for Tengboche which is located in the open main valley of the Dudh Kosi-Imja Khola at 3,867 m). More significantly, from a geomorphic point of view, rainfall intensities were low (mean value of 29 measured 30-minute maximum rainfall intensities for Khumjung was only 3.03 mm/hr; and between 1968 and 1984 24-hour intervals exceeded 25 mm on only thirteen occasions with an absolute maximum of 54 mm). Cumulative precipitation increases sharply during the summer monsoon but, at most of the study plots, sediment yield was slight (Figure 5.6). The one exception was the plot at Dingpoche (4,412 m, in an overgrazed pasture) that demonstrated a high sediment yield. This was representative of an alpine meadow–juniper shrub area with a sandy soil subjected to heavy human disturbance, including up-rooting of juniper shrubs. Partial explanation for the low sediment yields is provided by the vegetation cover data. Many areas had 50 percent or less ground cover in the pre-monsoon season, which partly explains why trekking-season visitors have

Figure 5.6 Cumulative precipitation at Khumjung (3,900 m) increases sharply with onset of summer monsoon, but produces surprisingly little sediment yield. In contrast, the lower precipitation total at Dingpoche (4,412 m) produces very high soil loss (after Byers, 1987: 86, Figure 1).

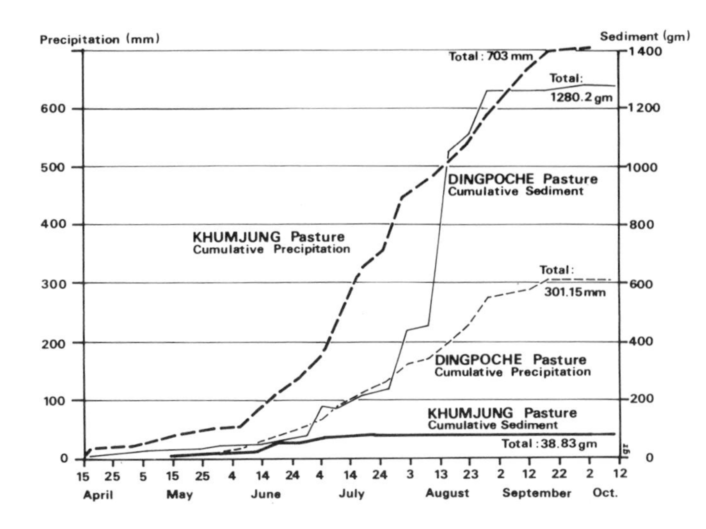

made alarmist reports of environmental degradation and anticipated severe erosion during the summer monsoon that they did not stay to witness. In contrast, Byers observed that pre-monsoon showers and light rains early in the summer monsoon season produce a rapid spread in ground vegetation cover and hence considerable protection against rain-drop impact (Figure 5.7).

Although this study monitored only surface erosion and did not consider such processes as debris flows, talus creep, and the higher-altitude glacial and fluvial systems, when the findings are coupled with the hazards mapping results it suggests that one of the world's highest mountain regions is far less active geomorphically speaking than might be suspected and than has been generally assumed. Similar results are anticipated from a parallel installation and data-collection programme for the 1985 summer monsoon season in the Yulongxue Shan area of the Hengduan Mountains, northwestern Yunnan (Ives, 1985).

The remaining topic for consideration in Khumbu Himal is the catastrophic outburst (Icelandic: *jökulhlaup*) of glacial lakes. A moraine-dammed lake drained catastrophically on 5 August 1985 in a tributary valley of the Bhote Kosi (Langmoche Glacier: Galay, 1986; Ives, 1986; Vuichard and Zimmermann, 1986, 1987). This provided a remarkable opportunity for before-and-after assessments and prompted a preliminary enquiry into the recurrence interval of such events. Hagen (1963), Hewitt (1964, 1982), Gansser (1966), Xu (1985), and others, have commented on such phenomena in the Himalaya-Karakorum region. In Khumbu Himal, at least, three

Figure 5.7 a) The summer monsoon retains its relative importance at high elevations in eastern Nepal, but the absolute magnitude is greatly reduced. b) Much of the anticipated sediment loss (based on dry season impressions) was apparently offset by sharp increases in grass/forb/shrub cover. Percentage cover of grass/forb, bare ground, litter, shrub, and rock were monitored throughout the study period (after Byers, 1987: 85, Figures 2 and 3).

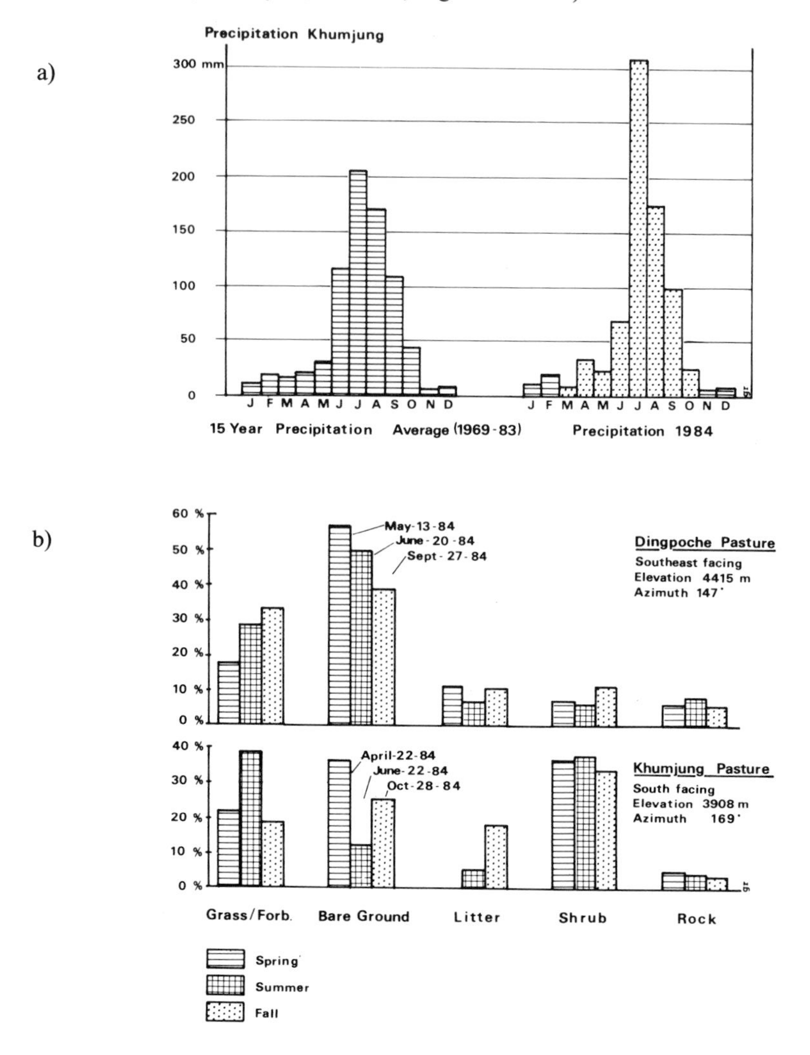

jökulhlaup, and possibly five, have occurred within living memory (Vuichard and Zimmermann, 1986; Brower, personal communication, 1986).

While these catastrophic events, with their point sources in the Greater Himalaya, are just being recognized as serious threats to water resource development (Ives, 1986) they may be of considerable importance also in terms of sediment transfer onto the Ganges Plain over the longer geological period. The largest-known glacier lake outburst, which deposited 5.5 km^3 of material in the vicinity of Pokhara, Nepal, some 600 years ago, has been referred to above (Fort and Freytet, 1982). The progressive retreat and thinning of most Himalayan region glaciers during the present century is resulting in the formation of new moraine-dammed lakes, or the enlargement of pre-existing ones between the retreating glacier front and the most recent Neoglacial moraines. Small ponds on the surface of lower glacier tongues are also enlarging and coalescing. A dramatic example of this process is a new lake that has formed since 1956 on the lower part of the Imja Glacier below Lhotse. It is now about 0.5 km^2 in extent; should a sudden drainage occur, the geomorphic, and human-destructive, impacts could be of catastrophic proportions (Ives, 1987, unpublished; Hammond, 1988). A cursory inspection of the metric camera imagery of the Dudh Kosi and Arun catchments in eastern Nepal–southeastern Xizang, revealed the presence of at least fifty ice-dammed and moraine-dammed lakes (Ives, 1986). While prediction of the frequency of collapse of their ice and moraine dams must await detailed field survey, their sheer number is impressive.

Vuichard and Zimmermann (1987) developed a sediment budget for the Khumbu outburst of 5 August 1985. They concluded that most of the material moved was redeposited within about 25 km in the stream channel and only about 10–15 percent (finer fraction) was transported out of the area. Their data indicate that 900,000 m^3 material was removed from the moraine dam but most of this was redeposited within the first 2 km below the breach. Much more material was picked up from the stream channel and valley sides further downstream. The peak discharge was calculated at 1,600 m^3/sec some 3 kw below the source and attenuated downstream.

It is extremely difficult to develop long-term sediment-transfer averages from the scanty information available. Nevertheless, the magnitudes of these events are sufficiently high and their recurrence interval, on an areal basis, quite small (5–15 years) so that they may prove a significant source of sediment for deposition at lower altitudes. Also, even if only a fraction of the material is far-travelled during the actual event, rivers subject to periodic *jökulhlaup* may well prove major source areas for subsequent sediment flux during peak monsoon rainstorms as the coarse material dumped in their channels is carried further downstream. The spectacular dynamics of the (Sapta) Kosi, for instance, and the development of its vast alluvial fan which occupies much of Bihar State, may be influenced to a considerable degree by the occurrence of *jökulhlaup* in several of its headstreams (Dudh Kosi, Sun Kosi, and Arun).

Soil Conservation and Sedimentation of Reservoirs

India's massive programme to develop Himalayan water resources for hydroelectricity as well as irrigation, and concurrent soil conservation measures, has produced a large amount of data on rates of sedimentation of reservoirs. These actual rates almost invariably have been much higher than hydroelectric engineers had predicted prior to dam construction. The discrepancies between predicted and actual sedimentation rates are so great that serious concern has been expressed in view of the multi-million dollar expenditures on river channel and discharge modifications on the Ganges Plain and in the foothills; even more critical from this point of view is the current trend to accelerate high dam construction rates in Nepal and India. Before discussing sedimentation rates in the context of the assumed causes (deforestation and mountain land mismanagement) some of the actual data will be presented.

Dhruva Narayana (1987), Director of the Central Soil and Water Conservation Research and Training Institute, Dehra Dun, has provided a valuable overview of the downstream impacts of soil conservation in the Himalayan region. He indicates that of India's 328 million ha of land area, approximately 175 million ha are suffering from intense soil erosion. He maintains that the Himalayan and lower Himalayan regions have deteriorated extensively because of deforestation, large-scale road construction, mining, and cultivation on steep slopes. In the northeastern Himalaya alone he attributes the serious degradation of 3 million ha to shifting (slash-and-burn) cultivation. Siltation rates in twenty-one Indian river valley projects were higher by 182 percent than the originally projected rates. He cites the Tehri catchment in the Garhwal Himalaya (total area 7,511 km^2, of which 2,328 km^2 are snow- and ice-covered) as producing 14.6 million tonnes of silt annually. The annual silt load from the Karnali catchment is estimated at 75 million m^3, equivalent to a denudation rate of 1.7 mm/yr. Singh and Gupta (1982) have determined that 28.2 tonnes/ha/yr are removed from the entire area of the Indian Himalaya. These are colossal figures and merit careful examination. Nevertheless, these figures also lie within the upper limits of derived denudation rates that have been attributed to geological, or natural, causes, as discussed above (pp. 105 to 109). This is an extremely important point. It implies that, assuming both sets of figures are within reasonable limits of accuracy, at least at a reconnaissance level, when considered on the regional scale (as distinct from that of a specific small catchment), the natural processes are so predominant that there is no requirement to seek human intervention as the cause of siltation. In other words, artificial reservoirs will silt up rapidly in this very dynamic region regardless of human influences – negative or positive.

On a more site-specific scale, Dhruva Narayana (1987) shows that soil losses of 80–156 tonnes/ha/yr have been recorded from the Chandigarh-Dehra Dun region (Lesser Himalaya and Siwalik small watersheds). These are absolutely unsustainable and destructive losses that can be drastically reduced by soil conservation measures. Such measures include contour cultivation, inter-cropping in maize, and planting of slopes with grass species.

In the Doon Valley, watershed gauging has demonstrated that transformation of naturally forested small watersheds to agricultural use results in a 72 percent increase in peak river discharge rates. Narayana goes

Table 5.2 *Sedimentation rates for major Indian Himalayan reservoirs (after Tejwani, 1986, unpublished).*

Reservoir and date[1]	*Annual Rates of Sedimentation*		
	Assumed	*Observed and year of observation*	*% increase in observed value*
Sutlej (Bhakra) (1971)	4.29	6.00 (1975)	39.9
		6.20 (1979)	45.0
Beas (1974)	4.29	15.10 (1975)	251.0
		23.59 (1981)	449.9
Ramganga (1974)	4.29	17.30 (1973)	303.3
Pohru[2]		7.41 (1973)	—
Giribata[2]		11.60 (1973)	—
Teesta[2]		98.20 (1973)	—
Gumti[2]		3.56 (1973)	—

[1] *Year of completion.*
[2] *No reservoir constructed.*

on to show that significant reductions can be achieved through conservation practices, especially terracing and bunding. Tables 5.2 and 5.3 give some of the Indian experimental watershed data. Bunding, in particular, is claimed to have reduced soil losses in experimental watersheds by 94 percent. Similarly, the construction of brushwood check dams in a forested watershed reduced soil loss from 4.7 tonnes/ha/yr to 2.8 tonnes/ha/yr (a 54 percent reduction).

Table 5.3 *Sediment load calculations for selected major Indian rivers (after Tejwani, 1986, unpublished).*

River	*Site*	*Sediment yield in acre feet/mi²/yr*
Ganga	Farraka	85
Arun	Tribeni	242
Karnali	Chisapani	600
Saptakosi	Sunarambh	330
Sunkosi	Tribeni	574
Tamur	Tribeni	1,255
Trisuli	T-Bridge	585
Brahmaputra	Pandu	160
Buridhirang	Rhowang	350
Dibang	Tiagaon	564
Dikrang	Dikrang Ghat	545
Lohit	Digaru Ghat	710
Manas	Mathauguri	145
Noadhing	Namsai	310
Ranga Nadi	Ranganadighat	180
Teesta	Anderson Bridge	2,070
Chenab	Ranihan	525
Ujh	Panchtirthi	1,650

Table 5.4 *Ranking of selected major rivers by sediment load:*

Table 5.4a) *Major rivers of the tropics (after Tejwani, K.C., 1986, unpublished).*

River	*Drainage basin (sq. km 000)*	*Average annual suspended load (million tonnes)*	*(tonnes/sq. km)*	*Estimated annual soil erosion from field (tonnes/sq. km)*	*(tonnes/ha)*	*Rank*
Congo	4,014	65	16	320	3	13
Niger	1,114	5	4	80	0.8	14
Nile	2,978	111	37	740	8	12
Chao Praya	106	11	107	2,140	21	9
Ganges	1,076	1,455	1,352	27,040	270	3
Damodar	20	28	1,420	28,400	284	2
Irrawaddy	430	299	695	13,900	139	5
Kosi	62	172	2,774	55,480	555	1
Mahanadi	132	62	466	9,320	93	7
Mekong	795	170	214	4,280	43	8
Red	120	130	1,083	21,660	217	4
Caroni	91	48	523	10,460	105	6
Amazon	5,776	363	63	1,260	13	11
Orinoco	950	87	91	1,820	18	10

(From Tejwani, 1980, unpublished)

Table 5.4b *Major rivers world wide (after Qiang and Dai, 1980).*

River (other)	*Drainage area (km^2)*	*Annual run-off (10^9 m^3)*	*Annual sediment load (million tonnes)*	*Average sediment concentration (kg/m^3)*
Colorado	637,000	4.9	135	27.5
Ganges	955,000	371.0	1,451	3.92
Missouri	1,370,000	616.0	218	3.54
Indus	969,000	175.0	435	2.49
Brahmaputra	666,000	384.0	726	1.89
Nile	2,978,000	89.2	111	1.25
Red	119,000	123.0	130	1.06
Irrawaddy	430,000	427.0	299	0.70

River (China)	*Drainage area (km^2)*	*Name of the gauging station*	*Average sediment concentration (kg/m^3)*	*Maximum sediment concentration (kg/m^3)*	*Erosion modulus ($t/km^2/yr$)*
Yellow	752,400	Sanmenxia	37.6	666	2,480
Yangtze	1,807,200	Datong	0.54	3.24	280
Haihe	50,800	Guanting	60.8	436	1,944
Huaihe	261,500	Bangbu	0.46	11.0	153
Liaohe	166,300	Tieling	6.86	46.6	240
	23,200	Dalinghe	21.9	142	1,490
Pearl	355,000	Wuzhou	0.35	4.08	260

Finally, when fast-growing Eucalyptus trees were raised in a denuded watershed (for fuel and paper pulp in a short rotation of ten years) total runoff was reduced by 28 percent and peak discharge by 73 percent. Similar advantages were obtained with coppiced plantations of eucalypts indicating the importance of a protective ground vegetation cover under a light crown of introduced trees.

From these data it is apparent that human intervention, both positive and negative, can result in large changes in both soil loss and river discharge on a small scale.

Table 5.2 provides sedimentation rates for some of the major Indian Himalayan reservoirs (Tejwani, 1986, unpublished, 1987). Table 5.3 provides sediment load data for eighteen major rivers and tributaries with a Himalayan origin and similar data for fourteen non-Himalayan rivers (Tejwani, 1986, unpublished, 1987). The very high sediment loads of the Ganga and Brahmaputra main channels, and the (Sapta) Kosi, Dibang, Teesta, and Lohit tributaries, corroborate the proposition that these rivers reflect the high rates of uplift and denudation in the Himalaya (pp. 98–99). The generally much lower figures for the non-Himalayan rivers are striking. On a worldwide comparison, the Ganges, Kosi, Damodar (and Brahmaputra), in tonnes per km^2 of catchment (estimated annual soil loss from the field), rank extremely highly. Table 5.4 shows the ranking of a series of rivers in different parts of the world. Two observations are noteworthy: one is that the figures indicate that extremely serious problems of erosion, sediment transfer, and siltation are facing the Ganges–Brahmaputra system and large parts of China; the other is that there are wide discrepancies in the different data sets (Qian and Dai, 1980, also included in Table 5.4). While the cause for alarm is real enough, this issue also falls squarely within our theme of 'uncertainty on a Himalayan scale' (Thompson *et al.*, 1986).

Table 5.5 shows loss of Indian reservoir capacity as percentage of dead

Table 5.5 *Loss of Indian reservoir capacity as a percentage of dead storage to live storage in ten reservoirs (after Tejwani, 1986, unpublished).*

Name of reservoir	*River system*	*Dead storage (%)*	*Live storage (%)*	*Remarks*
Sutlej (Bhakra)	Indus	16.0	2.5	up to 1974
Maithon	Ganga	26.0	12.0	up to 1974
Panchet	Ganga	35.2	17.8	up to 1974
Mayurakshi	Ganga	21.8	7.7	up to 1974
Nizamsagar	Godavari	100.0	60.4	up to 1974
Lower Bhawani	Cauveri	100.0	3.8	no date
Matatila	Yamuna/Ganga	15.0	8.5	no date
Gandhisagar	Yamuna/Ganga	27.2	11.5	in 14 years
Girna		33.8	7.2	in 14 years
Hirakund	Mahanadi	29.9	9.0	in 22 years

storage to live storage in ten reservoirs. While the data are incomplete and little is available after 1974 the development of a very serious situation is apparent.

Two points must be made concerning reservoir siltation, estimated suspended sediment load, and small river basin conservation response data. First, there are inadequate long-term data. Second, the data that are available do not give any reliable information about the relative importance of human land-use interventions and natural processes, except on a scale of tiny experimental watersheds. All that can be concluded is that the Ganges and Brahmaputra systems carry a higher sediment load than most other major rivers world wide. Some of the Himalayan tributaries, for example, the Tamur, (Sapta) Kosi, and Teesta, produce still higher rates in proportion to the area of their catchments.

Kattelmann (1987) has produced a valuable overview of hydrological problems in terms of development of Himalayan water resources. He stipulates that 'many watershed management projects have failed to moderate reservoir sedimentation except where there was little problem to begin with and [where] upstream cultivation and forestry were eliminated.' The former is irrelevant to our discussion of the highland–lowland sediment transfer linkage; reducing upstream cultivation and forest utilization is virtually impractical where a growing highland population is facing severe land shortage.

Impact of Road Construction on Sediment Production

Tejwani (1987), quoting data from Bansal and Mathur (1976), indicates that in the Indian Himalaya for each linear kilometre of mountain road, ten small to medium landslides occur as the direct result of slope instability caused by the road construction. Prior to the 1962 border war with China, the Indian Himalaya were, for the most part, accessible only on foot along trails; the only roads led to famous hill stations such as Mussoorie, Simla, Nainital, and Darjeeling, products of the British Raj. The shock of the Chinese military presence on the Himalayan frontier prompted a massive road construction programme that was put into effect after 1962 in great haste, for military expedience outweighed concern for careful planning and sound engineering.

According to Tejwani, post-1962 road construction has produced more than 10,000 km of highways in the Indian Himalaya. He estimates that poor alignments and ill-considered design are resulting in a total 'soil loss' of 1.99 million tonnes annually. This he equates to slope movement of 1.99 tonnes of sediment per linear metre of road per annum (recalculation indicates a misplacement of the decimal point, so this figure should read 0.199 tonnes/m/yr – authors' note). Valdiya (1985, 1987) uses a figure of 44,000 km and calculates that during the construction phase an average kilometre of road required the removal of 40,000–80,000 m^3 debris.[2] After construction, the extensive slope instability resulted in the production of enormous volumes of debris, usually dumped on the road bed and further downslope during heavy monsoon rainstorms in the form of debris flows, rockfalls, rockslides, and mudflows. Valdiya provides the following estimates of annual debris production per linear kilometre of road bed for three specific highways:

Western Himalaya: Jammu-Srinagar	724 m^3
Central Himalaya: Tanakpur-Tawaghat	411 m^3
Eastern Himalaya: Arunachal Pradesh	719 m^3

From these figures he derives an average annual debris production of 550 m^3 per km and a total debris production of 24×10^5 m^3 for the 44,000 km road network.

In practical terms, these circumstances, regardless of the somewhat different data derivations of Tejwani and Valdiya, have induced frequent highway blockages and enormous maintenance costs, especially during the summer monsoon periods. And since the standard method of road clearance is to dump the debris over the road side and down the slope below, this in turn further extends the area of instability. It causes destruction of downslope vegetation cover as well as the agricultural terraces of local subsistence farmers, who are usually not compensated for their losses.

Narayana and Rambabu (1983) accredit the enormous annual production of debris to unsatisfactory highway alignment and poor design. Valdiya (1985:24) states that 'The damage to the ecological balance [of the Himalaya] is mostly man-made or is caused by human negligence' and cites three main causes: road construction, overgrazing, and reckless deforestation. He, like Narayana and Rambabu, believes that road construction in seismically and tectonically unstable bedrock is 'the most important factor.' Haigh (1982a and b, 1984a and b) has undertaken detailed investigation of landslides along the Mussoorie-Tehri road in Uttarakhand. He measured 470 debris outfalls in 1977 and 1978 and concluded that the frequency of movement was related to depth of road cut, steepness of slope, degree of forest cover, and geological structure and lithology. He showed that annual costs for road clearance could be calculated from a simple measure of outfall width of the twenty largest landslides for this particular highway, data that could be obtained readily from air photographs.

Each of the above-cited researchers assumes that landslides produced by road construction are responsible for recent massive increases of suspended load in the local headstreams of the Ganges. Furthermore, a significant proportion of this is also assumed to be carried downstream to the Bay of Bengal, contributing to the development of New Moore Island (24 × 11 km) offshore of the Sundarban delta (Valdiya, 1985: 20–4).

Regardless of the weight of argumentation and the relative precision of the estimates of debris dumped onto specific sectors of the road bed each year, no accurate assessment of the proportion of the debris which enters the local rivers is available. Many of the landslides run part-way down the slope below the road bed and do not even reach the third-order stream channels. We feel obliged to insist, therefore, that while road construction and road maintenance problems in many parts of the Himalaya, and the attendant destruction of local slopes, are economically catastrophic, no data are available to support the claim that any of the debris so produced reaches the plains, let alone the Bay of Bengal. However, the landslide scars associated with the Himalayan roads are very evident even to the casual observer, and it is reasonable to conclude that sediment yield will be proportional to the density of the road network. The fact that Laban (1979) concludes that only 5 percent of Nepalese landslides are attributable to highway construction is presumably a reflection of the much lower mountain highway density in Nepal.

The data on road construction and induced slope stability introduced above are significant. The broader interpretations of the effects of roadworks on flooding and sedimentation on the Ganges Plain and on the seaward portion of the delta in the Bay of Bengal are supported by many workers (Haigh, 1982a and b, 1984a and b; Tejwani, 1984a and b, 1987; Valdiya, 1985; Narayana, 1987). We believe, nevertheless, that our general conclusions to the effect that man-made erosion (accelerated erosion) is negligible on a regional scale are not invalidated. At issue, however, is the 'uncertainty theme.' The seriousness of reckless road construction from a physical point of view cannot be doubted. Much avoidable damage has been, and is being, perpetrated. The costs of keeping the roads open are significant in themselves, as are the economic losses of repeated and widespread road closures. The development of between 10,000 and 50,000 km of road has also produced a major socio-economic impact on the Himalaya which includes greater accessibility of hitherto remote forests to commercial logging, ease of movement of people both from the mountains to the cities of the neighbouring plains and from the plains to the mountains.

GEOMORPHIC PROCESSES AND HUMAN INTERVENTIONS: SUMMING UP

We have tried to demonstrate in this chapter that mountain geomorphology faces many problems in its attempts to identify and to rank specific slope processes in terms of their relative importance or absolute amounts of 'work' accomplished. These problems confound both determination of the manner of evolution, over geological time, of the mountain landscape as we see it today. They also limit understanding of what is currently happening in natural, or near-natural, situations. The crux of this issue is the representativeness (or lack of) of collected data in both space and time. Especially difficult is assessment of the relative and absolute importance of high magnitude events with very long recurrence intervals compared with the continuous slow-moving downslope material transfer. Regardless of these problems a broad and extremely valuable understanding of denudation rates, orogenesis, and sediment transfer on a regional scale and within a geological time-frame has emerged.

We conclude from the foregoing discussion that natural processes in this dynamic region virtually obliterate the effects of human intervention, in so far as they can be gauged from the available data. Human intervention takes many forms; we have emphasized, within the Himalayan context, deforestation and general changes in land use as influenced primarily by the needs of a rapidly growing subsistence mountain farming population to sustain themselves. Other interventions have been introduced, including road construction and the installation of hydroelectricity facilities.

In taking our discussion from a rather academic review of geomorphology as a science into considerations of soil erosion and landslides that are widely presumed to be caused by bad land-use practices, we have inferred that the two sets of processes, geophysical and human, are probably several orders of magnitude apart. Furthermore, we believe we have demonstrated that the adherents of a theory stipulating that human intervention is the primary culprit of the large-scale environmental damage (as perceived from a human point of view) have failed to prove their case. Landslides and gullies, bare soil, floods, and silted reservoirs and irrigation works are visually graphic. At a local, site specific, scale they are serious and need to be corrected or reduced. However, we have found no reliable data to indicate that a specific mountain road, for instance, however badly constructed, is contributing a single grain of silt to the Ganges–Brahmaputra delta. We believe that, until there is very strong evidence to the contrary, the geophysical and climatic processes of our region, as reflected in some of the most rapid uplift estimates (orogenesis) and high rates of mountain denudation and concomitant accumulation of vast thicknesses of sediments, in the form of the Ganges and Brahmaputra plains, give adequate explanation for the workings of this extremely active landscape. The next major issue – large-scale processes on the plains – is the focus of Chapter Six.

NOTE

[1]In August 1988, northern India and eastern Nepal experienced a seismic shock of 6.4 on the Richter scale. Early reports indicated almost 1,000 deaths and thousands of injuries in the towns, due to the direct effects of the earthquake. At time of reading the page proofs (September 1988), only the first tentative accounts were available to the authors concerning damage in the mountains. It appears likely that thousands of landslides will have occurred, causing extensive damage in remote areas so far inadequately reported.

[2]The wide discrepancy between Tajwani's and Valdiya's assumptions of total highway length, and the apparent slippage of a decimal point in Tejwani's calculations, while not necessarily measuring up to Thompson *et al.* 's 67 factor, do provide us with yet another reinforcement of the 'uncertainty' theme.

6 THE HIMALAYAN-LOWLAND INTERACTIVE SYSTEM: DO LAND-USE CHANGES IN THE MOUNTAINS AFFECT THE PLAINS?

INTRODUCTION

The previous three chapters have discussed the linkages between those segments of the Theory of Himalayan Environmental Degradation that relate primarily to the physical processes occurring, or that are assumed to be occurring, within the mountains. This chapter focuses on the larger question: assuming that accelerated erosion due to land-use changes *is* taking place on a vast scale in the mountains, what is the evidence in support of claims for dramatic downstream impacts? We will now attempt to estimate the regional distribution of watershed degradation in the central Himalaya and its hydrological and geomorphic effects downstream.

Much of this chapter depends upon a major data collection and preliminary analysis by Andreas Lautherburg of the University of Berne (Lauterburg and Messerli, 1986, unpublished). During several visits to the region he collected as many data as possible on soil erosion and watershed degradation in Nepal, India, and Pakistan. In addition he collected all available data on stream flow (runoff) and sediment load of the main Himalayan rivers. Much of the critical information was found in publications such as Laban (1978, 1979), Zollinger (1978, 1979), and Carson (1984, 1985), and in the files of four government and international organizations: the Department of Soil Conservation and Watershed Management, HMG, Kathmandu; the Remote Sensing Centre, HMG, Kathmandu; the Central Soil and Water Conservation Research and Training Centre, Dehra Dun, Uttar Pradesh; and the International Centre for Integrated Mountain Development (ICIMOD), Kathmandu.

The data-acquisition phase of the study is by no means complete. This is partly due to time and funding limitations, but also to the fact that many data on hydrology and sediment transfer were not accessible.

The available data can be divided into two groups: direct and indirect. An example of the former is the direct measurement of soil loss from controlled test plots. This kind of information, as indicated in the preceding chapters, is both rare and unreliable (unreliable partly in the sense of lack of representativeness). Nor is it a major contribution for estimating the spatial distribution of degradation on a regional scale. Thus, of more value is the indirect information; examples of this are time series on the suspended sediment load in rivers and the density of landslides per unit area.

Much of Lauterburg's study of erosion records of test plots and small catchments parallels the discussion in Chapter 5. He concludes that, at the local (micro) level, soil erosion is highly influenced by human impact and that corrective measures could reduce this dramatically. We also wish to re-emphasize the positive aspects of certain forms of human intervention. Lauterburg also supports our earlier contention that the conversion of mountain lands under natural vegetation to an agricultural landscape does not automatically result in an increase in soil erosion (accelerated erosion) since soil loss is not dependent upon natural versus domesticated soil cover but on a conservation factor; an extreme case, for instance, is the positive influence of carefully tended agricultural terraces. This part of Lauterburg's analysis is taken no further here and we will discuss now the indirect data and its assessment.

REGIONAL ASSESSMENT OF WATERSHED DEGRADATION

Lauterburg analysed the work in Nepal by Nelson (1980), Laban (1978, 1979), and others. Nelson, for instance, attempted to determine the general status of degradation of all Nepalese watersheds by visually estimating a 'watershed condition index.' This 'relates the current state of soil erosion in an area in comparison with the soil erosion estimated for that area under natural or well managed conditions' (Nelson, 1980:2). This study shows a significant concentration of heavily degraded areas (accelerated erosion) in the central and lower Siwaliks and in the Kathmandu area.

Lauterburg reassesses Laban's landslide count from a light aircraft (Laban, 1979, and see above, pp. 106–7). This provides us with a *Natural Landslide Index*, which is obtained by dividing the number of landslides occurring within forested areas by the total number of landslides. This approach suffers from the uncertainty of being able to distinguish between 'natural' forests and degraded forests (lopped, partly grazed, etc.).

A complementary approach is the use of data on suspended sediment load in rivers. By assuming that the sediment load of several rivers is measured accurately over a significant number of years, the total (average) amount of sediment yield per annum divided into the total area of the watershed will provide the so-called *Specific Suspended Sediment Delivery* (SSD) of a catchment. This, as discussed in Chapter 5, does not include all material eroded within a watershed, since much is left in temporary storage and does not enter the fluvial system. In addition, there were no available data for bedload or solutes. However, as long as the watersheds under comparison are of approximately the same size and form (that is, their *Sediment Delivery Ratios* are similar) the SSD does provide a useful comparative unit. Table 5.1 gave the SSD rates (actually denudation rates) calculated in this way for Nepal. Figure 6.1 shows the Suspended Sediment Delivery for most of the Himalayan Region. However, this map has been compiled from very different

Figure 6.1 Suspended Sediment Delivery of some Himalayan rivers. Prepared by Andreas Lauterburg, Geographical Institute, University of Berne, with data from Ahmad (1960 – Indus), Goswami (1983, 1985 – Brahmaputra), and many sources (1974–85).

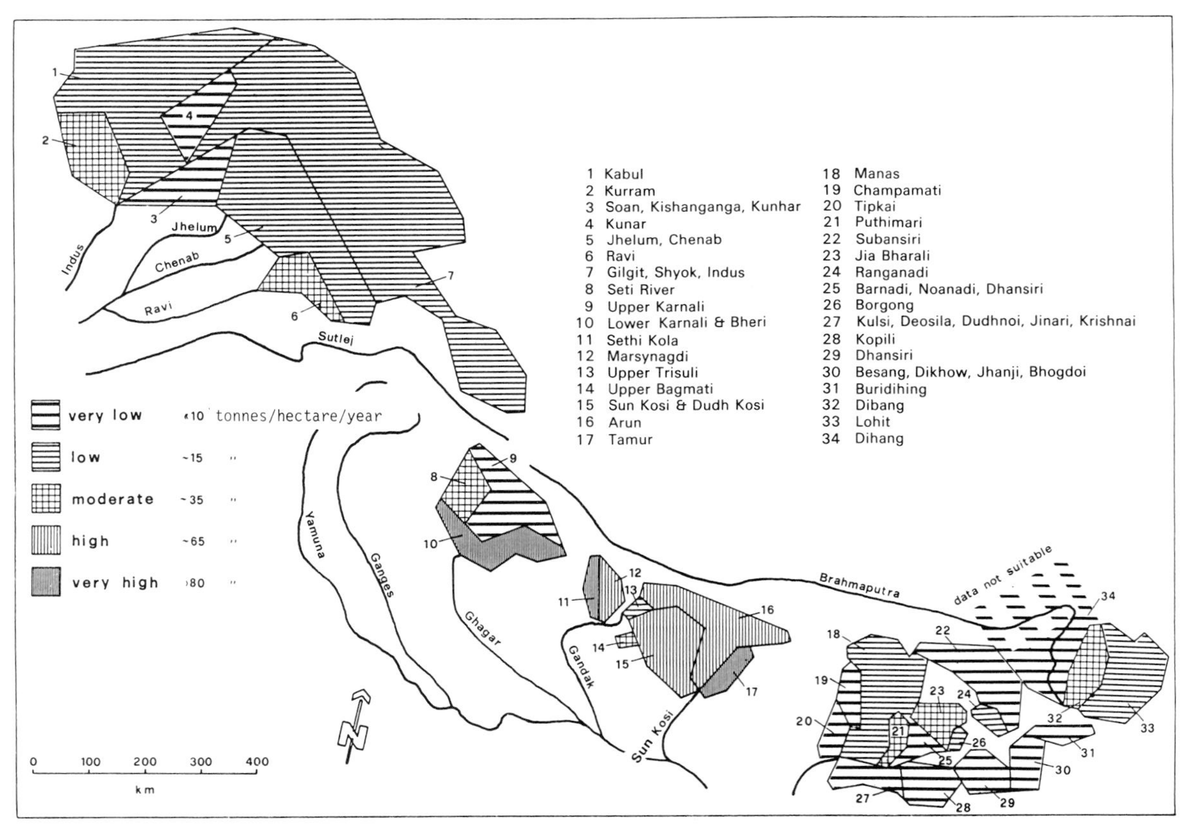

sources which cover different time periods. Thus the SSD data for the Indus watershed was collected in the late-1950s (Ahmad, 1960) and may be higher today. The somewhat higher denudation rates than those given in Table 5.1 are due to the use of estimated suspension data for Figure 6.1, especially for eastern Nepal.

Despite these obvious limitations, together with actual gaps in data availability, a broad-brush comparison may still be useful. Examination of the three macro-regions, the Indus, Ganges, and Brahmaputra watersheds, demonstrates immediately that SSD rates in the Nepalese Himalaya are much higher than the other two regions. Such a map (Figure 6.1), if updated with dependable information, would be extremely valuable. Unfortunately, this cannot be done at the present time. Another problem with this approach is that, whatever we learn about the current erosion rates, there is no time perspective. If an attempt is to be made to assess the relative, or absolute, proportion of total SSD due to human rather than natural (geophysical) processes, we would need to know the change in sediment load over the past hundred years or so. Moreover, no attempt has been made to estimate the *stabilizing* input by man through terrace construction and their effective maintenance.

As indicated in Chapter 1, the conservationist and scientific literature is replete with qualitative estimates of landscape degradation in specific small areas – this is largely what set in motion the Theory of Himalayan Environmental Degradation – but no attempt has been made to produce quantitative data nor, especially, to link them with the large-scale processes occurring on the flood plains and deltas of the major trunk streams.

In contrast, Figure 6.2 shows the *lithological erodibility* (specifically, the susceptibility to weathering and erosion of the bedrock) of the Karnali watershed in Nepal divided simply into low, medium, and high erodibility. The close relation between lithology and susceptibility to erosion is apparent. Of particular significance, the largest area of 'high erodibility' is situated in the high Himalayan zone which has a very low population density. However, this map shows susceptibility to erosion and not actual sediment yield. Rambabu *et al.*'s (1978) map of annual erosivity in northern India and Nepal (Figure 6.3) is more useful. This is based on a regression equation which calculates erosivity from monthly or annual rainfall data, incorporating 30-minute maximum rainfall intensities and total kinetic energy.

The construction of iso-erosivity maps normally depends on availability of an extensive network of stations that record short-term rainfall intensities as well as the catastrophic climatic events with very long recurrence intervals. As mentioned above there is a great shortage of these kinds of data for the Himalaya and the problems associated with this approach have been discussed in Chapter 5. Nevertheless, we concur with Lauterburg's conclusion that, compared with the other great mountain systems of the world, the Himalaya experience very intense erosivity and high probability of catastrophic high-intensity rains, and are at a very severe risk of climatic erosion. Even in making this very general statement, however, we must once

Figure 6.2 Lithological erodibility of the Karnali watershed, Nepal. Prepared by Andreas Lauterburg, Geographical Institute, University of Berne (data from Shrestha, 1980).

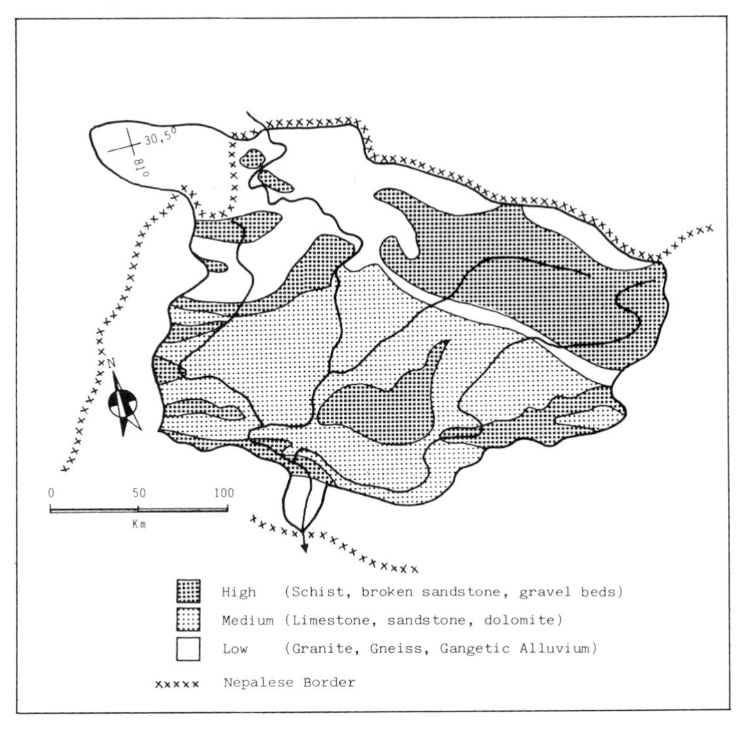

again emphasize the gaps in available data, as well as poor data quality.

The discussion of the ratio of natural (geophysical) erosion to that caused by human intervention (accelerated erosion) is taken a step further by Lauterburg. He compares maps of *natural erosion risk* (lithology and climatic factors) with maps indicating the state of watershed degradation and landslide frequency in Nepal (Figures 6.4, 6.5 and 6.6). Figure 6.4 indicates susceptibility to erosion according to a combination of lithology and climatic factors with data assembled on a grid. Figure 6.5 demonstrates watershed condition, incorporating Nelson's (1980) and Laban's (1979) determinations of landslide incidence induced by human intervention. Finally, Figure 6.6, which combines data from Figures 6.4 and 6.5, identifies those areas in Nepal where human activities have had very high, high, and moderate impacts on watershed degradation.

The data sets used in the compilation of Figures 6.4–6.6 include quantitative and semi-quantitative information as well as subjective estimates. There are also significant gaps in the data base. Figure 6.6 is principally a qualitative estimation of risk of watershed degradation. As anticipated in the previous discussion, there is a heavy concentration of 'very high impact,'

Figure 6.3 Iso-erosivity map for northern India and Nepal. *Annual erosivity (joules per square metre per millimetre per hour). Prepared by Andreas Lauterburg, Geographical Institute, University of Berne (data from Narayana and Rambabu, 1983; Mohns, 1981).

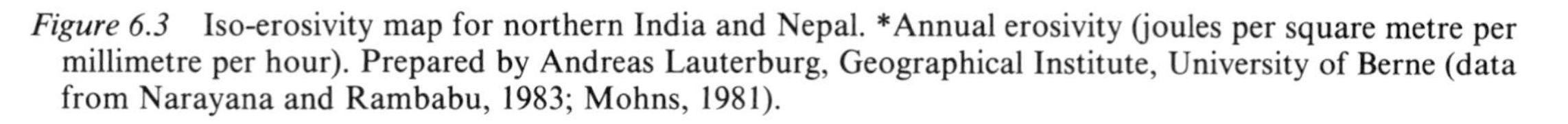

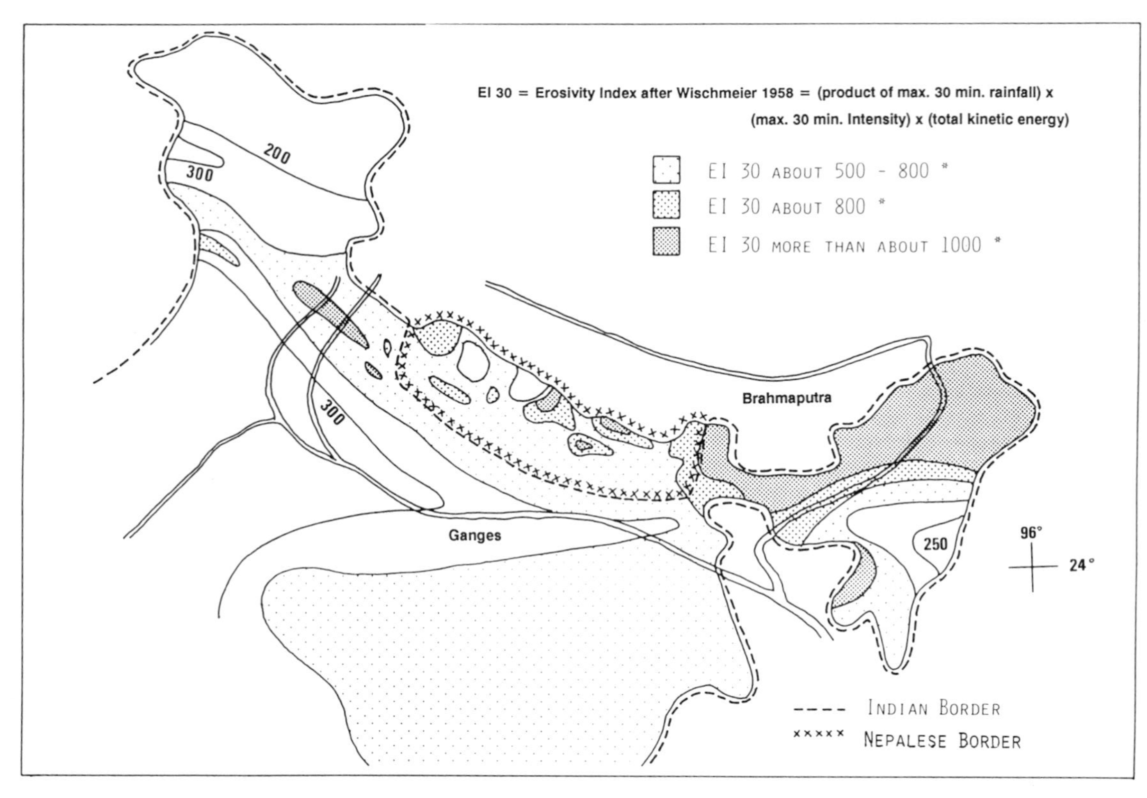

Figure 6.4 Climatic and lithological erosion hazard in Nepal. Prepared by Andreas Lauterburg, Geographical Institute, University of Berne (data from many sources including Shrestha, 1980; Mohns, 1981).

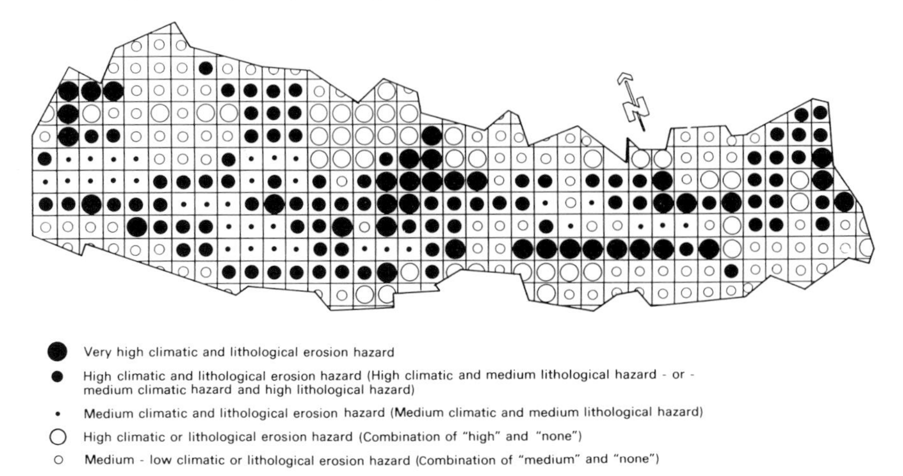

which is defined as low natural risk and high actual watershed degradation, in the central Siwaliks and the Kathmandu Valley and adjacent Middle Mountains. The much-discussed heavy damage in the Tamur and Arun watersheds in eastern Nepal does not appear and the Middle Mountains west of

Figure 6.5 Watershed condition and human-induced landslide increase in Nepal. Prepared by Andreas Lauterburg, Geographical Institute, University of Berne (data from Nelson, 1978; Laban, 1979).

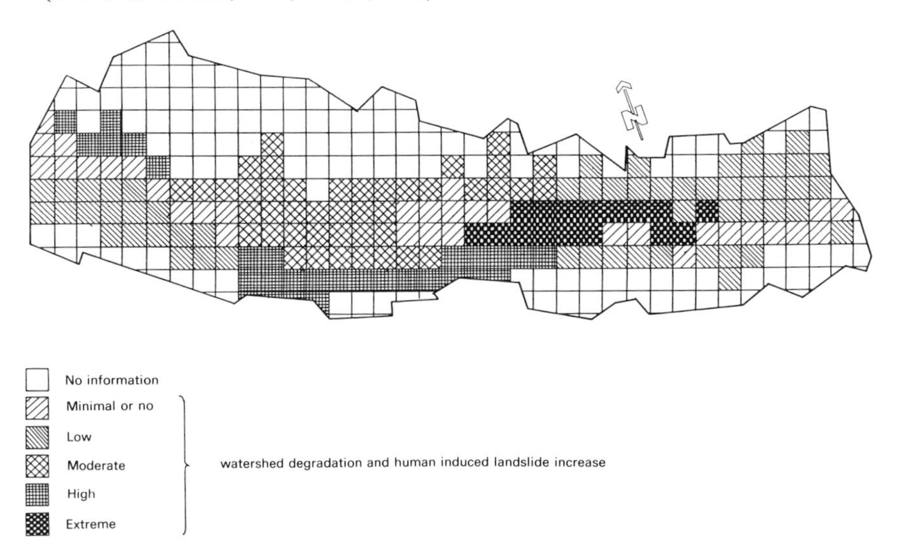

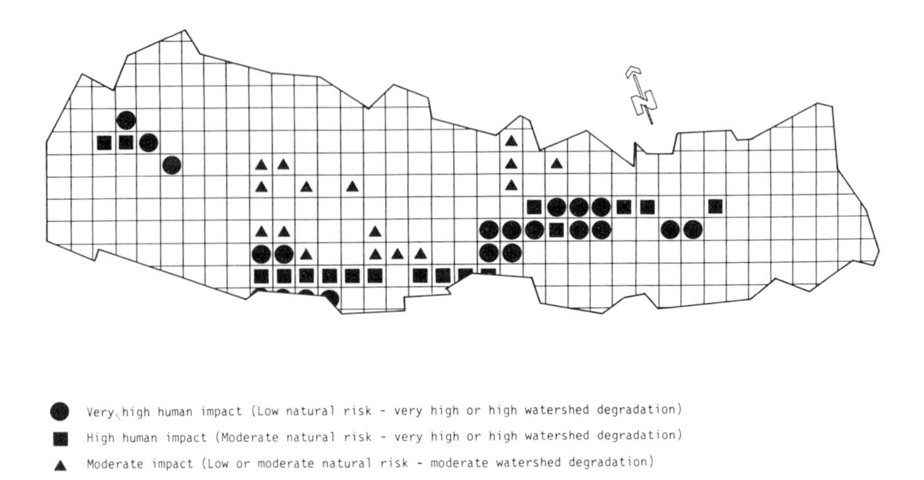

Figure 6.6 Human impact on watershed degradation in Nepal. Prepared by Andreas Lauterburg, Geographical Institute, University of Berne (data from Nelson, 1978; Laban, 1979; Shrestha, 1980; Mohns, 1981).

Pokhara are largely blank. The implication is, of course, that according to this approach, human impact (negative) must be quite low.

The approach of Lauterburg is interesting and, if more data can be accumulated, future construction of such maps could provide a valuable guide to land reclamation and watershed protection policy development. The present attempt, however, while giving some useful indicators, is not only limited because of data availability and accuracy, but cannot take into account the indigenous land-reclamation efforts of the farmers. Nevertheless, and regardless of gaps in data availability, Figure 6.6 creates the impression that the areas of high human impact in Nepal are very restricted. We are still left with the impression, however, that considerable areas of Nepal are in a condition of potential instability whereby heightened subsistence-farming pressures, or reduced maintenance of agricultural terraces, could lead to a rapid and dramatic increase in watershed degradation.

DOWNSTREAM EFFECTS OF WATERSHED DEGRADATION

Natural erosion in the Himalaya has been shown to be an important phenomenon and is probably higher than in most other major mountain systems. This implies that the Himalaya as a region is experiencing some of the highest, if not the highest, denudation rates in the world. This is due to the monsoonal character of the climate with high annual total precipitation concentrated in a three- to four-month period on an area of very high relief, susceptible lithologies and structures, and high seismic incidence. Furthermore, there is great annual variability in rainfall totals and occasional dangerously high rainfall intensities. And it is clear that this condition has existed for several million years.

It has been widely claimed in the literature that the devastating annual floods in the Ganges and Brahmaputra lowlands are influenced by extensive deforestation and intensified land use in the mountains. The human component of the total streamflow cannot be identified from any available data. Nor do the existing publications demonstrate any significant recent increase, either in sediment load of the larger rivers and tributaries, or in the magnitude of the annual flooding and levels of river discharge. Nor has any attempt been made to determine quantitatively the human impact on sedimentation and flooding on a large scale. Despite this it has been shown that, in very small watersheds, erosion and streamflow are highly influenced by man (Tejwani, 1984b, 1987). This prompts us to differentiate three scales, or sizes, of watershed to further our assessment of the downstream impacts of watershed degradation: the micro-scale; the meso-scale; and the macro-scale.

The somewhat arbitrary differentiation into three scales is illustrated schematically in Figure 6.7. The entire watersheds of the Brahmaputra,

Figure 6.7 Schematic representation of the relationship between human and natural processes at three different scales.

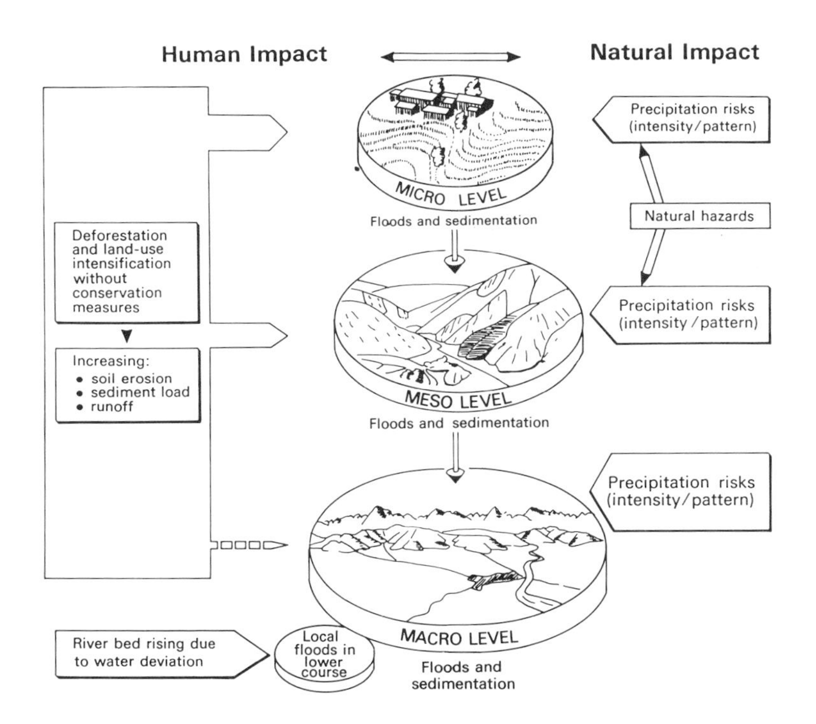

Ganges, and Indus, each represent the macro-scale. The meso-scale is illustrated by major Ganges or Brahmaputra tributaries, such as the Teesta, the Kali Gandaki, and the Karnali. The micro-scale can be illustrated by the Thulo Khola and Ghatte Khola in the UNU Kakani test area close to Kathmandu (Caine and Mool, 1981). Thus an approximate size class is as follows, bearing in mind that we will fudge this by including the (Sapta) Kosi in the macro-scale class for reasons explained below:

Micro-scale	< 50	km^2
Meso-scale	50–20,000	km^2
Macro-scale	$> 20,000$	km^2

Figure 6.7 demonstrates hypothetically how successful soil-conservation measures, traditional or introduced, can be in a degraded micro-watershed. Conversely, it also indicates the dangers of inherent slope instability and implies that adverse human impact can produce high-magnitude degradation. The value of soil conservation and watershed management practices on this scale have been amply demonstrated for watersheds of a few hectares (Garg, 1971; Chatra Research Centre, 1976; Mathur, 1976; Kollmannspererger, 1978/9; Impat, 1981; Christiansen, 1982; Tejwani, 1982, 1987; Narayana and Rambabu, 1983; Rambabu, 1984; and CSWCRTI, Dehra Dun, Annual Reports).

The lower and upper limits of the different watershed size classes are arbitrary, partly because there are insufficient data available for the Himalayan region to warrant a more precise approach. However, we wish to emphasize that this differentiation into three rough size scales of watershed is useful because the relative importance of human interventions within a watershed, and the downstream effects, change with watershed size. Because of limited data, however, critical sizes cannot be determined exactly.

There are good reasons for the foregoing statement. In small watersheds, as we have demonstrated, streamflow is less influenced by human activities than is the actual sediment load of the rivers. It is also reasoned that exceptional climatic events (for example, very high intensity rainfalls) will reduce the relative importance of human activities (accelerated erosion) compared to overall natural processes. The larger the size of the watershed, the greater will be the probability for local heavy rainfalls and these will also more likely influence the meso-scale watersheds downstream of the rainfall locality. Furthermore, large watersheds tend to have a smaller proportion of agricultural land to total area than small watersheds. This characterization only applies, of course, to micro-watersheds within the intensely settled Middle Mountains and lower attitudinal belts in comparison with the larger watersheds of which they are a part. For instance, micro-watersheds at high altitudes will contain little or no agricultural land. We therefore wish to emphasize the condition of meso-scale watersheds which contain a range of altitudinal belts with different degrees of land-use intensity, in comparison with a typical micro-watershed, such as the upper Bagmati (Kathmandu

Valley), or the Ghatte Khola, in Nepal. These are almost totally transformed into an agricultural landscape: the natural vegetation and even the original natural slopes, have been virtually eliminated (see Figure 5.4). A typical meso-scale watershed, such as that of the Trisuli, or Karnali (Nepal), embraces a high mountain belt with areas of steep rock, glaciers, and steep forested slopes where human subsistence activities are reduced to minute levels. Also, the larger watershed will possess a greater range of natural retention basins (lakes) that will reduce the downstream effects of processes occurring in the upper watershed (see also Vuichard and Zimmermann, 1987).

The Micro-watershed

Figure 6.8 is a graphic display of the rainfall, runoff, erosion potential, and actual soil loss for a small watershed near Chandigarh, Uttar Pradesh, that has been subjected to soil-conservation measures. Hydrologic response to treatment after 1965 as well as the reduction in soil loss are impressive. Comparison of the soil loss curve with the 'percent runoff to precipitation' (runoff = streamflow) curve leads to the conclusion that soil loss is much more

Figure 6.8 Annual rainfall, runoff erosion, and actual soil loss from small treated watershed near Chandigarh, India. Prepared by Andreas Lauterburg, Geographical Institute, University of Berne (data from Central Soil, Water Conservation, Research and Training Institute, Dehra Dun: Annual Report 1976).

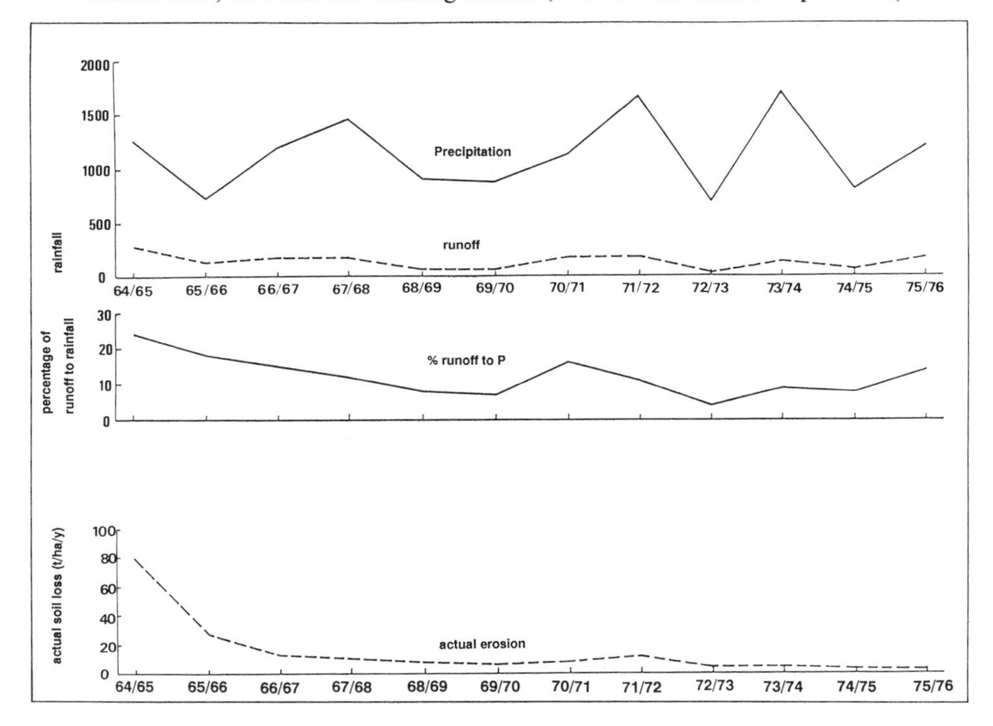

effectively influenced by land-use practices (in this case, soil-conservation measures) than is streamflow. From this it follows that the high variability of rainfall (Parthasarathy and Mooley, 1978), especially in the western and lower ranges of the Himalaya, has a much more important influence on measured streamflow regime than on soil loss. An important exception is the type of catastrophic rainfall event, such as described by Starkel for the Darjeeling area (1972a and b), which influences both streamflow and sediment transfer. Nevertheless, we can conclude that soil-conservation measures are more useful in preventing, or reducing, soil loss than in modifying the hydrological characteristics of the treated areas.

In terms of reservoir sedimentation it also follows from the foregoing discussion of the micro-watershed that soil-conservation measures can be extremely valuable in correcting a damaging situation for reservoirs in small catchments. However, the next sections will demonstrate that for meso-scale and macro-scale watersheds high reservoir sedimentation rates probably must be considered as an inevitable natural phenomenon.

The Meso-watershed

Lauterburg's search for data series over a long period illustrative of meso-scale watersheds was unsuccessful. But since we can assume that watershed degradation theoretically should influence streamflow characteristics an attempt will be made to interpret the few sets of streamflow data that are available for meso- and macro-scale rivers.

Figure 6.9 shows the annual peak discharge of the River Teesta from 1956 to 1975. The Teesta originates in Sikkim and is tributary to the Brahmaputra. The curve graphically illustrates the impact on peak discharge of the catastrophic rainstorms of 1968 and 1974 in the Darjeeling Himalaya. However, during the period for which streamflow data are available there is no demonstrable tendency for a trend toward higher or lower peak discharges. A full interpretation of the Teesta streamflow curve also requires access to watershed precipitation data as well as information on land-use changes over the same period. These are not available. In general, we do know that the Teesta watershed has been extensively modified by the spread of tea plantations and subsistence and market-gardening plots and that the area under these uses has increased since the early 1960s (Starkel, 1972a). It is also significant that catastrophic floods occurred in 1950, 1968, and 1974. The eighteen years between the first two events is perhaps long enough for local people largely to forget the impacts of the first event. The interval between the second and third events, however, is quite short, which can create an impression of increasing flood frequency. Nevertheless, the streamflow curve clearly demonstrates that the average streamflow has not increased; consequently we must regard the 1974 flood as a purely natural phenomenon. Because of the lack of the ancillary land-use and precipitation data and the short period of river flow record (twenty years) this statement is made as a working hypothesis which should be tested as more data become available.

Figure 6.9 Annual peak discharge and recurrence interval of River Teesta, northern West Bengal, India (Teesta Bridge gauging station). Prepared by Andreas Lauterburg, Geographical Institute, University of Berne (data from Gosh, 1983, in Central Board for Irrigation and Power).

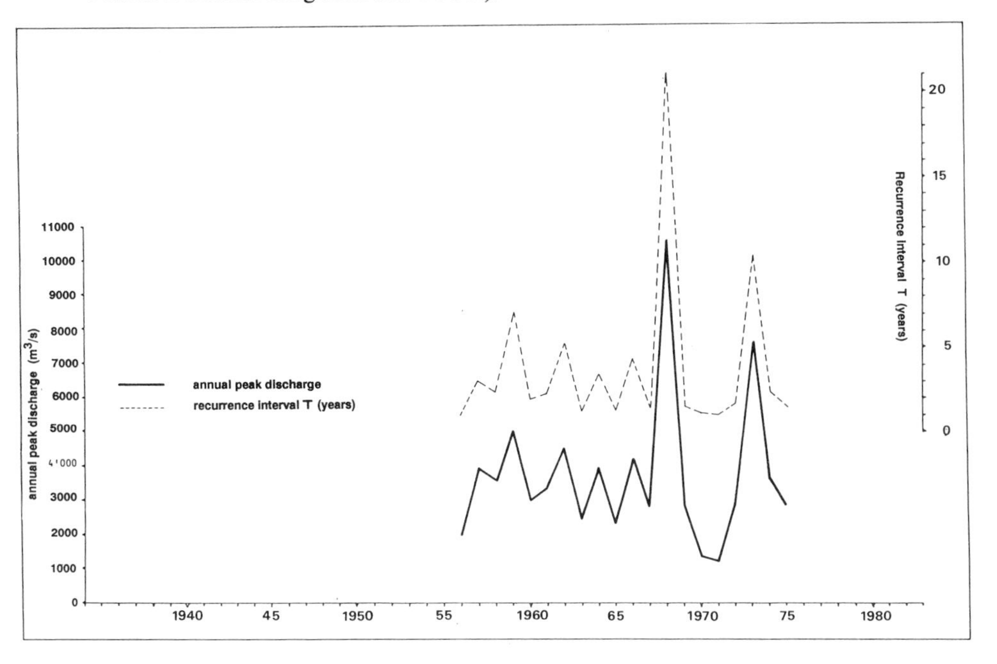

For the meso-scale watershed, therefore, downstream effects of landscape degradation are frequently stipulated but seldom have been demonstrated, and then only at the scale of a test plot (10 × 10 m^2) or a micro-scale watershed. Moreover, the dimension of anthropogenic downstream effects is not understood quantitatively. There is little information presently available with which to approach this problem. We conclude that the high intensity and high variability of natural events obliterate the effects of human interventions.

The Macro-scale Watershed

Macro-scale rivers in India and Nepal are much better documented than those of meso-scale watersheds. Nevertheless, because of the very recent formation of Nepal's Hydrologic Service (Shanker, 1983) only the big rivers in India have long-term data sets. Even here, however, many of the data collections of streamflow and sediment load are 'classified' and not available for scientific analysis – one of the especially unfortunate aspects of 'uncertainty on a Himalayan scale.' Some information is available for the Ganges, Brahmaputra, and (Sapta) Kosi and it will now be discussed.

Figure 6.10 provides data on streamflow, sediment load, and high- and low-flow hydrographs for the Brahmaputra between 1955 and 1979. This

Figure 6.10 Brahmaputra annual runoff, sediment load, and high- and low-flow hydrographs, 1955–79 (Pandu gauging station). Prepared by Andreas Lauterburg, Geographical Institute, University of Berne (data from Goswami, 1983).

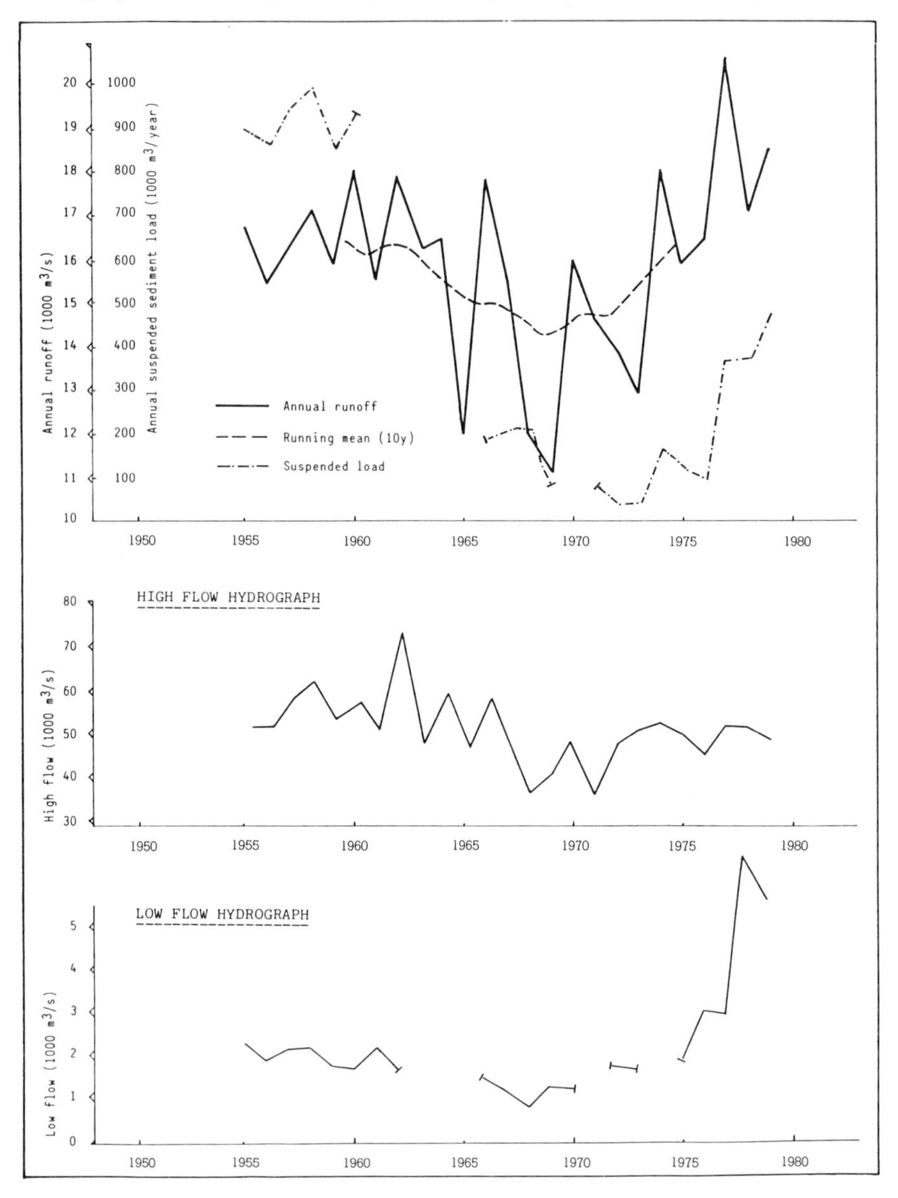

three-part figure provides an outstanding example of the risk of misinterpretation of relatively long-term data. The annual streamflow curve shows a significant increase in streamflow volume between 1969 and 1979. If we disregard the antecedent decrease between 1955 and 1969 we could interpret the increase over the 1969–79 period as evidence for the impact of human land-use change in the watershed. However, even for such a large river system as the Yarlungtsangpo–Brahmaputra, annual streamflow variability is extremely high (up to 100 percent). In addition the fluctuations in average streamflow over periods of approximately ten years are remarkable. Zollinger (1979), for instance, has indicated that major Himalayan rivers sometimes behave like mountain torrents. It follows, therefore, that long-term streamflow data series of less than about twenty years should not be used for calculating trends.

Figure 6.10 also shows variations in sediment load. This was extremely high in the late 1950s, decreased to less than 100,000 m^3 per annum in the early 1970s, and increased very rapidly betwen 1972 and 1979. Goswami (1985) and Rogers (personal communication, 1984) have interpreted the very high sediment loads of the 1950s as the consequence of frequent earthquakes in Assam between 1951 and 1956. Rogers maintains that the level of annual floods along the Brahmaputra has actually decreased since 1975 at an annual rate of about 15 cm. He ascribes this to the river having completed adjustment to earthquake-induced disturbance of its channel (including a throw of about 4 m). The reason for the steep increase in both annual streamflow and sediment load after 1979, however, is not known. It seems that earthquake frequency has not increased again (Goswami, 1985) and we dispute the assumption that deforestation is the cause.

The very rapid increase in suspended load of the Brahmaputra after 1976–77 (more than 300 percent) parallels the streamflow curve. Thus both increases, streamflow and sediment load, would have to be interpreted as resulting from the same causes – that is, deforestation or other human activitity. This is not a realistic conclusion. For instance, the high-flow hydrograph shows a decrease while the low-flow hydrograph shows an increase during the late 1970s. If human influence is to be considered, theoretically this would be demonstrated by a reduction in the water retention capacity of the watershed and, therefore, a reduction in low-flow and an increase in high-flow river discharge. This argument is based upon the standard predictions of the Theory of Himalayan Environmental Degradation. Thus, if increased human pressures on the landscape are resulting in land degradation – through accelerated deforestation, poorly maintained agricultural terraces, soil erosion, and landsliding, for instance – the anticipated hydrological responses of these processes would be increased streamflow during the summer monsoon (that is, an increase in high-flow discharge) and reduced availability of water during the subsequent dry winter and spring (that is, reduced low-flow discharge). On this basis, and in conjunction with the Brahmaputra streamflow data that are available, we must exclude changes in land use as a significant factor for explaining the streamflow and sediment load variations.

Having discussed the available data on the annual flow and sediment load of the Brahmaputra, let us now turn briefly to the Ganges and (Sapta) Kosi

Figure 6.11 (Sapta) Kosi at Tribeni: annual runoff 1948–76 (Tribeni gauging station). Prepared by Andreas Lauterburg, Geographical Institute, University of Berne (data from Zollinger, 1979).

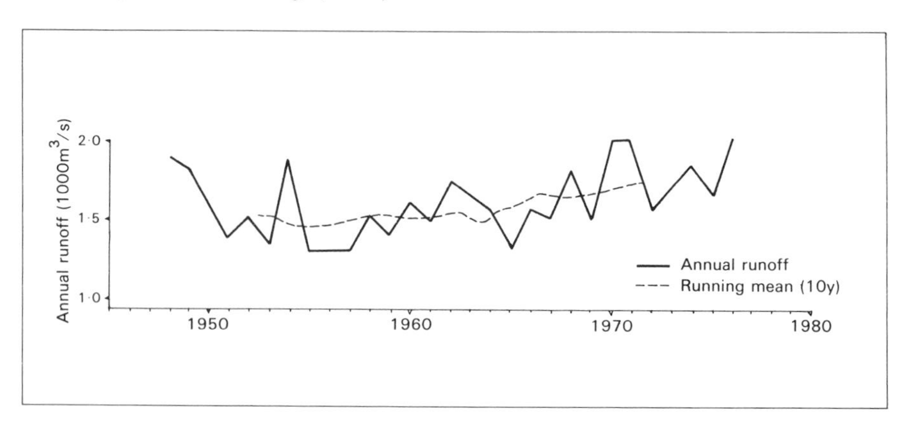

(Figures 6.11 and 6.12). We are including the (Sapta) Kosi in the macro-scale class along with the Brahmaputra, Ganges, and Indus, because of the large size of its watershed (30,000 km²) and its international position (it flows through China, Nepal, and India); as a tributary of the Ganges, of course, its watershed is much smaller than those of the other three. The (Sapta) Kosi is

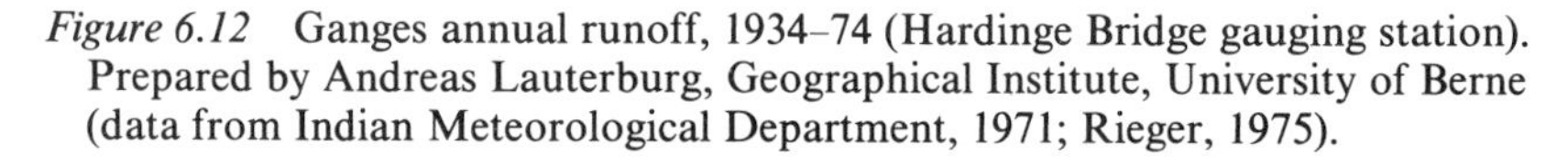

Figure 6.12 Ganges annual runoff, 1934–74 (Hardinge Bridge gauging station). Prepared by Andreas Lauterburg, Geographical Institute, University of Berne (data from Indian Meteorological Department, 1971; Rieger, 1975).

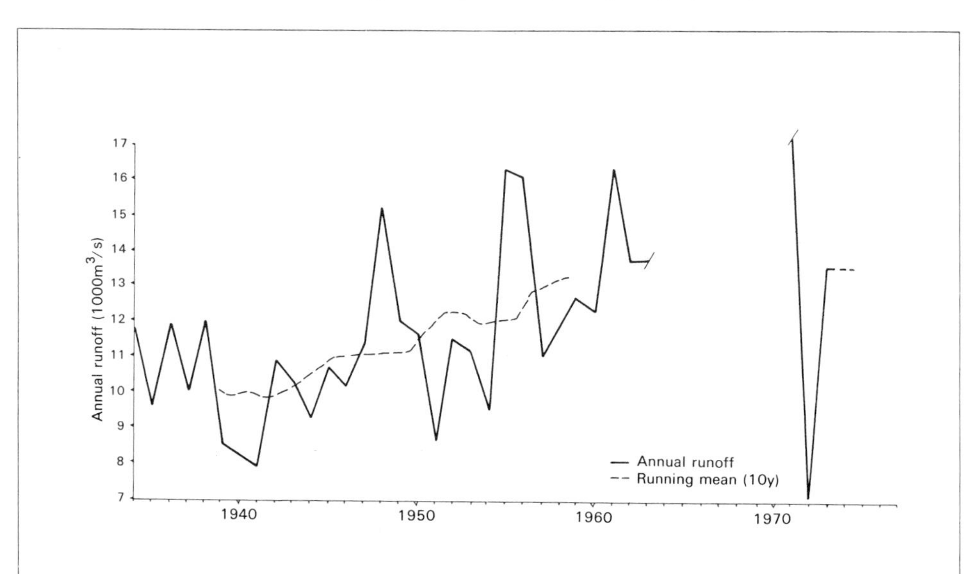

regarded as an extremely problematic river in terms of land use in Bihar State, and the attempts to control the river rank high amongst Indian hydraulic engineering projects in terms of total expenditures and duration of the work.

Figures 6.11 and 6.12 respectively show the annual streamflow of the (Sapta) Kosi at Tribeni (1946–76) and of the Ganges at Hardinge Bridge, about a hundred kilometres from the Bangladesh frontier (1934–74: 1963–70 missing), together with the 10-year running means. Both hydrographs display very high annual variability, especially that of the Ganges. The 10-year running means also demonstrate an increase in streamflow, for the (Sapta) Kosi from about 1954 to 1976, and for the Ganges from about 1940 onward.

The (Sapta) Kosi hydrograph is particularly interesting because of the widely acclaimed degradation of its upper watershed. While the information is qualitative, it is generally accepted that the Tamur watershed and the lower Arun valley (both main tributaries of the (Sapta) Kosi) are locally heavily degraded.

Whether or not the increased flow in the (Sapta) Kosi was caused by the increase in watershed degradation cannot be evaluated. This is because much of the upper course of the river and its main tributaries are located in Nepal and China (Tibet) and no long-term precipitation data are available. In addition, as we have seen in Chapter 4, the Dudh Kosi and Sun Kosi (also tributaries of the (Sapta) Kosi) are subject to periodic catastrophic outbursts of moraine-dammed and ice-dammed lakes. These natural events not only temporarily augment streamflow but add vast amounts of sediment to the river channels. Similarly, the large number of glacial lakes on the Chinese side of the Arun watershed would suggest that the Arun is also subject to such catastrophic and abnormal disturbances. Similarly, as has been demonstrated in the glaciological literature (for example, Østrem, 1974; Young, 1985) for watersheds with a significant proportion of their area under permanent snow and ice, the twentieth-century climatic warming has accounted for a significant increase in streamflow by accelerating glacier and high altitude snow melt. While no adequate supporting data are available for the entire (Sapta) Kosi watershed (but see Ikegami and Inoue, 1978; Fushimi and Ohata, 1980), that the glaciers in the upper reaches have been thinning and retreating since about AD 1900 is well known (Mayewski *et al.*, 1980). Nevertheless, we must conclude that a proportion of the increase in (Sapta) Kosi discharge *may* be due to human interference in the watershed. However, long-term changes in precipitation patterns and rates of snow and ice ablation could account for most, if not all, of the increase. Despite this it has been claimed that the (Sapta) Kosi is responsible for a massive increase in sedimentation and that the upper watershed is contributing 172 million tonnes/yr to the formation of islands in the Bay of Bengal (Fleming, 1978). Rogers (personal communication, 1984), however, counters that the Kosi barrage, close to the Nepalese border, effectively checks much of the downstream sediment transfer from the mountains onto the plain; the major problem below the barrage appears to be one of entrenching of the river

Figure 6.13 Channels of the (Sapta) Kosi River over the past 250 years showing the progressive westward shift of the river across northern Bihar State, India (after Carson, 1985).

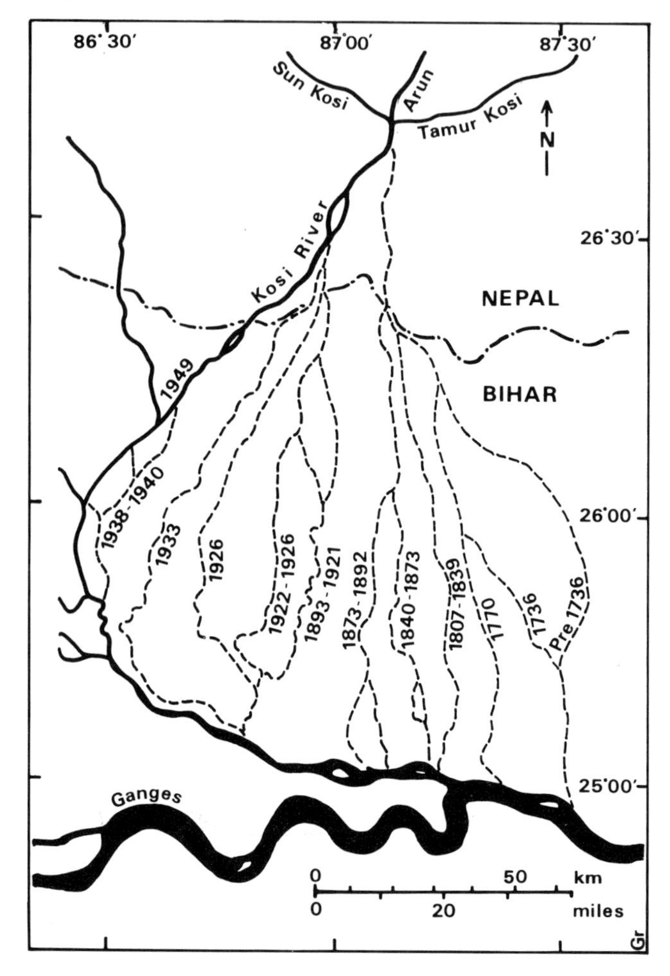

which has been relieved of much of its sediment load.

Figure 6.13 is a sketch map of the (Sapta) Kosi alluvial fan and shows the dynamic nature of the river's distributaries over the past 250 years. This indicates a 100-kilometre westward shift, and twelve distinct mainstream channels, during this period. This should be ample evidence to support the contention that the (Sapta) Kosi has been depositing vast amounts of sediment on its fan for a much longer period than that of recent (post-1950) human watershed intervention. This lends further weight to our supposition that the ratio of human (accelerated) erosion to natural erosion in the (Sapta) Kosi watershed could be very low indeed.

Figure 6.14 Actual and filtered rainfall for India from 1866 to 1970. Ten-year running means and periods of higher than average rainfall are indicated. Prepared by Andreas Lauterburg, Geographical Institute, University of Berne (data from Parthasarathy and Mooley, 1978).

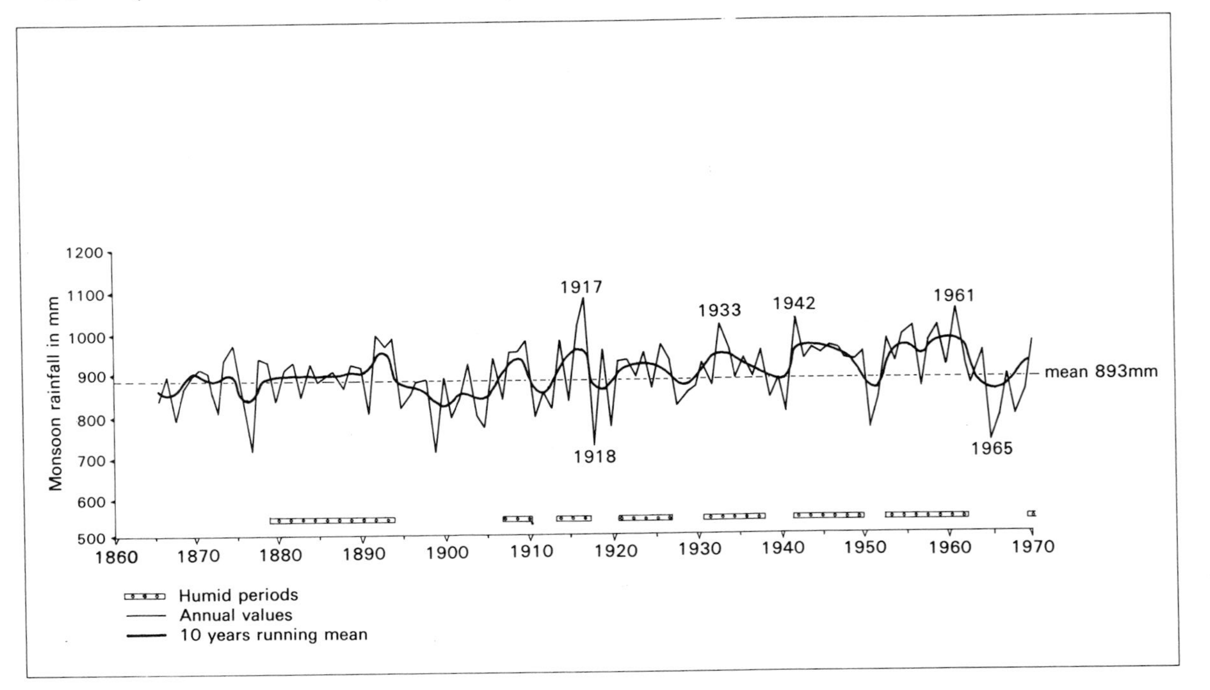

Figure 6.12 shows that most of the increase in streamflow of the Ganges (1940–74) is due to a decrease in the number of years with low discharge rather than the increase in high discharge years. The reduction in the number of low discharge years is certainly not due to anthropogenic impacts in the mountains, but must be the result of long-term precipitation fluctuations, with allowance for the impact of human intervention on the plains. To test this contention we have introduced Figure 6.14 which is an amalgamation of all the available monsoonal rainfall data for India for the period from 1866 to 1970 (Parthasarathy and Mooley, 1978). This long homogeneous series of rainfall data displays a remarkably high annual variability. Superimposed upon this, however, are the equally remarkable short-term fluctuations, particularly apparent between 1918 and 1970. This section of the curve can be broken into periods of ten to thirteen years. There are four periods each of about ten years' duration with higher rainfall totals than the long-term average. These are interrupted by periods of one to four years which are significantly drier than the long-term average (Parthasarathy and Mooley, 1978).

A comparison of Figures 6.12 and 6.14 shows a partial correspondence between variations in the streamflow of the Ganges and the aggregated annual monsoonal precipitation amounts. Lack of a strong correspondence may be due to the fact that the precipitation curve, constructed by Parthasarathy and Mooley (1978), is an integration of all Indian monsoon rainfall data. Despite this qualification, comparison of the two curves does suggest that human influence on changes in the Ganges discharge during the period 1944–72 is not demonstrable. It is emphasized, therefore, that streamflow curves of Himalayan rivers, especially meso-scale and macro-scale rivers, for periods of less than about twenty years should not be interpreted without taking rainfall data into account. Lauterburg's study failed to locate any macro-scale river that showed an increasing streamflow tendency over a long-term period. Nevertheless, there are short-term periods with increasing amounts or other features, such as variations in the high- and low-flow hydrographs, which give the overall impression that natural factors have dominated streamflow regime up to the present. From this it follows that, on the macro-scale, sedimentation rates are also dominated by natural processes.

The arguments presented above on variations in streamflow and precipitation of four major rivers of the Himalayan region do not lead to the inference that damage and loss of life from flooding has not increased during the course of the past fifty years or so. The scale and tragedy of these losses have been well documented. The large-scale flooding on the plains of Bihar, West Bengal, and Bangladesh, however, are most likely the result of the increasing number of human beings and livestock and the increasing intensity of agriculture in these areas in recent decades. Furthermore, monetary damage figures are often presented that have not allowed for inflation; thus there is the appearance of increase in damage even if the actual damage value had been the same in constant rupees or dollars. Nevertheless, the progressively increasing scale of loss is obvious, yet it cannot be demonstrated that this is due to human intervention in the mountain watersheds. Because long-term data are not available we are compelled, therefore, to draw the following conclusions:

1. The information network has improved rapidly since about 1940 and has promoted a much fuller awareness of the basically natural phenomenon of large-scale flooding on the lower Ganges and Brahmaputra.
2. The rapidly increasing population of northeastern India and Bangladesh has led to both intensification of agriculture and its extension into areas that probably were always affected by flooding. However, these areas could not be used for permanent agriculture until recently, or were not required when population pressure was much lower.
3. Hydro-technical modifications to the main river channels of the North Indian Plain (including causeways, barrages, and canals, for irrigation purposes, hydroelectric power plants, and spillways for diversions out of the macro-watershed of the Ganges) may themselves play an important role in changing the water–sediment ratio. Consequently this would affect the downstream sediment-carrying capacity below major infrastructures, and therefore may lead to local increases in the level of the river bed and to outbreaks of water from the main channels.
4. The fact that water is withdrawn from the river for irrigation purposes, which will also result in overloading the rivers with sediment further downstream, has not been taken into account, again because of the difficulty of obtaining adequate data. Nevertheless, responsible Indian ministries (personal communication to Lauterburg, 1985) are well aware of this kind of problem.

The overall message of this chapter, therefore, is that the fluctuations in annual streamflow and high sediment loads of the macro-scale rivers are the consequences of natural (especially climatic) processes. Locally, flooding and excessive sedimentation on the plains may be due to human intervention *on the plains*, rather than to the activities of subsistence farmers in the mountains. As attention is focused on progressively smaller watersheds, down to a few hectares, the potential for human impact undoubtedly increases. Even at the micro-scale, however, periodic catastrophic rainfalls and lithology, or specialized processes such as the outburst of glacier lakes, may heavily outweigh in importance the negative impacts of human activities.

We do not contend that soil-conservation practices are either unnecessary or not effective for the specific purposes for which they are undertaken. This is an entirely separate issue. It should not be confused by the macro-scale claims that a few million subsistence hill farmers are undermining the life support of several hundred million people on the plains (WRI, 1985). It follows that forestation of mountain watersheds, and extensive soil-conservation measures, are valuable for their own sake and, if appropriately carried out, are vital for the well-being of the hill farmer. It is potentially disastrous, however, for foreign-aid agencies and national government authorities to undertake such activities with the conviction that they will solve problems on the plains. Incorrect identification of causes of large-scale problems and attempts to treat them would appear to constitute a major factor in the steepening of the perceived downward spiral into environmental and socio-economic supercrisis. At the very least, considering the enormous costs and energy requirements needed for forestation of large areas in the mountains, if such were undertaken with the major objective of modifying conditions on the plains, an expensive disappointment is a likely result.

7 THE HUMAN DIMENSION: WHAT ARE THE FACTS?

INTRODUCTION

The preceding chapters have been concerned primarily with the physical components of the Theory of Himalayan Environmental Degradation. While the discussion has repeatedly expanded to questions of human intervention (it must be admitted, for instance, that deforestation is essentially a human process), we have concentrated on the physical facts and pseudo-facts of progressive reduction in forest cover and the assumed and real consequences of that reduction in terms of landscape degradation and far-distant downstream effects. At issue has been the relationship between natural, or geophysical, processes and human intervention (accelerated erosion, sediment transfer, and downstream flooding). We have concluded that human intervention has the potential for significant landscape changes at the scale of the micro-watershed or individual mountain slope. Yet even here we point out that periodic catastrophic rainstorms, for example, will tend to over-ride the effects of human activity. At the macro-scale, however, natural processes will reduce the role of human interventions to insignificant proportions, although this is not to claim that they are unimportant in specific locations.

Our approach to this discussion has relied heavily upon geomorphic theory, documentation of actual deforestation for those few areas for which information is available, and the few data sets that we can accept as reliable. This has resulted in our sweeping attack on conventional wisdom by exposing its lack of solid factual support and its dependence upon intuition, emotion, and the dramatic reactions of short-term visiting 'experts' whose *opinions* have been reiterated so often in the recent literature that they have become 'facts.' These 'facts' have then been popularized by a group of writers and journalists (cf. Sterling, 1976; Eckholm, 1976; Myers, 1986) and have permeated the mass media and the conservationist literature. This, we believe, is how the latter-day myths, the 'sacred cows,' of the Theory of Himalayan Environmental Degradation have come to life.

Despite our insistence that much available information is inadequate, or even invented, we have attempted to sift out facts that we do believe to be reliable, to use the theoretical base of our own discipline, and to conclude that the physical process components of the Theory are either invalid or unprovable. In this case, therefore, we submit that the Theory as an entity cannot be defended.

We persist in our belief, however, that the Himalayan region is facing a serious and accelerating crisis, both in terms of its degrading environment and in the increasingly desperate poverty of its rapidly growing population. In this chapter we will attempt a review of the information that leads to the definition of this growing poverty, and of the deductions drawn from it. We must also ask what are the standards by which a country, or community, or an individual human family is judged to be in a condition of poverty. This leads to an assessment of the data base as well as of the perspectives of those who set the standards.

We have not yet tackled the *causes* of this increasing poverty nor the relations between poverty and the apparently uncontrollable population growth. Perhaps we would do better to admit that, while we can make some tentative suggestions, we cannot produce a definitive answer, and simply put out a call for more research. This approach would be nothing less than an abrogation of moral responsibility; a giving way to the criticism that, when a university professor does not know what to do he collects more data; vast sums are already being spent on 'pure' and 'applied' research, and even vaster sums on development projects throughout the region. But how can research and 'development' proceed and succeed if they are based upon unreliable conventional 'scientific' wisdom and when the Himalayan Problem remains shrouded in uncertainty. Thompson and Warburton (1985a) and Thompson (1987) have asked the unthinkable question, unthinkable, that is, in terms of conventional science: if you don't know what the facts are, you can ask, what would you like the facts to be? This leads to their advice that, since the prevailing uncertainty is not likely to go away, then we should explore how we can use the uncertainty to good effect. This will be taken up more fully in the concluding chapters.

This preamble to a discussion of the human dimension indicates the necessity for a change of pace, in comparison with Chapters 1 to 6. The subject matter is very different and much more diffuse. We believe that the foregoing chapters hang together, in that they move, link by link, down the chain of the Theory of Himalayan Environmental Degradation.

To illustrate this need for a change of pace, we are prompted to raise the question: is there such a theory in relation to the human dimension? Our answer is that, most definitely, there is: it embraces the whole congeries of assumptions that leads providers of aid to believe that they are bestowing development on a series of 'target' populations. This is the idea of pulling Least Developed Countries (and communities within LDCs) up to the scientist's bench, or the surgeon's operating table, and doing the necessary to them. It is that aspect of the Theory that Griffin (1987, and see Chapter 10) is attacking so effectively and that the Bahugunas and Bandyopadhyays are resisting so strongly. A major consideration is that, contrary to all these mindless statistics and indicators (which we will go on to review) those 'ignorant and irrational peasants' do not just sit there waiting for development to hit them. (This paragraph is largely based on a personal communication, dated February 1987, by Michael Thompson.)

We wish to emphasize, therefore, that the so-called ignorant and overly fecund subsistence farmer is the central element of the Theory. We will try to arrange this chapter around this statistic, who, after all, is a human being with sensitivity, knowledge, and desire, like anyone else. The positive contributions of the subsistence farmer, in converting steep mountain slopes to productive agricultural terraces, in maintaining the stability of those slopes by dint of the muscles of his and her back, arms, and legs, aided by a great body of accumulated environmental lore, is as much at issue as the negative impacts, some of which are real, some assumed. Similarly, the villager's, or subsistence farmer's, attitudes toward forests, for instance, often in the face of negative governmental intervention, will largely determine whether or not remaining forests will be preserved and whether or not afforestation projects, large or small, will succeed. To quote Thompson *et al.* (1986) once again, it is necessary that we overthrow the conviction that the subsistence farmer is part of the problem and argue that he is part, a very large part, of the solution. This view has also been championed by Macfarlane (1976), Campbell (1979), Dani and Campbell (1988), and Messerschmidt (1978, 1981, 1987), and many other development anthropologists; this will become apparent later in this chapter.

To enter the complex issue of the human dimension we will concentrate initially on a discussion of the meaningfulness and reliability of the socio-economic and demographic data base and on the manner in which it is used. It is reasonable to ask: if, as we think we have shown, the physical 'facts' are largely unreliable, why should the human 'facts' be any better?

POVERTY IN THE HIMALAYA

This section draws heavily on one of the Mohonk Mountain Conference background papers which discusses poverty, women, and young people in the Himalaya (Pitt, 1986). It analyses information available in the files of major agencies of the United Nations Organization and the World Bank in an attempt to lay a foundation for answering the question: what is poverty in the Himalayan context and how is it determined?

Any large agency that has to face up to responsibility for a major section of humankind, will have to address the problem of data collection, storage, retrieval, and analysis. This is especially formidable when dealing from scratch with such issues as health, poverty, human welfare, in Developing Countries that were initially without any centralized bureaucracy comparable to those that inhabitants of the industrialized countries accept as given. Thus problems of data reliability, even of what kinds of data to collect, are a part of the process from the beginning. It is appropriate, therefore, that we attempt a review of this process in the context of the Theory of Himalayan Environmental Degradation.

For some decades the World Bank has ranked the countries of the Third

World using a variety of socio-economic indices. From this we are informed that, amongst the large number of very poor countries (Least Developed Countries = LDCs), Nepal and Bhutan were ranked fifth-to-last and last, respectively, in the 1981 *World Bank Atlas*. We cannot assess the comparable standing of other sections of the Himalaya (for example, the Kumaun and Garhwal Himalaya of Uttar Pradesh State, India, relevant parts of Xizang Autonomous Region, People's Republic of China, and northern Pakistan) because the relevant data are aggregated within those of the respective nation state. We will assume that, if specific data had been available, these mountain regions would compare closely with Nepal and Bhutan.

The ranking of Nepal and Bhutan is based upon a comparison of the Gross National Product (GNP) per annum per capita. The five bottom states on the World Bank LDC list are regarded as being in a condition termed 'supercrisis' (we intend to use the term 'supercrisis' on a regional scale here, rather than at the level of a single country, and will discuss it in more detail in Chapter 9). Since the other three lowest rank positions (in addition to Nepal and Bhutan) were taken in 1981 by Bangladesh, Chad, and Ethiopia, this alone should indicate the seriousness of the Himalayan situation if we assume that the listing has any significance.

Relying for the moment solely on the inferences that have been drawn from ranking by GNP per capita, it appears that extreme poverty characterizes sub-Saharan Africa and the land-locked areas of the Himalayan region, plus Bangladesh, and several Andean and other countries. Furthermore, the most recent figures from the *World Development Report* (World Bank, 1984a:218) show little relative change from the 1981 listing. There is perhaps one significant exception: Bhutan has dropped off the list entirely. Nepal, after showing an increase in the growth rate of GNP per annum per capita in the 1981 *Atlas*, registered in 1986 an annual decline of 0.1 percent for the period 1960–1982 – a rate of decline only exceeded by Chad amongst the last five LDCs.

Pitt (1986) makes the mildly cynical suggestion that most of the new Himalayan data on poverty refer to Nepal since it appears to have become very fashionable to study Nepal, both for academics and for agencies. This, if justified, raises another issue relating to data reliability, comparability, and representativeness, and is worthy of a fuller discussion. The reasons for going to Nepal, of course, may in part be due to the beauty, romance, and general attractions of the country and the friendliness and hospitality of its people. Pitt insinuates, however, that it may be due also, at least in the case of some agencies, to the fact that opportunities for intervention elsewhere have been significantly reduced. The favourite countries for technical aid in the 1960s have since turned sour. For instance, at one point the United Nations was even evicted from that South Seas Paradise, Western Samoa, while in many African countries military coups and dictatorships have made life very uncomfortable for development and research agencies: on one occasion a top-level UN mission on rural development was detained in a hotel for several weeks. Within the Himalayan region itself university scholars certainly find access into Nepal much easier than into most other areas: the northern frontier zones of India are especially restricted.

This openness of Nepal, however, may be as illusory as the welcome to those outsiders who thought that they had grasped the Samoan situation. The waters of Nepal may be even more muddy and the cultural mosaic more intricate. Certainly, the presence in Kathmandu of a large expatriate subculture living on comparatively enormous salaries, and the lack of co-ordination amongst a bewildering array of UN and bi-lateral agencies and competing projects, have caused a development problem of their own. And as we implied earlier, with certain exceptions, much of the rest of the Himalaya are virtually a closed book, and even the new UN reports on Bhutan are not very informative.

The ranking of LDCs, therefore, and especially the very low ranking of the Himalayan region as a whole, must be viewed critically in terms of data reliability. The indicators of poverty that are normally used, GNP per capita as indicated above, and social indicators, such as infant mortality rates, do not vary significantly over time, nor among the different sources in which they appear. But this may be because the sources are all quoting the same primary data so that the appearance of corroboration is spurious. Nevertheless, even if the data do need to be handled with scepticism, some relative points can be made.

There are two groups of comparative indicators of poverty which are in wide use, the one economic, the other social (notably health and education). In the 1984 *World Development Report*, two economic indicators are termed *basic*: the GNP per capita and the inflation rate. The Nepal score in the first is US $170 per annum per capita for 1982. This shows an increase of US $40 over the 1979 figure (*World Bank Atlas*, 1981). That this is regarded as a *decline* by the economic wizards is apparently not sleight-of-hand but the product of the method of calculation of GNP (World Bank, 1984a: 274), and the fluctuating rate of the purchasing power of the US dollar.

It is generally agreed that there are many things wrong with using GNP as an indicator of poverty, or indeed of development. This is especially true when we descend from consideration of a country as a whole to that of an ethnic minority, or of an individual family. GNP is a measure of the total domestic and foreign output claimed by residents, without any reduction for depreciation, divided simply into the total population. No distinction is made for distribution of income. It is not possible to include the so-called 'black economy' or the activities of big, and especially multi-national, companies, nor the extent of foreign-exchange transactions. Only quantified and quantifiable elements are included, so that in addition to the other limitations, the entire subsistence economy is omitted from consideration. In Bhutan, the 1981 *World Bank Atlas* claimed a GNP per annum per capita of US $80 for 1979. This figure is quite meaningless because it is estimated that traditional subsistence agriculture accounted for more than 90 percent of total 1980/81 production (World Bank, 1984b:18). Mr. D. N. S. Dhakal (personal communication, 1985), a UNU Fellow working with us, made the telling point that in his village of birth in Bhutan the average farmer, who is the faceless statistic accounting for the average US $80 GNP per capita/yr, would not know what to do with the money if he had it – 'it is GNH, gross national happiness, that counts, not GNP!' A similar situation prevails in Nepal and presumably throughout the Himalayan region. Gross National Product per capita, in terms of access to cash, hardly indicates wealth or poverty in a non-monetary economy judged on local standards rather than from a Eurocentric perspective.

An interesting issue is the starting point for the entire process since the GNP per capita for Year I, the first year that a country is ranked, provides a benchmark for evaluating its economic progress or decline. The case of Nepal is of particular interest and it may serve as a caution when viewing the ranking progress of several other countries. In the early 1950s Nepal was virtually bereft of normal government files and accumulated information relating to the national economy. Stiller and Yadav (1979: 35–36) make the point that: 'It is common knowledge, for instance, that the base-line for economic growth, the *per capita* income for Nepal during the early years of the planning period was arrived at by just such a method of approximation.' They go on to recount Dr. Y. P. Pant's account of this amazing story. It is quoted in full, not only for its own sake, but because it also explains how he, as Secretary of Finance, Secretary of Planning, and Governor of the Rastra Bank, chose to qualify the deed.

> In Nepal no serious attempt could be made earlier to calculate national income due to the dearth of statistics. The first attempt was made by Prof. W. O. Thweatt, an economist in the Ford Foundation, for the year 1954. For 1954 he estimated the national income . . . at $40 (Rs 304) [per capita]. He derived these figures simply by comparing the *per capita* income of Pakistan, which he guessed to be almost similar to Nepal's with respect to economic development during the period. . . . Prof. Thweatt's estimate was almost an *ad hoc* approach, which, no doubt, has helped to create some basis for further investigation. He himself calls it only a rough approximation.
>
> (Y. P. Pant (1970) quoted in Stiller and Yadav, 1979:36)

The commentary by Stiller and Yadav that follows this quotation is also worth citing in full:

> Besides explaining the origin of this statistic that is often quoted for establishing growth curves in Nepal, Dr. Pant has provided some interesting examples of how even among economists statistics slip from 'rough approximations' to harder facts. He calls Thweatt's estimate *almost* an *ad hoc* approach, even though he knew very well that it was sheer fabrication.
>
> (Stiller and Yadav, 1979:36)

Stiller and Yadav also explain that Thweatt's knowledge of Nepal was 'good for a man in his position but limited nonetheless.' He had absolutely no observational data and so was prompted to make 'this hilarious jump to Pakistan . . . ,' and Dr. Pant, after sombre reflection on this 'almost *ad hoc* approach,' believed that it assisted in creating some basis for further investigation. The most telling point, however, in this passage from Stiller and Yadav (1979) is their concluding remark: 'Even the softest of statistics are treated progressively as harder and harder data until the statistics become accepted as fact.' This at least draws the useful comparison with our earlier concern over the physical 'facts': the arena we have now entered should not be entirely unfamiliar even though we are facing a different species of lion.

To conclude the excursion into establishment of Nepal's benchmark GNP, if we accept Dr. Pant's approach, then we have demonstrated that by 1982 Nepal's GNP had improved by a factor of more than four. If we refuse to accept the approach, what else is available? This is the very problem that faced us in the preparation of this chapter: if we leave our readers confused over which conclusions or inferences to accept and which to reject, at least they will sense the essence of the dilemma that faces all decision makers and most scholars.

To be fair, the World Bank has drawn attention to the fact that GNP per capita figures should be used with extreme caution. And in general they are supplemented by a battery of other economic and social indices. For example, the 1981 Conference on Least Developed Countries (UNCTAD, 1981) omitted GNP per capita completely and placed more weight on agricultural indicators (such as percentage of labour force in production, agricultural production growth rates, output per worker, and so on). The agricultural production growth rates for Nepal showed an actual 1 percent per capita decline for the period 1960–79. While the Bhutan growth rate was zero, this is better than the average rate for all LDCs which showed a decline of 0.8 percent.

One indicator, however, is quite remarkable for Nepal: that is the 1976–77 income distribution figure quoted in the *World Development Report* for 1984 (World Bank, 1984a: 272). According to this, the highest 10 percent of households accounted for 46.5 percent of the total income. This is very nearly a world record, being exceeded in the World Bank table only by Brazil with 50.6 percent. This means that the distribution of wealth in Nepal is highly stratified and that poverty is heavily concentrated in the lower-income groups, creating a large gap between them and the upper-income groups. Of course, it is another reflection of the predominantly subsistence economy; and it would be instructive if comparable data could be obtained for various Indian states which have a lowland and a Himalayan division, such as Uttar Pradesh, Himachal Pradesh, Jammu and Kashmir.

Inflation is a much more difficult indicator to use in the Himalayan situation although there are certainly figures available. For example, the World Bank 1984 *Report* for Nepal gives a rate of 8.9 percent per annum for the period 1970–82, compared with 7.7 percent for 1960–79, and 11.7 percent for the thirty-two low-income countries (excluding India and China). Nevertheless, the rate for Nepal had exceeded 10 percent by the end of 1983

(ESCAP, 1984). The 1984 World Bank *Bhutan Country Report* suggests a rate of about 11 percent for consumer prices. The figures for Nepal and Bhutan are relatively low compared with some African countries where inflation is running away at 30 percent or more per annum.

We must ask, however, what do inflation figures indicate in terms of purchasing power or, more precisely, the ability of an ordinary family to feed itself and provide for basic needs? The lowest average inflation reported by the World Bank for any country in the world for the period 1970–82 was Ethiopia! By the end of this period, as is well known, Ethiopia was facing a national calamity and earned the descriptor of 'supercrisis' state.

The inflation rates quoted refer only to quantified elements in the commercial economy and tell us little about the subsistence economy. In the case of Bhutan, with over 90 percent of the economy estimated as subsistence, the inflation rate is merely a mirror image of the cost of imports from India. The subsistence sector in Nepal is probably between 70 and 85 percent, although it is declining (Acharya and Bennett, 1983, Table 2).

Inflation, of course, can take place under subsistence conditions, as anthropologists have shown in discussing the Kwakiutl potlatch, a traditional feast, and similar conditions. More relevant to the Himalayan region, however, is the high and rapidly increasing cost of marriage (*biha*) in Hindu society, and of the *arghun* funerary activities and expenses amongst the Gurungs (Messerschmidt, 1976a). But in these situations there are ritualized means of stopping an upward spiral of inflation and of mitigating any effects, notably through gift exchanges. The basic problem facing subsistence farmers, however, is that people cannot transfer their subsistence incomes (farm and forest products) into cash, nor can they easily obtain cash. This becomes more serious when taxes and other financial obligations must be paid in cash rather than in kind. Therefore, the inflation rate is effectively increasing the difficulties of obtaining any cash, and thus enhancing the processes which divorce them from their land, livelihood, and means of self-reliance. If it could be calculated, this inflation rate likely would be very high indeed.

The difficulties in drawing meaningful conclusions from income figures and other economic indicators have led to an emphasis on social indices – the so-called physical quality of life indices (PQLI) – a step closer to the Bhutanese gross national happiness evaluation. For example, some countries which rank very low in the World Bank's (1984) GNP per capita table (Western Samoa, Sri Lanka, the state of Kerala, India) show a much higher standing when the indicators used are infant mortality rates, literacy (notably female literacy), and life expectancy. Health without wealth has been seen as a desirable goal for development and a situation where poverty (as defined by the numbers game being discussed) is not necessarily a problem at all. Such a circumstance has been variously explained (Defence for Children, 1982). The status of women and a relatively egalitarian social structure seem important prerequisites. For example, in income distribution tables, Sri Lanka scores very well (28.2 percent of the household income being obtained by the highest

10 percent of households, an egalitarian rate compared to some of the other LDCs).

The PQLI figures, however, are also suspect, especially infant mortality rates, since many people do not register births and certainly not the infant deaths, even discussion of which is often taboo. Without these figures, expectation of life can only be a guess. Literacy, also, is an ethno-centric concept, and is based upon criteria obtained by written cultures (usually the industrialized countries). Most people, especially those of the LDCs, are part of the world of oral cultures.

With all of these reservations, let us consider the PQLI for Nepal and Bhutan. They still do not indicate an acceptable standard of living, although they are not apparently declining. For example, infant mortality rates in Nepal for 1980 were given as 150 per thousand (UNCTAD, 1981:32ff) whereas the World Health Organisation (WHO) goal, for comparison, is 50, and the average for all low-income countries is 87 per thousand. The 1982 figure for Nepal was 145 per thousand (World Bank, 1984a:262). For Bhutan it was given as 147 for 1982 (World Bank, 1984b:18). The adult literacy rate in Nepal in 1975 was 19 percent, although the most recent government figures (1981) claim 24 percent (Ministry of Education and Culture, 1984), which is the LDC average. The adult literacy rate in Bhutan for 1982 was set at 10 percent (World Bank, 1984b:18).

Access to safe water has been a major problem, and probably underlies the high infant mortality rates. In Nepal in 1975 it was calculated that only 9 percent of the population had access to safe water (UNCTAD, 1981:32), and the experience of tourist visitors to Kathmandu might serve to challenge even this low figure as optimistic. In Bhutan the figure is probably lower, as only 6 percent of the rural population was rated to have access to safe water in 1982 (World Bank, 1984b:18) compared with a 35 percent average for all LDCs. In contrast, both Nepal and Bhutan had a higher than average daily calorie intake as a proportion of estimated minimum requirement (91 percent and 88 percent, in 1977, respectively), compared with 86 percent for all LDCs (UNCTAD, 1981:32). An overall assessment of these PQLI figures does indicate that there is some improvement, provided we can assume that the margins of error are consistent, and this is not necessarily a safe assumption.

The conclusion may be drawn that, by available measurement standards, Nepal and Bhutan are very poor countries though in some areas (notably water power) there is a great potential for improvement (World Bank, 1979). The assessment of great poverty, however, does not necessarily infer that they will become supercrisis states like Ethiopia and Chad. It could be argued that in those latter two countries there are additional catastrophic elements that are not present in the Himalaya at the moment. In Ethiopia disaster, including a rapid rise in mortality, is due, in part, to political unrest and conflict in the midst of drought, which not only takes its own toll of lives but also disrupts subsistence agriculture and severely curtails the effective distribution of food aid. Nor is drought a particularly pervasive and significant phenomenon in the Himalaya, as it is in Ethiopia and Chad and other parts of Africa.

In order to understand better the relationships between the environmental problems in the Himalaya, however defined, and poverty, and the effects of recent changes, it may be necessary to disaggregate, to look in greater detail at the poorest groups within the region, and at changes over time. Here the UN and World Bank statistics are not particularly helpful, partly because they consistently aggregate, and partly because quantitative figures may be lacking for certain areas and for time series, and because they tell us little about causes and dynamics. Much more reliance, therefore, has to be placed on field studies including, where available, anthropological work, even if this work is scanty, and may not be at all statistically representative.

What, then, can be said about recent changes in the poverty situation and the links to the broader environmental problems? What light is thrown on the key causal patterns? The simple explanation proposed by many outside agencies has been population pressure, and indeed indicators based on per capita estimations suggest that poverty deepens with increased population. One might presume, however, from one statistic that this is not the way women, at least, interpret the situation. The World Bank's 1984 *Report* (p. 256) states that in Nepal only 7 percent of married women of child-bearing age are practising contraception (itself one of the more dubious statistics). The population appears to be increasing at an unexpectedly rapid rate. Nor is everyone blaming overly fecund peasants. One recent International Labour Organization (ILO) study has boldly stated that 'the conclusion often drawn – that it is population growth that is *largely responsible* (their emphasis) for the growing poverty of the mass of the Nepalese people – is a gross over-simplification and ignores the crucial fact that demographic change, like material deprivation or poverty, is a social product, conditioned and determined in the last analysis by the economic and social structures of the state' (Seddon, 1983:1). Population growth, then, may be regarded as one of the symptoms, not one of the causes, itself to be explained by what is happening in other sectors.

The Seddon ILO (1983:32) study is of exceptional interest since it is based on detailed fieldwork of household behaviour and economic status. A major emphasis is food production – not the overall quantities but the access that households have to food supplies. The conclusion is that grain availability for Nepal had moved from surplus in 1971 to deficit in 1981 (from +66,921 to –108, 278 metric tonnes). In 1981 the only region with a surplus was the Terai and it was predicted that this would be eliminated by the late 1980s. The Terai surplus so far has been earning foreign currency rather than being used to maintain minimal nutritional standards. Messerschmidt (personal communication, March 1987) sees the surplus grain of the Terai as earning two things: (1) foreign currency (initially); and (2) foreign debt (incurred later when Nepal re-imports rice to meet deficits, usually in the Spring, when local stocks have been depleted). The situation in the Middle Mountains is very much worse, and the 70,697 metric tonnes deficit of 1971 increased to 123,755 metric tonnes in 1981. In the Terai both production and yield, according to the Ministry of Food, Agriculture, and Irrigation, seem to have declined between

1974 and 1978, despite the increase in land under cultivation (60,000 ha) through irrigation, the rise in fertilizer use of about 40 percent, and the enormous increase in area under high-yielding crop varieties (Seddon, 1983:34). The reasons given are poor weather, failure to extend irrigation further, and inability to apply new agricultural technology. All crops seem to be in decline, except possibly rice (Seddon, 1983).

These conclusions of Seddon, based upon a historical perspective, are now somewhat out of date as well as being open to challenge because they were derived from fieldwork in a limited area. Since the 1970s wheat production in the Middle Mountains and in the Kathmandu Valley has increased significantly, due in large measure to the introduction of new varieties and the rapid growth in cultivation of winter wheat. Potato production has also risen, although its contribution to the total food consumption is rather insignificant (Horton, 1987). To cite one specific instance (Messerschmidt, personal communication, March 1987), Marsiangdi, Lamjung district, became a winter-wheat producing centre of some note in the ten years following Messerschmidt's introduction of new varieties in 1972. Today the entire valley is green with wheat in winter when previously it was brown with fallow.

Such improvements are to be found in many parts of the Himalaya. Nevertheless, food shortage crises, especially in the Middle Mountains, occur with increasing frequency (it appears that the improvements do not keep pace with population growth). As a result, not only are foreign earnings from export crops declining, but international food aid is required. A special committee was established in 1980 and agreements were signed with the World Food Programme, India, USA, and others. Nevertheless, food has continued to be exported and new agreements have been reached with China, for instance, for the export of flour, beans, and pulses. Only some mountain regions maintain grain surplus, but even here food aid and subsidization has been necessary.

The implications of this combination of local food deficit and increased aid are considerable. It has been shown that the effects of such aid in Africa, where there have been many negative development effects, such as corruption and internal distribution problems, a dramatic decline in subsistent, self-reliant, agriculture, and a decline in maternal and child health, have occurred (Defence for Children, 1982). A similar scenario seems to be building up in Nepal.

The aid-giving agencies, however, have fought vigorously against the notion that introduced aid may be counterproductive. The crisis of food production in Nepal has been blamed on the Nepalese farmers who, according to several agencies, have not accepted quickly enough the introduced technology, but have preferred to stay with what is called their 'archaic' and 'traditional' agriculture. Such a stereotype needs careful examination. Recent studies (Seddon, 1983) have shown that the larger farmers with more land (especially in the Terai), more financial resources and capacity, do invest in the new technology. Tractor-owning farms were an average size of 17.7 ha, compared to a general average 3 ha according to one study in Parsa and Chitwan. But even here, investment was modest by international standards. One can argue, however, that machines have a limited utility on slopes, and the cases cited here refer to the Terai.

Fertilizer use is limited, according to Seddon (1983), not because of lack of finance, access, nor the weak development of agricultural extension services, but by the farmers' own choice. Nevertheless, the large increases in wheat production, referred to above, result primarily from the introduction of new varieties and the adoption of winter cultivation and reduction in winter fallow, rather than from increased use of artificial fertilizers.

The lessons from the history of agricultural production are relatively simple. The retention of traditional technology has not only a long-term logic in the sense of preserving a delicately balanced agro-ecosystem, but is probably the only available course of action when incomes, notably from the agricultural sector, fall below a certain level (Pitt, 1970, 1976; McNeely and Pitt, 1985). At this point, population increases because traditional agricultural methods are labour intensive (this is only one of a number of possible causes of population growth). It is human hands and energy that provide the capital. In this respect, there appears to be a striking difference between Nepal and Bhutan, the so-called Himalaya poor states, on the one hand, and the supercrisis states of Africa, on the other. Traditional African agriculture or pastoralism have been seriously undermined and in places virtually destroyed, in some cases by deliberate government policies which have discriminated against marginal social groups, and have, in the interests of centralized political control insisted on sedentarization (see also Blaikie, 1985; Blaikie and Brookfield, 1987). The traditional Himalayan social system cannot be broken so easily in this way, although one might hypothesize that governments may have a vested interest in encouraging, or at least in not discouraging, the recent urban drift, because it provides a reserve army of labour and may facilitate political control.

It is already obvious that any simple label of poverty does not adequately define a situation where there are very considerable sociological differences in the degree of poverty. What are these sociological dimensions in Nepal and other areas of the Himalaya? There are some interesting recent materials from which a more detailed mosaic can be constructed, although these deal mainly with Nepal.

A first fact is land ownership. Generally speaking, those who own land are better off than those who do not, and the larger the landholding the higher the standard of living. The *World Development Report*, World Bank (1984a), as mentioned above, indicates a high degree of concentration in the distribution of Nepalese wealth. In 1981, in the rural sector, 2 percent of all rural households were said to cultivate about 27 percent of the land (Seddon, 1983:93). The land-reform measures had obviously not worked, partly because the official ceiling for landholdings was quite high, 25 *bighas* (about 18 ha). Inequalities in size were most marked in the Terai, but in the Middle Mountains the problems continued to relate to the small, often uneconomic size of holdings. However, the proportion of landless (continuing at about 8–10 percent) is not high by Asian standards, but most of the rural poor who are landowners have too little land on which to eke out a viable living. They are therefore forced to rent additional land, to practise sharecropping on larger estates, or to seek waged labour. These options may be precluded, at least locally, by the high rentals and the scarcity of employment opportunities. As a result, according to Seddon (1983:93), 63 percent of family labour days were underemployed. This has led to a further stimulation of migration.

The size of landholdings also relates directly to nutrition and calorie intake. A 1978 survey of households showed that farms of less than 1 ha had an average calorie intake of 1,500, or less, per person, whilst those over 1.01 ha produced 2,250 calories per person, or more. This depends to some degree on types of crops and the annual yield but overall, according to Colin and Falk (Kosi Hills Rural Development Project – KHARDEP, 1979), small farms are deficient, and farms under 1 ha produce only 60 to 70 percent of household requirements while those over 1.01 ha produce 86 percent.

The Agricultural Projects Services Centre (APROSC, 1978) report that annual per capita food consumption (in kg) varies significantly between regions (297 in the far west, 480 in the centre, and an average of 373 for all Nepal). In the far western region 59.2 percent of households were considered to be below the poverty line (defined in Nepal as an earning capacity of 2 N rupees/day on 1977 figures). What is not clear from these admittedly inconsistent figures is how quickly the situation may be deteriorating. A decline is not necessarily corroborated by the steeply rising migration rate because there are many reasons, other than economic, for migration (Pitt, 1983), but for the Himalaya, the two are probably linked (Gurung, 1973; Elder, 1974; Manzardo *et al.*, 1975).

Landholding, in certain circumstances, may be a sign of poverty. Nevertheless, out-migration from the Middle Mountains is usually the result of abandonment of marginal land which cannot provide the minimum sustenance for enlarged families. However, since there is no base-line it may be that the numbers of those below the poverty line (which is more of an ethnocentric concept, rather variously applied from outside) have not changed greatly.

What is more interesting perhaps in the most recent figures is the indication of the concentration of poverty. But it is significant that the figures do not pinpoint the social groups which are most deprived nor explain why. Individual regions contain many ethnic and caste groups, households vary in sex and age structures, and the size of landholding tends to become irrelevant when the vital extra income has to be earned by other means. Seddon (1983) did not find an answer in any feudal-type obligations since, presumably, even the larger landowners had few resources. More significant may be the differential access which different castes and ethnic groups have to wage-earning possibilities: for example, Gurungs, Magars, Limbus, and Rais are noted in military service, high-caste Brahmins and Chhetris are preferred by employers, and 'untouchables'[1] encounter employment barriers by virtue of their low or outcaste social status. These conclusions of Seddon (1983) must also be handled with extreme caution, as the whole issue of employment opportunity is much more complex than one of caste and status. Status plays a role, along with the related factors of education and political influence. It could be argued that 'untouchables' encounter no greater employment barriers than others in the sense that (a) the upper castes have cultural barriers to many categories of employment also, while (b) lower castes have access to manual labour, portering, etc, from which the higher castes are barred. The argument could even be turned on its head: high castes have reduced latitude for employment because of the restrictions of their caste status (Messerschmidt, personal communication, March 1987). Nevertheless, there are widespread indications of a relatively steep recent fall in standards of living because of the rural people's declining purchasing power (for rice or maize, particularly) (ARTEP, 1982).

This fall in purchasing power seems to have had a number of major effects in terms of the satisfaction of basic needs. First, nutrition: in the Middle Mountains and high Himalaya total available calories per person per day dropped from 1,569 in 1977–77 to 1,426 in 1983–84, and they were projected to drop to 1,299 by 1989–90 (Seddon, 1983:124). Though the situation in the Terai was not as critical as this in terms of averages, most people had less money, especially the smallest farmers and the landless. Nutritional problems were accentuated amongst certain ethnic and caste groups because of food taboos. The fact that demand concentrates on certain foods raises the price for these foods, although caste and ethnicity may be less important in some areas. Sacherer (1979) has shown clearly in her studies in Dolakha district how this factor affects food sufficiency. Here Brahmins had from 7 to 9 months of food sufficiency whilst 'low castes' had only 4.6 to 5 months (other figures were Chhetris: 5.9 to 6.9; Newars: 6.8; Tamangs: 8.4; Sherpas: 4.8).

This example indicates how the problem of malnutrition, at least, is in part a result of a deleterious change from traditional customs where taboos were in many senses related to seasonal supply and demand, and the 'modern' system which has superseded it, where supply and demand are manipulated or otherwise work against the consumer. The pressures on traditional society, deriving in part from progressive intrusion of the world monetary economy, therefore, may have accentuated the problems.

The unwise change from traditional management also applies in time as well as space. The months prior to the monsoon are months of hunger. To bridge the food-gap, people such as the Magars, Gurungs, and Tamangs, for example, have traditionally foraged for forest products at this time: tubers, berries, and such like, and have taken game birds, animals, and fish. The deteriorating quality of the forests and the difficulties of common-land rights, have further curtailed these possibilities. People may even eat the next year's seeds at this time. More significantly, this is the time they fall into debt. The question of indebtedness, however, is also very complicated and will be discussed further in Chapter 9. Here, nevertheless, it is necessary to indicate that changing (rising) aspirations play a large role. For instance, increasing desires to own "luxury" goods such as radios, wrist-watches, and to enable at least one child to go to school, all place an increasing strain on very limited monetary means, leading to indebtedness.

There are, in fact, estimates that 90 percent of all debt in Nepal is for consumption items (Seddon, 1983:132). As much as 80 percent of land (especially irrigated land) may be pledged (Caplan, 1970). Interest rates may be extremely high. In the Middle Mountains the problem is particularly acute, with reports that 35 percent of households are in debt. On the other hand, the debts may be rolling ones, where the principal is not repaid and foreclosure rare. Debt, of course, also occurs in the traditional sector, notably for life-cycle ceremonies. Here, however, it was a redistributive mechanism, part of the Maussian gift-exchange structure of reciprocity, where giving (and lending) was a prerogative of rank. This traditional ethos may have carried over into the modern situation, so that debt is perhaps not the enormous problem it appears to be from the available figures. Rotating credit associations are other local mechanisms that have been established to ameliorate this problem (Messerschmidt, 1978).

The effects of malnutrition, however, are cumulative. For example, there is a vicious circle between decreases in production and increasing malnutrition. An outcome is ill-health, particularly the incidence of debilitating disease and, notably in Nepal, diarrhoea. Blindness may be a result of malnutrition, especially Vitamin A deficiency. On the other hand, most health indicators, from infant mortality rates on, show a gradual improvement, as the number of physicians per capita has more than doubled in the period 1960–80 (World Bank, 1984:267). But even this may be mainly an artefact of collected statistics, such as the number of hospital admissions, which refers overwhelmingly to the Kathmandu region. There is much less information for other regions, or social groups, and what is available is often contradictory.

The available evidence suggests that any discussion of poverty and basic needs in relation to wealth, income, and land, needs careful interpretation. For example, Sacherer (1979), in her studies in Nandu Kabre, has pointed out that whilst wealth may increase as one ascends the caste ladder, health, or at least nutrition, may decline. The main reason is that 'the higher the caste group, the greater the number of restrictions on diet' (Sacherer, 1979:31). These restrictions constrain what is eaten by which member of the family, what is eaten at what age, by women or men at what time of the month, year, and so on. Moreover, the more able and ambitious people, who wish to acquire wealth and status, will restrict their diet to move up the social ladder. High-caste children, Sacherer concludes, may therefore suffer the greatest malnutrition. Brahmin food prohibitions, for example, include beef, buffalo, pigs, chicken, and eggs. Children eat after adults in high-caste families, and girls after boys. If there is a shortage of food the child at the end of the line suffers most. There is a Nepalese saying that 'boys are raised up, girls are starved up' (Messerschmidt, personal communication, March 1987). It is possible then to have wealth without health as well as health without wealth.

But even these social correlations should be handled with care. For example, many young Brahmins told Sacherer that they did not believe in caste rules, restrictions, or privileges. They had travelled to Kathmandu and, indeed, often to India and, whilst orthodox in their home villages, wanted changes – especially for their children. The situation was complicated by inter-caste and inter-ethnic marriage (for example, a Brahmin married to two wives, one Brahmin and one Chhetri). It is uncertain, however, whether social distance affects inter-marriage; in other words, do proximate caste people inter-marry more readily, and how often is caste inter-marriage part of wider Maussian gift exchange where things cancel each other out? Nor is there an adequate explanation of the 'success' stories that Sacherer (1980) encountered among the mid-ranking Sherpas, Jirels, and Newars. Finally, Sacherer's study is only of two panchayats and it is not known how representative these are.

Some speculation is in order here. As an example, there is the curious paradox in Sacherer's figures that the groups which made the greatest use of the health posts – the Brahmins, Tamangs, and Chhetris – had the highest rate of child malnutrition, although the relationship to mortality and morbidity rates is unknown. The health posts also may not be visited by lower-caste people and ethnic minorities where they are typically staffed by high-caste people, based in part on their easier access to education. The hypothesis proposed elsewhere (Defence for Children, 1982) that health posts may do more harm than good (for example, by distributing inappropriate baby foods) should be carefully examined in Nepal. Another interesting point is that female child malnutrition rates were higher than male in all caste and ethnic groups except the Sherpas and Jirels where the rate for males was higher. This would seem to indicate that females are not prejudiced against amongst the Sherpas and Jirels, as they appear to be in many caste and ethnic groups. This raises several questions. Once again, is the traditional context important in improving health and nutrition? Is the introduced system of health, in some senses, counter-productive?

Discussion of other basic needs, such as education, housing, and water, is restricted by lack of evidence. Sacherer's (1980:18) studies indicate that Brahmins in one panchayat were considerably more literate (23 percent) than Chhetris (12 percent), or Newars (9 percent), but not Tamangs (22 percent). But in the second panchayat Brahmin literacy was both low (8 percent) and not significantly different from that of Sherpas and Jirels (7 percent). Overall in the Dolakha district males were much more literate (18.3 percent) than females (2.3 percent). The national averages at that time (presumably 1977) were 23.6 percent and 3.9 percent, though the most recent survey of education (Ministry of Education and Culture, 1984) gives males 34 percent (six years and over) and females 12 percent. Brahmins and Tamangs were also over-represented in the primary-school population of the two panchayats, whilst Sherpa males (though not Sherpa females) were under-represented.

These data raise additional questions. How much do literacy rates reflect local abilities or central inabilities? For example, it has been suggested that teachers are not of a high quality and both teachers and students often play truant. Dropping-out is common. But how important is literacy? In an age of cassette-records, oral cultures may find a new importance and greater relevance to the problems of solving poverty.

Literacy has been linked especially to health, through family planning. Again, the Sacherer (1980:43) evidence shows that Brahmins were more likely to practise contraception than, for example, Sherpas who, in her study, did not practise contraception at all. But the overall level of use was so low that these figures may have no significance. Nor is the evidence on numbers of living children necessarily indicative even though, in one panchayat, it is the 'untouchables,' and lower castes who have the greatest number of living children (5.5) compared to Brahmins (3.3). However, if this discussion is broadened beyond the confines of Sacherer's two panchayats, the conclusions are once more turned on their head. For instance, the military groups (Gurungs, Magars, Rais) and the Sherpas, and a few other groups, have greater exposure to introduced institutions. This is particularly true of the Newars and all the inhabitants of the Kathmandu Valley. The difficulty facing any such discussion as this is not only the unreliability of available data, as has already been emphasized repeatedly, but also the lack of representativeness, so that conclusions based on the detailed study of one or two panchayats may be totally inapplicable elsewhere.

Despite these problems, some inferences are warranted. The tentative suggestion proposed by Pitt (1986) is that one contributing cause of poverty is the decline of traditional culture and that introduced institutions, while attempting to work against poverty, may even be exacerbating it. This helps explain phenomena such as relative health without wealth amongst some ethnic groups. In health and nutritional matters, in education, in family life, and so on, some caste and ethnic groups may insulate themselves against outside intrusion. But we must also emphasize Macfarlane's (1976) conclusion which associates increase in population with resource degradation, and hence a breakdown of traditional patterns in all aspects of life, including agriculture, consumption, and so on. As suggested elsewhere, independence and identity are not correlates of poverty, but rather an alternative means of achieving significant life values, wants, and aspirations (Pitt, 1983). Ultimately a firm base of self-reliance may become the best vehicle for promoting economic development. When the independence and identity of a social group is broken down, severe problems may result – not simply the breakdown of social rules and order, but also a continuing dependence, which may or may not be the intention of those newly introduced institutions. At certain times traditional societies are more exposed to intrusions, and perhaps least able to resist, at least in certain sectors. Famines, epidemics, and related disasters, or military disorder, have been the classic examples of outside 'intrusion,' notably on the food or health front. Once involved it seems difficult for outside forces to disengage. The saving grace of Nepal, as opposed to sub-Saharan Africa, may be the ability to prevent the label of disaster being attached so often and so widely. In this context the concept of the preservation of cultural diversity, and hence dignity and self-reliance, must be placed on an equal footing with the ecologists' and environmentalists' demand for bio-diversity and preservation of the gene pool. There is a good argument for stipulating that bio-diversity and cultural diversity are complementary.

WOMEN AND YOUNG PEOPLE

Since the mid-1970s many UN reports about women, children, and youth have appeared, including reports from Nepal, even if the UN has never defined (or recognized other definitions of) what is a child and has accepted a purely chronological limit for youth (up to 26 years!). The argument is being developed that women, in the past, to use the World Bank's phrase, have been 'invisible,' and that all three groups are the most deprived in poverty, especially in extreme poverty situations.

How does this fit in Nepal and the Himalaya in general? Are women/children/youth 'problem' groups? What roles do they play in the developing 'crisis'? What roles do they, can they, play in the dynamics of development (local innovation versus centric, or outside agency, intervention – Messerschmidt, 1987)?

At first glance women seem to have received the worst of the poverty situation. Female infant mortality rates (1976) are higher; women's incomes, life expectancies, literacy rates, and so on, are lower in the Himalaya, though, as we have indicated, this does not hold true across all caste and ethnic divisions, nor for life expectancy, which in some instances, at least for 1982, is nearly in balance (females – 45; males – 46 years).

The World Bank in its latest report (1984a:198) also has a battery of indicators showing the 'low' international status of Nepalese women. In 1980 they had a ratio of adult male/adult female literacy worse than any other country in the world except the Yemen Arab Republic. They also had the fourth lowest percentage (aged 15–44) in the world enrolled in a primary school (just ahead of Somalia, Burkina Faso, and Mauritania, and equal with Ethiopia). In 1977 they had the lowest mean age at marriage (17), sharing lowest place with the Yemen Arab Republic.

The Child Welfare Coordinating Committee (CWCC, of UNICEF, 1980) have also collected very interesting documentation on the deprived status of the child in Nepal. The death rate of children aged one to four years old was high (35.3 percent) and this was highest in rural areas. Morbidity rates were high, and the CWCC gives a rate of 23.2 percent for intestinal infections, the most common illness, with pneumonia second (16.7 percent). One survey (CWCC, 1980:15) reports a high incidence of stunting (48.1 percent) amongst rural children, especially males. Dr. Bagchi, of WHO, reports that 70 percent of pre-school children are malnourished (WHO, no date). Primary school enrolment has risen (Ministry of Education and Culture, 1984:14), although it is much lower in remote regions, and the drop-out rate (Grade I) has been put at 53.3 percent, mainly due to examination failures. In 1977 68.8 percent of girls, but less than 10 percent of boys, had no access to education. Many children are working, a figure of 13.5 percent (for the late-1970s) of the total work force has been officially given, which is very high internationally when compared with the global average (cf. Bangladesh, 10.2 percent). Unofficially as many as 75 percent of children of school age may be working or are what the ILO euphemistically calls 'unemployed.' The effects of child labour on health and life opportunities are enormous. Nevertheless, it is difficult to interpret these statements without the ability to determine what is meant by 'working.' And again, many of these figures need to be broken down into ethnic, caste, or regional components. Infant mortality rates are much lower in urban areas, for instance; fewer children in the Terai (CWCC, 1980:46) are below height for age (45.5 percent) than in the Middle Mountains (55.5 percent). Rates for Terai Brahmins were lower (35.9 percent) than for Terai Chhetri (43.3 percent) and both were better than their mountain caste peers.

Women and children certainly do suffer from many forms of deprivation or exploitation. First, they work relatively long hours. The women's sphere in Nepal is still mainly in the household, including subsistence farm work. There are exceptions: for example, Thakali women are involved in the hotel and catering business, educated Gurung and Chhetri women are employed in clerical positions in the private and public sectors, women from occupational castes may work as labourers or porters (Seddon, 1983:139). Women who are in the household sector have a major role in subsistence work. Acharya and Bennett (1983) estimate that women account for 57 percent of adult input time into subsistence agriculture and contribute half of household income (compared with 44 percent for men and 6 percent for children). These patterns varied to some degree in different communities, but always the women put in more hours than the men. For example, Acharya and Bennett again report that in Parbatiya Bakundel the women work 12.5 hours per day as opposed to men 8.16. The lowest number of hours worked by women was among the Tamang, 8.46 hours, compared with Tamang men, 7.65 hours. In several groups men did more work in what Acharya and Bennett (1983) call sphere one (cooking, serving, cleaning, shopping, and child care); notably the Tharu-Sukraware men worked 7.29 hours per day, the women 4.97 hours.

The long hours worked by women were not well rewarded in money terms, even if the value of the subsistence sector was relatively high, up to 80.8 percent in the Acharya and Bennett (1983) studies, and even if women's work was more valuable than men's in the cash and subsistence sector combined. Nor did women have the major voice in the important decision-making processes, certainly not concerning labour allocation, though they did help to decide technical matters such as what crop to plant, what seed to use, and the amount and kind of fertilizer.

It is necessary, once again, to reflect on the risks of generalization, this time in terms of the claims for the systematic exploitation of women. That women work long hours and are not well rewarded in monetary terms is equally true of the developed world, capitalist and centralized economies alike, and the same qualification, 'except in pockets,' is likewise true. Such pockets do exist in the Himalaya. Sherpa women seem to experience a relatively egalitarian existence with their menfolk, and some of the most successful entrepreneurs in tourism and trade are Sherpa and Thakali women. To distinguish these pockets and encourage their growth and multiplication, however, requires disaggregation of a national or regional population according to sectoral strategies. This appears to be something of great importance, which is inhibited by the steadfast processes of data aggregation characteristic of many agencies (Thompson, personal communication, February 1987).

These studies, however, do provide data which may challenge ideas of how women spend their time, at least in some areas. The classic version of the deforestation segment of the Theory of Himalayan Environmental Degradation, for instance, has women walking further and further to the shrinking forests for fuelwood, or for water, as aquifers, wells, and other water supplies dry up on eroded terrain. As a result the women are worn out, and their own and their babies' health suffers. World Bank figures show that the most time any group of women spent on fuel collection was 0.69 hours per day (Newar women), or about 6 to 7 percent of their time. In a number of cases men spent more time than women collecting fuelwood, but this never exceeded 0.58 hours per day. Water collection involved more time in some communities, although the average was about the same as fuelwood collecting. The most time recorded was 1.12 hours per day in Pangma Panchayat, which was still less than 10 percent of the time spent on all activities. The biggest single activity involved agricultural work. However, in Ehrich's (1980) fieldwork in Kangra district, Himachal Pradesh, India, fuelwood and water collecting did account for 55 percent of women's time, again raising the danger of extrapolating conclusions drawn from one region to another; perhaps an even more fundamental question raised by this discussion is, once again, that of the reliability of the primary data.

There is also now documentation showing the low status and oppression of many women within this domestic–subsistence sector (CEDA, 1979). Seddon (1983) cites of a strong tendency in 'high-caste' Hindu families, and particularly amongst Brahmins, for women to be more oppressed than among the tribal groups, such as the Limbus, Magars, Gurungs, or the 'untouchables.' These restrictions result in severely reduced access to education and health facilities. Women, especially young women, in many areas, had few rights with regard to choosing marriage partners or the age of marriage, and the early arranged marriages may be a significant contribution to the population explosion, although this situation is not new.

Poor health and nutrition amongst women were reflected in the health and welfare of infants and young children, according to recent materials published on the child-care programme in Dhankuta, a "Save the Children" United Kingdom project, and in the Kosi Hill Area Rural Development Programme (Ministry of Local Development and Britain's Overseas Development Administration; Nabarro, 1984). These figures show both time and space variations and raise important issues. For example, wasting amongst children was highest at the end of the summer monsoon (Nepalese month of Bhadra –

Table 7.1 *Comparative average daily time allocation by sex and age-group, expressed in hours and decimals, in Nepal (after Acharya and Bennett, 1983).*

Activities	15 years and older			10 to 14 years			5 to 9 years		
	Male	Female	Both	Male	Female	Both	Male	Female	Both
Labourforce participation									
Animal husbandry	1.43	0.97	1.17	2.46	2.44	2.45	1.10	1.17	1.13
Agriculture	2.73	2.74	2.72	0.90	1.51	1.25	0.53	0.72	0.63
Manufacturing	0.42	0.45	0.44	0.03	0.08	0.06	0.01	0.01	0.01
Market activities (in village)	1.24	0.46	0.81	0.25	0.13	0.18	0.13	0.35	0.24
Subtotal	5.82	4.62	5.14	3.64	4.16	3.94	1.77	2.25	2.01
Subsistence economic									
Hunting and gathering	0.17	0.05	0.11	0.08	0.03	0.03	0.02	0.03	0.01
Fuel collection	0.24	0.38	0.32	0.15	0.43	0.31	0.06	0.14	0.10
Fetching water	0.07	0.67	0.40	0.23	0.70	0.50	0.11	0.21	0.16
Housing construction	0.25	0.08	0.16	0.06	0.02	0.04	0.03	0.03	0.03
Food processing	0.18	0.97	0.62	0.12	0.34	0.24	0.07	0.09	0.08
Subtotal	0.91	2.15	1.61	0.64	1.15	1.12	0.29	0.50	0.38
Domestic									
Cooking/serving	0.27	2.05	1.25	0.15	0.63	0.42	0.03	0.07	0.07
Cleaning dishes and pots	0.03	0.39	0.23	0.03	0.23	0.15	0.01	0.07	0.04
Cleaning house/mud	0.04	0.46	0.27	0.04	0.24	0.15	0.02	0.05	0.04
Laundry	0.02	0.15	0.09	0.01	0.06	0.04	0.01	0.03	0.02
Shopping	0.24	0.17	0.20	0.06	0.06	0.06	0.02	0.06	0.04
Other domestic	0.04	0.13	0.09	0.05	0.06	0.06	0.03	0.03	0.03
Child care and rearing	0.16	0.06	0.45	0.22	0.35	0.29	0.15	0.34	0.24
Subtotal	0.80	3.41	2.58	0.56	1.63	1.17	0.27	0.65	0.48
Total work burden	7.53	10.18	9.33	4.84	7.31	6.23	2.33	3.40	2.87
Education	0.43	0.10	0.25	1.72	0.83	1.22	0.96	0.14	0.54
Personal maintenance	1.45	1.12	1.27	1.35	1.33	1.33	1.40	1.45	1.42
Social activities	0.31	0.16	0.23	0.12	0.23	0.19	0.10	0.10	0.10
Leisure	6.30	3.81	4.93	7.98	6.30	7.03	11.21	10.92	11.07

August/September) and lowest in Mangsir (November/December), three months after the maize harvest. Food shortage was a reason for this but the prevalence of diseases, notably diarrhoeal, was an important contributing factor. Nabarro speculates that the incidence of diarrhoea was related to outbreaks of measles. But the causes of the wasting were multiple: food shortage; a moist environment which favoured faecal transmission; parents who could not afford time to go to a health post or to provide the child with extra ghee or sugar; and measles, which often may have been, as studies have shown elsewhere, the catastrophic straw which breaks the camel's back. But there were also many hidden factors: the time it takes to try to raise extra income or to negotiate loans, or to persuade children without appetites to eat when they are sick. An interesting point to emerge is that, although the research team could show that wasting or stunting vary with the amount of land a family held, it was less easy to explain morbidity and mortality. Much depends on the mother's behaviour and ability to cope. Some mothers 'innovate' because of superior education or wealth; other, poorer mothers out of desperation.

The status of older children seems to be bad also, and is compounded in the case of females. Child labour is apparently far more prevalent in Nepal (at least according to official figures) than in India or Bangladesh. In the Acharya–Bennett studies (1983:43) girls contributed more than twice as much as boys to household production (Table 7.1). The effects and consequences of child labour are not well understood, partly because in some sectors, notably the urban, it is a relatively new activity and the long-term effects (for instance, on health) may not be apparent for some years. The UN, ILO, and some non-government organizations (NGOs) (Pitt and Shah, 1982) have recently put together a catalogue of major health and other exploitative consequences which may contribute to low life expectancy and high morbidity. The UNICEF/CWCC investigation (CWCC, 1980) has provided case studies of child labour in Nepal. This includes work in restaurants, domestic service, and newspaper stalls. Working children are to be found also in the so-called 'black economy,' and in criminal and related activities, especially in Kathmandu. There is also a large amount of anecdotal information about the fear of girls and young women being kidnapped and taken to India and being forced into prostitution. This was headline news in the *Rising Nepal* (Kathmandu's English-language newspaper) in early 1986, for instance, which described a rash of kidnappings of young girls from Terai locales for sale to the prostitution markets of Bombay.

Most working children (well over 90 percent), however, recorded in the censuses are in the agricultural sector and this proportion has not changed much since the 1952 census. Some authorities have pointed out that the figures on health hazards are somewhat speculative and, in any case, if children did not work in the agricultural sector, families would starve. The force of this argument is somewhat diminished by the fact that some families would starve anyway and that the causes of this starvation are related to external factors, notably prices, and the associated difficulty for subsistence families to acquire cash.

Nonetheless, a distinction should be made, as Bouhdiba (1982) has made, between exploitative child labour and that labour which contributes to the child's education and his family's subsistence and well-being. Certainly the growth of and opportunities for child labour is one reason for a continuing high population growth.

One of the most serious consequences of child labour is simply that children, especially female children, do not receive even primary-school education. In 1977, for example, in a number of districts, notably in the far west, over 90 percent of female children had no access to primary school education (CWCC, 1980).

Little is known about the social dynamics of child labour, and even less about the 15–26-year-old group, which is generally called 'youth' in the UN nomenclature. This group has certainly a high rate of unemployment or underemployment. Many are married with children. In the 1971 Census, 7 percent of the population between 6 and 14 years old in the central and western Development Regions in Nepal were already married. Many of this group migrate to the cities, notably to Kathmandu, and Sacherer's studies (1979, 1980), for example, show that young people were more ready to change their attitudes and behaviour in response to outside cultural influences.

Nevertheless, it should not be assumed that Nepalese youth are simply a component of the global mass-culture youth model. For example, in Nepal migration has been regarded as a 'final' option for those in the most desperate category. According to a study quoted by Seddon (1983), there has been a change in the basic reasons for migration over the past twenty-five years. In the early period migration of households (and this should be distinguished from individual migration) was largely because of natural calamities, particularly the loss of hill land through erosion, or its progressive degradation until it could no longer support a family. It is argued that more recently the increasing pressure of debt has become the main cause. Both reasons involved loss of, or greatly reduced access to, suitable land. Young people were carried along on this tide with their families, but until recently such migration was to other farmland and primarily involved migration from the Middle Mountains to the Terai. In recent years, however, this option has gradually been closed by the government's move to control resettlement, as conflicts have developed between officials and other land-hungry people in the Terai, and as the stock of land has progressively decreased.

Desire for land has remained the fundamental motive, and when a sample of Middle Mountain people were asked why they wanted to migrate, most (over 60 percent) gave reasons relating to the difficulties of producing enough on land in the Middle Mountains and the prospects of more land elsewhere. A comparable proportion of those who had already migrated to the Terai gave similar reasons. It is interesting to note that a relatively small proportion (around 5 percent) of both samples gave the loss of land through foreclosure on outstanding debts as the reason, or indeed, total loss of land through erosion and natural calamities (3.9 percent of the Terai sample).

New strategies for households to cope with this situation seem to be emerging and in these young people play a major role. First, since there are, or are thought to be, wage-earning opportunities in the cities, permanent migration of households may be giving way to individual migration of young people who are remitting wages back to support older and younger members of the family. As the debt problem is one of structural relationships – however unequal – between lenders and borrowers, other forms of exchange may be taking place across this divide. For instance, Sacherer (1980) has referred to the marriage of low-caste young women to higher-caste men. However, the situation may not be stable since the higher-caste men are also, or have been, migrants. This may help to explain the sudden upsurge in urban population numbers in the 1981 Nepal Census (Goldstein *et al.*, 1983).

Overall, therefore, there appears to be a worsening situation in the status of Nepalese women and children in the general context of deepening poverty. This is not the place to examine the efforts of government and the various agencies who are trying to improve the situation. The literature that is available paints a picture in the 1960s and 1970s of a centralized monarchy which had introduced the so-called 'panchayat democracy' in order to preserve the traditional structure. There was much rhetoric about decentralization and 'back to the village' slogans, but popular participation was, according to most reports, limited. Shrestha and Mosin (1970), in their survey of village panchayats in the late 1960s, report that most people were confused about, or had little knowledge of, the panchayat system, though a popular vote in 1981 confirmed its continuation and political parties have been banned since 1961.

Since the beginning of the 1980s there have been changes stimulated by unrest amongst the students and intelligentsia and by pressure from leaders of banned parties, if not from the grass roots. The Royal Palace is still central, perhaps more so, but here, and in government and the bureaucracy, exists what Seddon (1983) has called 'a new cast of thinking.' The essence of this new thinking, which can be seen clearly in the sixth and seventh Five Year Plans (1981–85; 1986–90), is the so-called Basic Needs Strategy. This strategy, according to Seddon, initially owed much to the work of the international agencies and foreign donors. But it has been taken up by local politicians, bureaucrats, and the intelligentsia, and has already some concrete results, including specific legislation on decentralisation and increased participation. What was seen by most observers as a 'bottom-up' rhetoric early in the 1970s, may become a part of the dynamic structure of the society.

Notwithstanding the impacts of the new thinking, some doubts have been expressed. First, in recent interviews the King himself has favoured free-market economy forces: this is contradictory to the Basic Needs philosophy. The top ranks of the bureaucracy (including the military) are dominated by three castes, Brahmins, Chhetris, and Newars, and this has been true for over one hundred years (Beenhakker, 1973). Most come from the Kathmandu Valley (in one estimate, nearly half of the entire Civil Service comes from this region, which has only 5 percent of the total national population). However, increasingly a meritocracy is emerging, of what Blaikie *et al.* (1980) call 'modern bureaucrats' – those who have taken advantage of the new educational openings, including study abroad. These 'key' personnel have 'radical' new ideas which are already showing, for instance, in the Nepal National Planning Commission. There remain many obstacles to this new thinking and action, notably inter-departmental rivalries, lack of vertical co-ordination, the poor quality and training of the middle-level bureaucrat, the existence of a parallel foreign-manned bureaucracy in some sectors disguised as a 'counterpart system,' and so on. The Decentralization Act of 1983, however, is a particularly important initiative on the part of the government and is further evidence of the progressive shift from rhetoric to determined action.

Important as any new directions at the centre may be, the new decentralization processes put district-level officials in much closer contact with the panchayats and create new district organizations which group together workers, women, young people, and ex-servicemen. There are many problems, too, with the new decentralization measures; for instance, continuing feuds and corruption in the panchayats and the persistent influence of a rural elite. Most influential panchayat positions (for example, chairmen) were held by the larger upper-caste landowners in the 1960s and 1970s, except in those villages which are all low-caste. This power structure continues and has been cited as a major factor in inhibiting recent food distribution schemes in food-scarce areas.

In terms of meeting basic needs the greatest difficulties can be expected in the health and education fields. Although there have been significant improvements, Nepal still spends much less (as a proportion of total expenditure) on such sectors as education, health, housing, community amenities, and social welfare, than many other countries. According to the 1984 World Bank *Development Report* (p. 268), in 1981 9.7 percent of total expenditures was devoted to education (11.5 percent was the average for low-income countries), 4.1 percent to health (a decline from 1972 when 4.4 percent was the low-income country average), 1.5 percent to housing and community amenities, social security and welfare (6.1 percent for all low-income countries). All of these data have an important bearing on the situation of the poorest sectors, and especially women and young people within those sectors. It is too early to predict whether conditions will improve after the new legislation and whether there will be many ameliorating effects but certainly it will be an uphill struggle.

Most central government expenditure in Nepal during the 1970s and 1980s (around 57 percent) has gone into the economic sector, particularly agriculture, and in the views of many observers this has favoured the wealthier rural classes. Some Nepalese by the mid-1970s, however, were calling for 'an integrated rural development' and this idea was supported by a number of international agencies and foreign donors. The Agricultural Project Services Centre (APROSC) acquired a central role, and many new programmes were directed toward the small farmer. There was a national 'back to the village' campaign and the co-operative movement (Sajha) took on new momentum. Blaikie *et al.* (1980) have claimed that the more general campaigns were largely rhetoric but specific programmes may have been more successful.

But are there no indigenous collective social forms? What place do women and young people have in this structure and process? What roles can they play? The role and potential of traditional co-operatives have been described elsewhere (Messerschmidt, 1981). These may remain outside the official ambit but despite – or because of – this, may have a significant, if unrecognized, potential. In Nepal, co-operative groups become increasingly important in any switch from herding to cropping, where there may be more need for co-operative labour inputs. In such 'non-conventional' co-operative forms both women and young people play an important role. And women play a much greater role in health activities. Here, and in the analysis of the role of young people, there is a lack of data. There is certainly a growing literature on women in Nepal both from expatriate researchers and local scholars. But many of the foreign works, Pitt (1986) claims, at least in anthropology, reflect the biases of western scholarship. For example, there is the obsession with kinship, originally mainly male, but now concerned with the analysis of topics such as 'mothers' milk and mothers' blood,' that is, with women's symbolic roles.

It must also be emphasized, however, that kinship and women's symbolic roles are extremely important to an understanding of social organization, including co-operation in traditional forms. And there is a small but growing group of expatriate and local applied anthropologists who are playing an increasingly vital role in furthering this understanding. This group includes Manzardo (Manzardo *et al.*, 1975), Fisher (1978), Messerschmidt (1978, 1981, 1987), Campbell (1979; Dani and Campbell, 1988), Gurung (1981a and b), Goldstein (1981; Goldstein *et al.*, 1983), Acharya and Bennett (1983), and Dani *et al.* (1987). Further progress along these lines is essential if unwanted and unexpected social disruptions arising from development policies and their central implementation are to be avoided.

There is also literature on women's roles in agriculture (Schroeder and Schroeder, 1979), and in maternal and child health (CWCC, 1980). The CWCC has produced interesting materials on the role of women in child rearing in a number of castes and ethnic groups. For example, in the Sherpa community studied (Rolwaling) there was no diarrhoea, partly, it seems, because nobody defecates or urinates near the water sources, and parents take pains to clean up the children's defecations. Nutrition in this high-altitude community depends a good deal on potatoes, which are also fed at an early age to children. Infant and child mortality rates appear from both this and Sacherer's (1979, 1980) studies of this group to be low, and the level of nutrition relatively high. Other factors may be the somewhat later age at which women marry, the spacing of children, and breast-feeding.

In the Rolwaling case the role of women is important, but it is not particularly collective. Women perform tasks, even childbirth, without the help of others. Much of what they do is based on the traditional oral knowledge system. Twenty percent of the people, in fact, had learned to read and write from the lamas, but none of these are women, except for two *anis* (nuns). However, since 1972 there has been a school established by Sir Edmund Hillary's Himalaya Foundation and this has been enthusiastically received, and girls, who have apparently been impressed by the superiority of women tourists, are going to school in increasing numbers.

The role of women and young people in agricultural co-operatives shows how this kind of organization may form the base for an upward internal generation of development and, even more interestingly, provide a mechanism for crossing horizontal ethnic and caste lines. Some of the basic tenets of the official co-operative movement (formal government, as opposed to informal and traditional) in Nepal are questioned by Messerschmidt (1981). Compulsory savings were the problem in the 1960s, 'faltering' administration in the 1970s, and outright official hostility in the 1980s, as the economic situation worsened. But most forms of traditional co-operation in Nepal were neither officially inspired nor directed. They existed before, during, and after the official programmes and in them women and young people were prominent, if not the principal participants. Most commentators from western societies have discussed co-operatives in terms of spheres of male influence and power, but it is hardly surprising that women and youth should be the most active groups in the Nepalese co-operatives since most of the work is done by them. There are certainly some male-dominated co-operatives in Nepal, apart from the official ones; they include, for example, the temple associations (Greenwold, 1974).

Traditional forms of labour co-operations are rather different. A well-described and typical example is the Gurung *Nogar* (Messerschmidt, 1981), which is a temporary village-level association of Gurung youth. *Nogar*, in fact, are occasionally all female, but the shortage of males has been explained by their absence on military service with the Gurkhas. Most *Nogar* members are in their teens and twenties, usually drawn from the female *rodi* communal houses.[2] *Rodi* girls, in fact, invite boys to work with them. *Rodis*, it should be noted, act as a means of uniting people from different villages. They also play an important role in courtship, focusing on the *rodi-ghar* (girls' sleeping dormitories). *Nogars* may also cross ethnic and caste lines. There are examples in the literature of non-Gurung blacksmiths (Kami) joining. There are also records of exclusively 'untouchable' *Nogars* (Pignede, 1966), and Tamang membership (Macfarlane, 1976). The fact that there are not more records, according to Messerschmidt (1981), is because scholars have never really looked at this phenomenon and because caste groups live at some distance from each other. The poorest families have their own smaller co-operative activities.

Nogars seem to be very successful with great morale and bursts of energy. This positive, productive attitude, what Pignede (1966) calls the *ésprit de nogar*, is a major driving force in the subsistence economy. Because *Nogars* are 'part of life', especially part of the marriage cycle, they encourage motivation. All this is part of the Gurung success. Other groups, however, may be less likely to co-operate successfully. An example is Doherty's (1975) study of the Brahmin and Chhetri castes where the woman's role is much less dominant.

Finally, the long-standing existence and survival of many groups through the period of nationalization of the forests by the government, and recent re-activation of the traditional systems of natural resource control and protection have been and remain extremely important (Messerschmidt, 1987).

CONCLUSIONS

Despite all the complexity and scarcity of systematic data, it is possible to draw some general conclusions. Some of these raise doubts about much of the conventional wisdom on the Himalayan crisis.

1. This general discussion on the human dimension demonstrates a comparable degree of lack of data, confusion in the use of data, reliability and representativeness, to add to the earlier discussion on the physical components of the Theory of Himalayan Environmental Degradation. Poverty, population growth, and increasing demands on limited, if not finite, natural resources, are central to the Himalayan Problem.
2. In terms of the problems of subsistence and access to land, the role of debt may be a less disadvantageous factor (for local people) than burdens which are imposed from above and from outside, as the influence of the world market economy expands.
3. Women, children, and young people do suffer most in a situation where wealth, power, and prestige are to some degree male preserves and are assets which often are distributed through interventive mechanisms, such as foreign aid and development and government central planning without local input.
4. The low status of women and children varies across different sub-cultural groups and in different situations. Some caste/ethnic groups are more egalitarian, and some young people are achieving mobility. Both women and young people are involved in co-operative, traditional activities.
5. It may be that the more traditional caste groups, in some cases, are faring better, despite, or possibly because of, their traditionality which provides resilience against the more negative and disruptive aspects of intervention.
6. There is some evidence in Nepal, as in the Indian Himalaya, that women are the heralds and foot-soldiers of grass-roots movements in which environment and health are major sectors. Hugging the trees is not only to be found in the Chipko Movement (cf. Shiva and Bandyopadhyay, 1986b). But in the Nepalese setting, the woman's role may be rather in health, particularly child health, which remains quite traditional and sometimes very effective, especially amongst the sub-cultures of the high Himalaya. The recent improvements in infant mortality rates, therefore, may well be a sign of this growing role of women.

7. If women are a force in local developments, young people are more involved in the linkages between innovation at the local level and the processes of intervention because they are the migrants in body, and mind, and because they are prepared to change their attitudes and behaviours.

The seven concluding points are primarily applicable to Nepal. However, they are relevant to a certain extent to other parts of the Himalaya. For example, in Bhutan (World Bank, 1984b), though data is still very scanty, the traditional ways of life are still very strong. Urban development has been much slower than in Nepal, partly because of a government policy making sure that each district has its own town. Subsistence agriculture dominates, is labour intensive (so that population increases are absorbed), but also it is even more of a male preserve. Also land is apparently temporarily reassigned to equalize the land–labour ratio. According to the UN Food and Agricultural Organization (FAO) there is no significant pressure on fuelwood sources. The absence of labour and forest problems has been partly due to government controls through compulsory employment practices and severe restrictions on commercial felling (from 1979) amounting to nationalization. Tourism, too, has been greatly restricted, partly because there is a relatively small need for hard currency. Very little is known, however, about the status of women and young people in Bhutan although, even compared to Nepal, a very small proportion are in school.

Certainly the World Bank report paints a relatively glowing picture of Bhutan and expresses surprise at finding a 'well-managed' economy and a much higher than expected overall standard of living, despite the retention of the traditional way of life. This is a reflection on the view from the top! Nevertheless, not all indicators are so positive, of course – for instance, the infant mortality rate. But this rate itself may require closer scrutiny. If the total population of Bhutan is not known with any degree of precision, it is unlikely that a far more socially sensitive figure, such as infant mortality, will be accurate. In any case, such rates are themselves heavily influenced by cultural practices, as for example, in attitudes to handicapped babies. More important perhaps, infant mortality rates can be lowered dramatically when women (through village committees) have more power.

In the Indian Himalayan regions, women's movements of the Chipko type have been a marked and well-publicized feature of recent history. This, however, has been less a sub-cultural social fact, in the sense of the preservation of traditions, rather than an assertion of power at the grass roots. It is populism rather than traditionalism, with many political ramifications, notably a strategy to remain in the mountain regions and to confront the outside interventionist forces. Nevertheless, it is firmly based on a very old traditional philosophy which also gave birth to the Gandhian non-violent protest movement and the forest satyagrahas – organizations for resolving conflicts by non-violent refusal to co-operate – of the last half-century of British rule (Shiva and Bandyopadhyay, 1986b). As mentioned earlier, the current situation in Arunachal Pradesh, and the eastern Indian Himalaya in general, is unclear because of difficulty of access and lack of reliable data. It is clear, however, that the entire area is in a condition of serious political unrest, with demands in the Darjeeling Himalaya for autonomy from the State of West Bengal and a large and growing population movement from the Brahmaputra plains of Assam into the mountains of Arunachal Pradesh. This latter development is particularly interesting because it is the converse of migration patterns typical of most of the Himalaya where people are moving onto the lowlands from the Middle Mountain belt. Goswami (1985, and personal communication, October 1987) presents a strong case for recent increases in sedimentation and annual flooding in the Brahmaputra Valley in Assam compared to the long-term (past two to three million years) average. This he attributes to present-day rapid uplift of the mountain system, recent earthquake activity, and the high susceptibility of geological formations to erosion by running water under a monsoonal rainfall regime. Of particular relevance to the topic under discussion in this chapter, these geophysical processes are prompting a significant migration trend into the mountains, presumably with increased pressures on the mountain landscapes and accelerated socio-economic and political conflict.

We can only sum up by returning to our opening questions: (1) the subsistence farmer does, indeed, lie at the core of the Theory of Himalayan Environmental Degradation and also at the core of what we prefer to introduce as an entirely differently structured Himalayan Problem; (2) misconception and misunderstanding are rife, in large part because of absence or unreliability of data and the unrepresentative character of the data that are available – there are, indeed, as many 'sacred cows' that must be confronted as in the 'physical' sector of the Theory; and (3) resolution of the Himalayan Problem will depend heavily upon much better problem definition than hitherto has been achieved, together with the introduction of multiple approaches to solution identification and implementation to match local-felt needs.

NOTES

[1]'Untouchables' are legally and technically not recognized in Nepal and the term, in a strictly social sense, refers only to a few castes (cf. Höfer, 1979).

[2]*Rodi* are semi-permanent Gurung girls' associations organized as age-graded dormitory structures. By inviting boys to work with them, they assist in the formation of *Nogars*, or work associations of an ephemeral nature. The dormitories also provide the locale for social events, such as nightly singing and dancing (Messerschmidt, 1981).

8 TWO APPROACHES TO THE POPULATION PRESSURE/LAND PRODUCTIVITY DECLINE PROBLEM IN THE HIMALAYA

INTRODUCTION

Two case studies are introduced in this chapter, in part to illustrate the problems of the increasing population pressure on limited land resources in predominantly subsistence societies and, in part, to emphasize the extreme range of proposed responses to that problem. The first case study deals with the Himalayan foothills of Uttar Pradesh, India, concentrating on the Kumaun Himalaya. It is based on a paper prepared by M. J. Jackson (1983, unpublished) and presented at a conference on environmental strategies for the Himalaya, held at Nainital in October 1983. It involves a farmer-based strategy incorporating a minimum of 'outside' technology transfer. Its essence is the creation of local political will and self-help. The second case study deals with Nepal. It is essentially a large-scale technological 'fix' based upon careful survey of land, water, and vegetation-cover resources and population growth trends. It is taken from a paper prepared by J. P. Hrabovszky and K. Miyan (1987) which, in turn, is the synthesis of three related studies undertaken for the National Planning Commission, HMG Nepal. These are: land-use plan; agricultural plan; and long-term food plan.

We do not wish to imply that either approach necessarily constitutes the correct or only answer to one of the more pressing problems facing the Himalayan region – often viewed by development agency experts as 'too many people,' or by the villagers as 'not enough food' (Thompson *et al.*, 1986). The case of the Kumaun Himalaya provides a good illustration of the intricate inter-relationships between the three major sectors of a mixed subsistence hill farming economy, namely, the forests, cultivated land, and the livestock. It goes on to describe what happens when these inter-relationships fall apart, and how they might be set right. It also emphasizes that, in order to achieve the proposed solution, a new institutional framework must be developed which would lead to a radical change in access to natural resources, including *control* of land. Whether the existing social and political structures can be adjusted to achieve this goal is a moot point. The case for intensified agriculture in Nepal involves massive inputs of technology – large-scale irrigation, increases in artificial fertilizer application, and increased use of high-yielding varieties (of food crops and livestock). Even if we adopt an optimistic attitude toward this proposed solution, as Hrabovszky and Miyan (1987) demonstrate, the attempt to achieve a balance between increasing food production and continued population growth will be a very hard struggle.

LIVESTOCK AND LAND DEGRADATION IN THE KUMAUN HIMALAYAN FOOTHILLS

Livestock, as already emphasized, are an integral part of mixed subsistence hill farming. Figure 8.1 illustrates the critical energy flows between the livestock, forest, and arable sectors, expressed in a simplified form in terms of the dry weight of the materials involved – that is, fodder, dung, grain, milk, and draught power, in relation to human population – rather than their energy equivalents.

Jackson maintains that, because of the close integration of livestock, available forest resources, and crop production, this local subsistence system has several advantages over more 'modern systems in which livestock and crops are separated into watertight compartments.' The latter systems are illustrated in simplified form in Figure 8.2. As long as the subsistence system is in proper balance it is efficient in terms of the ratio of food and fibre output to total solar energy captured, and there is almost no wastage. It is also a self-reliant system, the importance of which cannot be over-emphasized for LDCs dependent upon imported petroleum products.

Jackson believes that the attempt by scientists and agricultural planners in India to visualize the goal of livestock development in terms of a 'western' technological model (Figure 8.2) rather than in terms of improving the efficiency of the existing system (Figure 8.1) has been responsible for the

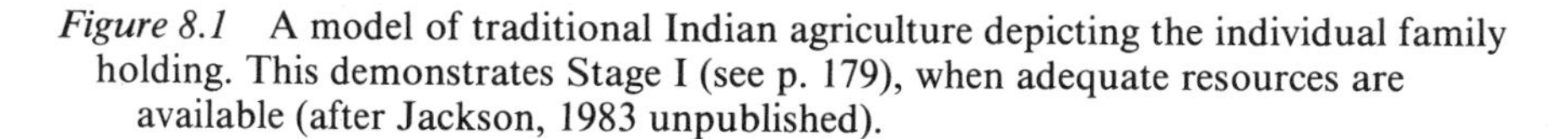

Figure 8.1 A model of traditional Indian agriculture depicting the individual family holding. This demonstrates Stage I (see p. 179), when adequate resources are available (after Jackson, 1983 unpublished).

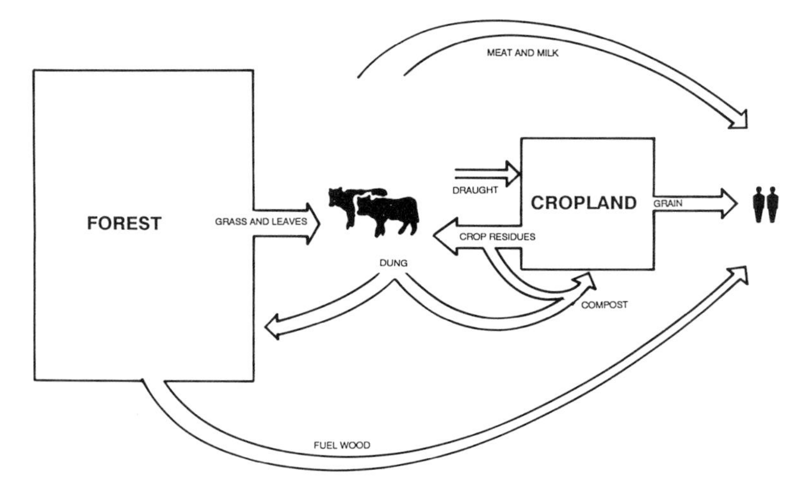

Figure 8.2 A model of present-day 'western' agriculture dependent upon high input of artificial fertilizers and pesticides (after Jackson, 1983, unpublished).

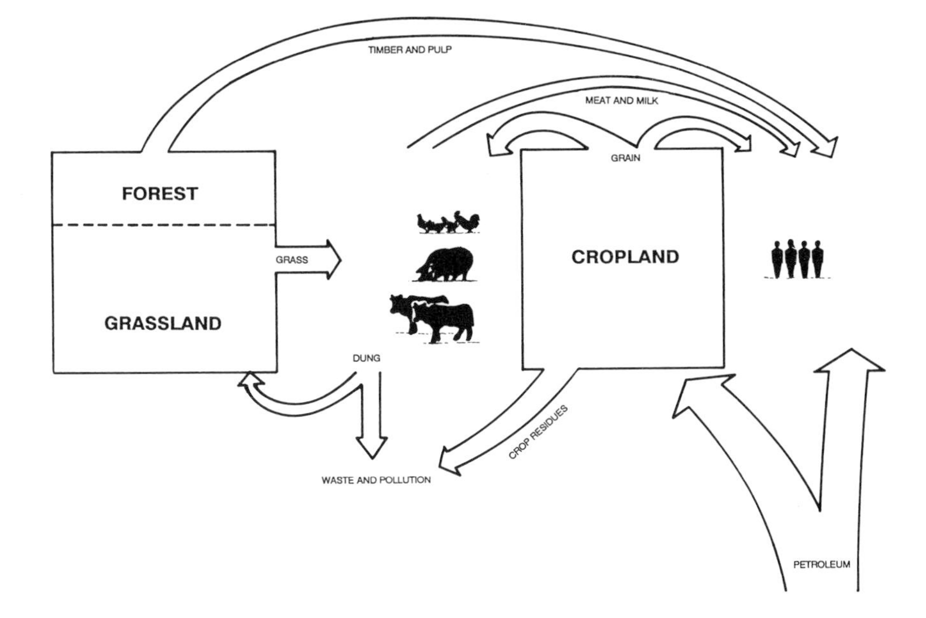

failure to increase livestock productivity. He argues that the most unsatisfactory feature of the subsistence system is the manner in which uncultivated land is managed (primarily the forest component of Figure 8.1). Unrestricted grazing of livestock, in a situation where both animal and human populations are growing rapidly, has led to progressive decline in the density of the vegetative cover and the severe depauperation of the forests. The species composition changes toward a predominance of useless species, while rates of precipitation runoff and soil erosion increase. The productivity of the livestock itself falls as the animals obtain less and less green fodder, which in turn reduces crop yields because there is less available animal manure. (This argument is reminiscent of one of the vicious circles of the Theory of Himalayan Environmental Degradation, but is relevant to the specific case and worthy of following further.)

Vegetative Regrowth, Grazing, and Burning

In order to retain a vigorous and sustainable level of growth, plants – both grasses and trees – must not be defoliated too frequently. After defoliation new leaves initially grow at the expense of food stored by the plant in its root system. At a certain point in their growth they cease being net consumers of previously stored energy and become net producers. This surplus energy is in turn stored in the root system: thus root reserves of energy must be recouped each time before defoliation if the plant is to remain healthy and survive.

Thus, in an ecosystem undisturbed by man, populations of wild grazing animals are usually naturally regulated in relation to the ability of the vegetation to withstand the grazing pressures. Domestic livestock, on the other hand, are stabled at night and thus graze a limited area constantly near the village. A careful management system, therefore, is necessary if a balance is to be maintained. Overgrazing may be defined as 'too-frequent defoliation,' so that net shoot productivity is reduced. This applies both to direct grazing by domestic livestock and to lopping of trees for leaf fodder. Animals and their owners gradually move further afield each day in search of more abundant vegetation, slowly enlarging the circle of degradation. As the human population increases, more villages come into existence and these circles of degradation begin to intersect.

Two additional processes also hasten the deterioration of the overgrazed forests. If seedlings of broad-leaved trees are grazed or lopped too often, they die. Tree seedlings, including the unpalatable species, such as chir pine, are also killed by being trampled. Thus forest areas that incur unrestricted grazing become depauperate and eventually are deforested because there are no new trees to replace those that are cut or that die. The rate at which trees are cut for timber and fuelwood may or may not be greater than the natural rate of growth of replacement trees, but this will be immaterial if grazing prevents the establishment of sufficient replacements (cf. Moench and Bandyopadhyay: 1986 – the 'nibble effect').

Finally, the effects of burning must be considered. Farmers often use fire during the dry season to clear dry grass stubble and dry tree leaves and needles. This is to induce the sprouting of new green shoots from the grasses and herbs to provide a 'green bite' for the animals. This practice, however, also further weakens the understorey plants, kills the seedlings, and even large trees.

Stages in the deterioration of the Kumaun foothills

The climax vegetation type of the lower Himalaya, and also of the Terai, is a broad-leaved deciduous forest dominated by sal (*Shorea robusta*); that of high altitudes consists of various species of conifers interspersed with oaks occupying the sites with better soil and moisture conditions (Gupta, 1979). The succession of grasses leading up to the forest stages in the Kumaun lower Himalaya between 500 and 2,000 m is described by Dabadghao and Shankaranarayan (1973). At the highest altitudes the grass species that occur are not only the most productive, but they also cover the ground most completely (they are bunch grasses). At lower altitudes the grass and herb species are both less productive and provide a less complete ground cover. Table 8.1 gives an estimate of the areas of the land categories for the eight 'hill districts' of Uttar Pradesh.

During the early historic stages of human occupation of this area people were few in number, crop and animal productivity were high, and little pressure overall was exerted on the mountain forests. (Large areas were also

Table 8.1 *Areas under various land-use categories in the eight hill districts of Uttar Pradesh (1979) (after Jackson, 1983, unpublished).*

Category of land	*Area, 1,000 km²*
Snow-covered	10
Forested*	22
Alpine pastures and blanks	4
Cultivated	12
Remaining (shrub and grass land)	9
Total land area	57

*This refers to the area actually under forest at present, and not to the area officially classified as 'forest,' which is much larger since it includes much of the land here shown as 'shrub and grass land.'

hunting reserves of the local rulers.) Lower-lying sites accessible for irrigation were the first chosen for arable land, and the fields received ample manure from the relatively highly productive domestic livestock. The forests provided a nutritious and adequate diet of green grasses and tree leaves, and many other products. Human diet was adequate and well balanced. Figure 8.1 depicts this situation, which is designated also as Stage I in a progression of deterioration.

Progressive increase in the human population results in an increase in clearing of the original forest cover for arable land. This reduces the

Figure 8.3 Traditional Indian agriculture, Stage II. The progressive increase in human population has resulted in the regression of nearby forests to grassland, or at least, shrubberies and depauperate forest. The reduced access to fuelwood results in an increasing amount of animal dung being used for fuel (after Jackson, 1983, unpublished).

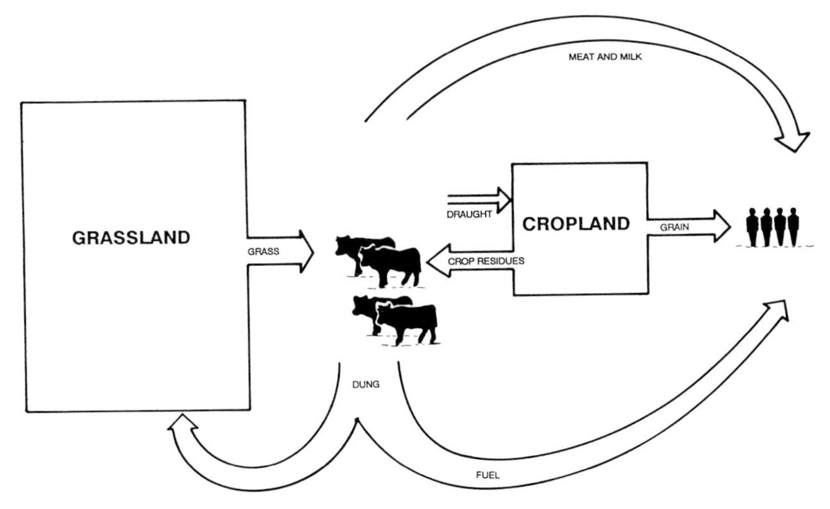

remaining forest land in relation to fodder, bedding, and fuel requirements. Regression to grassland, or at least shrubberies and depauperate forest, occurs. This is Stage II and is depicted in Figure 8.3. The system is still productive at this stage as grass yields are high. Two important changes are occurring, however. First, there is a reduced access to fuelwood and an increasing amount of animal dung is used for fuel. This lowers the productivity of the cultivated fields. Second, there is a reduction in the amount of green tree leaf fodder to feed the animals during the long dry season (from October to June). Since grass is only green and nutritious during the rainy season, the dry season fare for the animals is largely restricted to straw. Thus fertility of females is reduced and milk yields decline; the growth of calves is limited during the dry season. The productivity of the livestock, and of the entire system, has fallen; so has the well-being of the human population.

Continued increases in the human population result in further extension of cultivation. But now recourse is made to steeper slopes with much less fertile soils; farms become smaller. Yet each farmer maintains a traditional complement of livestock: a pair of bullocks, a cow and/or a buffalo, and two or three young animals as eventual replacements. Thus the number of grazing animals per hectare of uncultivated land increases, causing a further regression of the natural vegetation cover to sparse inferior grasses and unpalatable 'weeds.' Output declines further, as Stage III is reached (Figure 8.4).

Jackson (1983, unpublished) maintains that today the vast majority of the land area of the entire Indian subcontinent has been reduced to Stage III.

Figure 8.4 Traditional Indian agriculture, Stage III. Continued increase in human population results in further extension of cultivated land, but now recourse is made to farming on steep slopes with less fertile soils. The number of domestic animals per hectare of uncultivated land thereby increases and natural grassland cover is reduced to inferior species and unpalatable 'weeds'. Output declines further (after Jackson, 1983, unpublished).

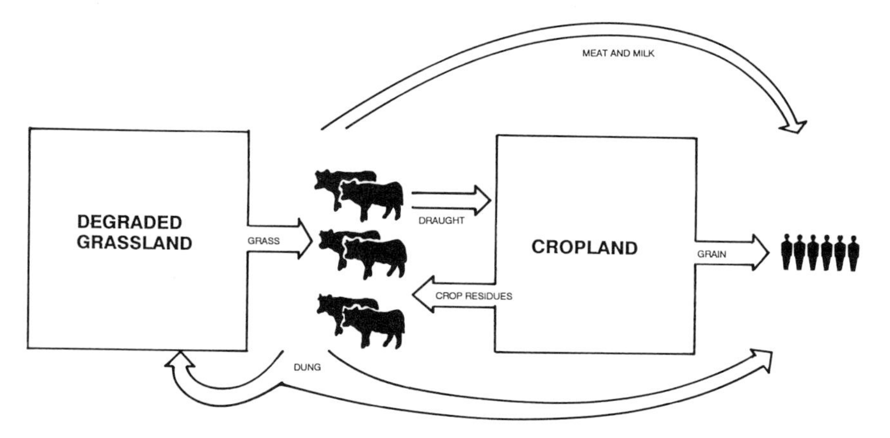

Figure 8.5 Traditional Indian agriculture, Stage IV. Further increases in grazing pressures reduce much of the grassland to barren wasteland where rainfall is low or where rainfall is high as in the Kumaun foothills, to shrubberies, sparse grasses, and unpalatable species (after Jackson, 1983, unpublished).

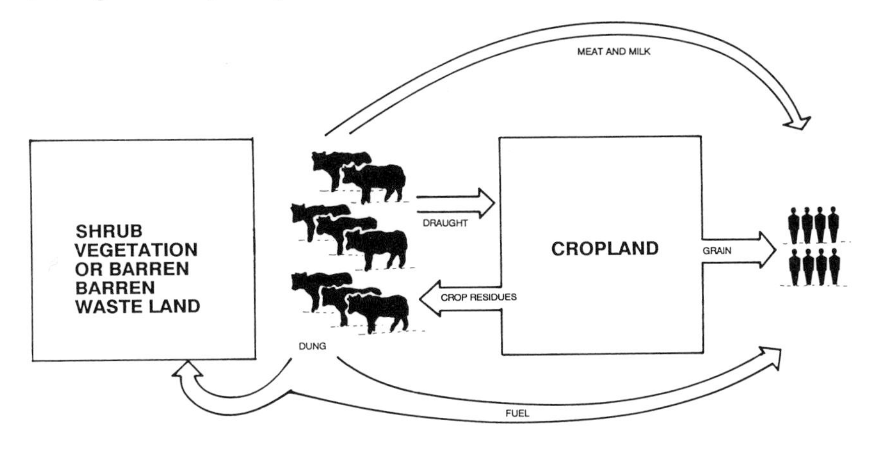

Spontaneous regeneration can still occur on most of this land if grazing pressures can be reduced. If it is not, deterioration will proceed at an accelerating rate, to Stage IV (Figure 8.5). Here, uncultivated land is reduced to barren waste where rainfall is low; where rainfall is fairly high, as in the Kumaun foothills, shrubberies predominate; grasses are scarce and unpalatable species common.

Even under Stage IV conditions there is often a tendency for the number of animals per hectare of cultivated land to increase if the number of farms continues to increase by subdivision of existing land. This trend, however, is ultimately replaced by one of falling animal numbers. When this happens preparation of land for sowing of crops is delayed and even hand digging may become necessary (Vaidyanathan *et al.*, 1979; *DANIDA, 1980).* Degeneration to Stage IV virtually precludes natural regeneration. Assisted regeneration by planting, however, can slowly bring the land back into productivity, as experience in West Asia and China has shown; the cost, nevertheless, is very high.

Why are there too many animals?

Since the obvious immediate cause of this progression, overgrazing, is too many animals, it is worthwhile looking into the reasons for animal population increase. Why, for instance, do farmers maintain animals that may be infertile and moribund? Even if the animals do not provide an adequate return, the farmer also has no costs under the present system of unrestricted grazing on village common land or on government land. The labour expended on grazing the animals cannot be put to alternate productive use since the family members are underemployed. The sale value of old and unproductive bovines is so low that it is usually worthwhile to keep them solely for their low level of dung production. The individual farmer is most likely well aware that the collective grazing of too many animals is a major cause of the progressive environmental deterioration, but as an individual he is helpless to do anything about it. If he reduces his animal stock, his plight may become worse than at present because he cannot be sure his example will be emulated. Again, he probably knows well enough that a change in the system of land management (especially of the uncultivated land) could enable a significant increase in productivity, but as an individual he is powerless to make any changes because everyone has rights of access to the uncultivated land. Social and economic disparities in the village have served to prevent effective collective action.

There is a widely held view that the continued existence of excessive numbers of animals is the result of religious stringencies. However, the problem of excessive livestock numbers (including cattle) and overgrazing is common in many non-Himalayan countries and the basic cause is lack of control over grazing. Jackson (1983, unpublished) also points out that a study of livestock census figures reveals that the growth in numbers in India is due primarily to an increase in buffaloes and goats, not cattle. Finally in this context, Vaidyanathan *et al.* (1979), by means of a critical analysis of bovine species and sex ratios, have concluded that religious sentiment is not an important factor in the decisions farmers make to keep or dispose of their animals. It is extremely important, therefore, for agricultural development planners to understand this point. Jackson believes that, if this were not the case, there would be no solution.

Proposal for the restoration of productivity in the Kumaun foothills

The present situation in the Kumaun Himalayan foothills can be described from the point of view of the individual family. In Figure 8.6 an attempt is made to do this using the local farming model. The data, provided in Table 8.2, have been obtained from a detailed survey of households in Dwarahat Block of Almora district, Uttar Pradesh. The biomass data are derived from actual monthly weighing. The survey is as yet incomplete and the data are thus tentative. However, no other set of data suitable for this purpose is known to exist and refinements will be made as more and better data accrue.

Dwarahat Block is in the more heavily populated belt that runs northwest to southeast through the mountains of Uttar Pradesh. Most of the cultivated mountain land is found in this belt. In satellite photographs the area shows up distinctly because of its thinner vegetative cover and higher incidence of soil erosion. Human population density in Dwarahat Block is about 170/km^2. One-third of the area of the block is cultivated. The Forest Department manages 50 percent of the uncultivated land for commercial forestry, though villagers have grazing and fuel-gathering rights on this land. A further 30 percent of the uncultivated land is managed by the Forest Department solely for the benefit of the local population. The remaining uncultivated area is in panchayat land or is privately owned. Nowhere, however, are there thick stands of trees; the general aspect of the landscape is of treeless hills with sparse grass cover, shrubberies, and individual trees. Altitude ranges from 1,000 to 2,000 m. Water scarcity is acute during the dry season. These characteristics are typical of the larger cultivated belt of the outer Himalayan ranges of Uttar Pradesh.

Figure 8.6 indicates that the net result of the present agricultural activity in Dwarahat Block is only enough food for a very inadequate diet for the residents for about two-thirds of the year. The output of forest products which earn government revenue is also meagre, because of the very poor condition of the forests. And as dismal as this picture is, the more important point to be grasped is that it will get worse. The uncultivated land will become completely barren, or, at best, covered with useless shrubs. And it cannot be argued that in the plains this stage was reached centuries ago and yet agriculture is still continuing there, so why should we be concerned about the mountains? Mountain soils are inherently poorer than plains soils – in structure, depth, composition, and nutrient content, as well as in water-holding capacity. If their productivity is low now, what would it be with no organic matter additions?

The deterioration in the productivity of the mountain environment has now been defined as a function of deterioration in the vegetative cover of uncultivated land. The basic strategy for development, therefore, must be to repair that vegetative cover and then to maintain it. Unless this is done nothing else that is being attempted (such as soil conservation, new cropping practices, orchards, roads) will alter the present trend of decay.

Figure 8.6 The situation of an individual family holding today in the Dwarahat Block, Almora district, Uttar Pradesh (after Jackson, 1983, unpublished).

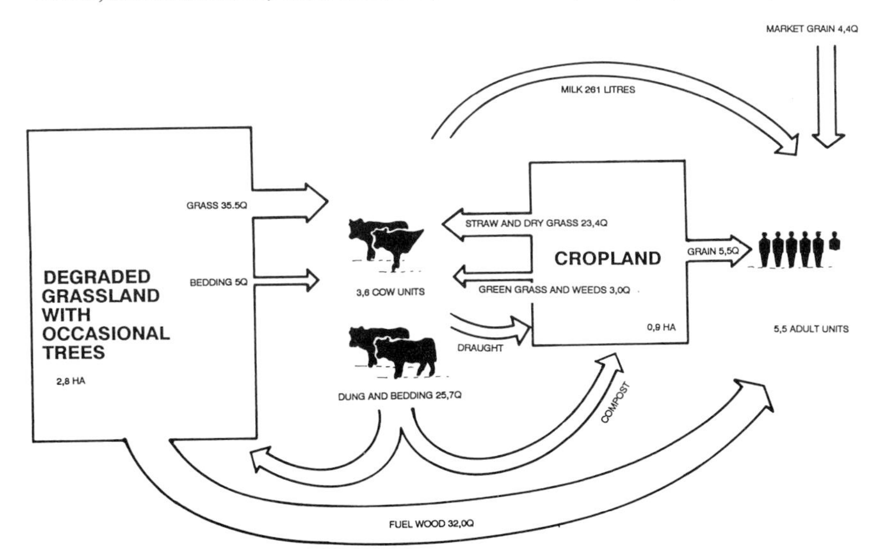

Degraded grassland – Stage III – will regenerate spontaneously if properly managed. 'Proper management' implies reducing the frequency of defoliation so that grass plants gain strength, spread, and set seed. To achieve this the management of livestock will have to be altered. If animals are to be grazed, grazing on any one parcel of land must be permitted for only a few days at regular, long intervals (for example, one month during the rainy season and six months during the dry season). The alternative is not to graze animals at all, but to cut the grass and stall-feed animals. The advantages of cutting and carrying are that: 1) it is easier to organize than rotational grazing; 2) young tree seedlings are not destroyed; 3) cattle tracks, which erode in the rain, are not created; 4) all dung goes to cultivated land; 5) livestock parasite loads are less; 6) damage to standing crops and terrace risers is reduced. Against this is the great labour requirement of cutting and carrying. However, if the objective of hill development is to create more employment, this should be acceptable, provided that the labour expended on cutting and carrying grass is remunerative. Figure 8.7 suggests that this can be achieved.

Much thought has been given to the mechanics of rehabilitating uncultivated land in the context of cutting and carrying grass, as part of the Tinau Watershed Management Project (1980). This is a major Swiss-aid supported project in similar terrain in southern Nepal. A start is made by enclosing (preferably with quick-growing hedges) approximately one tenth of the area to be rehabilitated. Grazing is eliminated from this area, but allowed to continue in the remaining nine-tenths. The grass in the enclosed area is not cut until after the rainy season so that it will shed seed. If the existing cover is too thin, seeding of higher-yielding indigenous bunch grasses can be done by hand. Within three to four years output should have increased two- to fourfold. At this point, the grass may be cut two to three times during the rainy season on a rotational basis to give a daily supply of green grass. Another section of the original area, two to four times larger than the first, should next be enclosed for rehabilitation. In this way even bigger portions of the original area can be rehabilitated without reducing total grass output during the process. A gradual change from grazing to cutting and carrying is also effected.

As has been pointed out earlier, grassland yields green, nutritious fodder only during and immediately after the rainy season. In the interest of effective feeding of livestock, therefore, use of trees, which provide green leaf fodder during the dry season, should be part of the rehabilitation programme. This will also serve the needs for timber and fuelwood. Here one is obviously not thinking of natural regeneration of forest, but of planting and intensively managing (for example, coppicing) specifically selected species. Trees also yield more fodder per hectare than grass. Again the Tinau Watershed Management authorities (1980) have identified species, collected information on growing them, and have organized seed and seedling production. Their programme of rehabilitating uncultivated land calls for the planting of trees for fodder production even while the grass cover is regenerating. Within eight to ten years these trees will produce the bulk of the fodder from the area so

Figure 8.7 Idealized representation of plan for rehabilitation of uncultivated land in Dwarahat Block, Almora district, Uttar Pradesh (after Jackson, 1983, unpublished).

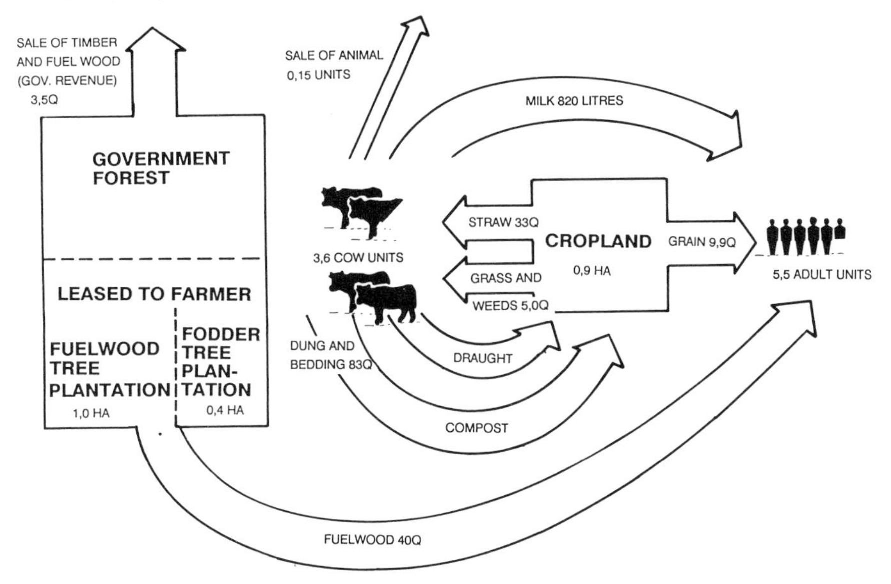

treated. It is estimated that yields of green forage of 120 quintals/ha on an air-dry basis are feasible (plus 15 quintals of grass from under the trees). Many of the fodder trees they have identified are leguminous. Intensively managed firewood plantations are estimated to be capable of yielding 35 quintals of dry wood/ha/yr. Of course, fodder trees also yield some fuelwood, and fuelwood trees some leaf fodder.

The impact of rehabilitating uncultivated land in Dwarahat Block along the lines described above is depicted in Figure 8.7 using data provided in Table 8.2. By following the proposed programme the productivity of the entire farming system will improve. The output now includes enough cereal grains to feed the family for the entire year, or 7.4 q/yr more than at present; 2.25 litres of milk per day, or 1.5 litres per day more than at present; 40 quintals fuelwood/yr, or 25 percent more than at present, and an occasional surplus animal for sale, where there is none at present. The monetary value of these gains in productivity is about Rs 2,300/family/yr (about US $250) at current market prices. Increased revenue from government-managed forests would be of the order of Rs 4 crore/yr (US $4 million) for the block as a whole (18,000 ha × Rs 2,700/ha), or a ten-fold increase over present revenue. If 20 percent of this revenue were to be paid out in wages and salaries to local people an average family would earn Rs 600/yr. If present income per family is taken to be Rs 4,000 (G.B. Pant University, 1976) average family income would increase by 75 percent (Rs 4,000 to Rs 7,000/year).

Further gains in productivity and income could also be obtained from ancillary measures. These include chopping fodder and stall-feeding it, chemical treatment of straw to improve its feeding value, providing mineral supplements, better health care for animals, better composting techniques, improvement of terraces to reduce erosion, the use of chemical fertilizers for crop production, the use of a smokeless *chula* (with 50 percent increased fuel efficiency), and improved means of grain storage. And, in addition to making life more pleasant, increased resources would be available for developing irrigation.

Still further gains in income might be made by farmers near the better-class roads and towns by using the increased productivity of the land differently. Some might find it more profitable to replace cereal crops with grass and fodder trees and produce milk for sale by keeping more animals. (They could also replace their bullock with a cow, producing still more milk on the same feed budget.) Similarly, fruit might be more attractive to some farmers than grain. A shift from grain to fruit would also permit more animals and more milk could be produced. These alternatives should be particularly attractive to farmers whose fields are marginal for cultivation.

Administrative and legal considerations

Having considered the technical aspects of rehabilitating uncultivated land, and the possible benefits, the administrative and legal aspects must be taken into account. In the Kumaun mountains uncultivated land was 'reserved' at the beginning of the present century, that is, it was closed to the local population who previously had had free access to it for grazing their livestock and collecting fuelwood. This caused much resentment which expressed itself in large-scale burning of forests. This led to the forest satyagrahas associated with the Gandhi resistance to British rule, from which the Chipko Movement has its inspiration (Shiva and Bandyopadhyay, 1986b). In 1921 about half of the closed area was reopened to the local people by the creation of civil and panchayat 'forests'. The government, however, retained ownership of the civil 'forests' and assumed the responsibility for their maintenance. The management of the panchayat 'forests' became the responsibility of the village panchayats. At present, even government land not redesignated civil or panchayat 'forests' is, for the most part, open to the local population. By an area being 'open' it is meant that the local people can graze their livestock, lop trees for fodder, collect fallen branches for fuel, and pick up forest litter for bedding livestock. They are also allowed timber from civil and panchayat forests. Rules exist for regulating all these activities but they are not now enforced. Indeed, they cannot be, in view of the dwindling availability of grass and wood and the increasing demand for them.

This is an extremely unsatisfactory state of affairs and is the root cause of the abuse of uncultivated land. Continued ownership of the land by the government and all the rules and regulations by which it seeks to control the use of this land has placed it in opposition to the people. The fact that the land is 'open' to them or earmarked only for their use has not mitigated against this antagonism. The mountain people, understandably, have developed an irresponsible attitude toward this land, one of getting as much out of it as they can with no feeling of responsibility for its maintenance. At the same time they are not unaware that their activities are causing the evident deterioration. But they feel helpless in a situation in which they do not have effective control of the land that sustains them. It is this pervasive attitude of helplessness that characterizes the mountain people's attitude to farming and to life in general. No 'development' whatsoever is possible under such circumstances.

The lesson to be learned from this situation is that people must own the land that sustains them if they are to use it responsibly and if their sense of self-confidence is to be maintained. It might be argued that some land *was* turned over to the people, the village panchayat land, and that its condition is as bad as that of the civil 'forest' land. However, the true significance of this fact is that collective ownership of land is not the solution to the problem either. As pointed out earlier, the prevailing social and economic disparities among people in the village are not conducive to a democratic and effective control over the use of community-owned land.

Table 8.2 *Statistics for the average farm in Dwarahat Block, Almora, district Uttar Pradesh, present and rehabilitated (after Jackson, 1983, unpublished).*

Item	*Present*	*Rehabilitated*	*Note reference numbers*[2]
Number of people	7.1	7.1	1
Number of adult units	5.5	5.5	2
Area of cultivated land, ha	0.9	0.9	3
Area of uncultivated land, ha	2.8	2.8	4
Number of cow units	3.6	3.6	5
Grain produced, q/yr[1]	9.2	16.6	6
Straw and dry grass produced, q/yr	23.4	33.2	7
Green grass, weeds, etc. from cultivated land, q/yr on air-dry basis	3.0	9.0	8
Green fodder from uncultivated land, q/yr on air-dry basis	35.5	74.0	9
Bedding material, q/yr	5.0	25.0	10
Fuelwood, q/yr	32.0	40.0	11
Timber, q/yr	1.0	—	12
Milk, litres/year (after feeding calf)	261.0	820.00	13
Saleable livestock, cow units/yr	—	0.17	14
Quantity of compost, q/yr (undecomposed, air-dry equivalent)	25.7	83.10	15
Nitrogen addition to cultivated land through compost, kg/yr	22.9	88.80	16

Source: G.B. Pant University (1980).

Notes: [1]q/yr = quintals/year.
[2]See notes to the table, overleaf.

The proposed remedy, therefore, is to parcel up and lease panchayat and civil land to individual families. Leases could be for a lifetime and should be non-transferable, except to one legal heir of the lessee. Leased parcels of land should be equal in area for all families, irrespective of how much cultivated land they own. This land could be used by the family only for grass and/or trees but not for crops or orchards. A provision could be made for revoking the lease for misuse of the land. The land area to be thus leased would be equal to the present combined area of civil and panchayat land, but should be that land nearest to the villages and least vulnerable to erosion. Distant land in a difficult terrain should be retained by the government as forest land and managed intensively for profit. Local people would no longer have any rights in this forest. The government could assist the farmers to rehabilitate the land leased to them by providing technical guidance and (for payment) inputs such as tree seedlings.

Notes on data in Table 8.2

1. The average number of people per household was obtained by random survey of sixty families in fifteen villages (G.B. Pant University, 1980). Data from a survey conducted in Dwarahat Block in June 1980 (Data Systems Inc., 1980) indicates that population has not increased since the last general census in 1971. Consequently, no increase in population is being assumed for the next decade.
2. An adult unit is defined here as one adult (individuals above eighteen years of age), or two children (individuals below eighteen years of age).
3. The cultivated land areas for individual holdings have been obtained by dividing the total cultivated area in a village by the number of its families. Data were collected from Patwaris in thirty randomly selected villages in the Block (G.B. Pant University, 1980). The term 'cultivated land' includes areas in permanent grass and orchards since such areas are privately owned and are not grazed. It is considered that the area of cultivated land should not increase in the future, keeping in view the very marginal nature of the remaining land.
4. The average area of uncultivated land per holding was calculated by dividing the total uncultivated area of the Block (excluding only land under 'non-agricultural uses') by the number of families in the Block (calculated from the thirty-village sample mentioned in note 3 above). This figure therefore includes all uncultivated land, irrespective of its official classification as to ownership or for revenue purposes. This is done because virtually all land not cultivated is subjected to grazing and tree lopping by the local people. In the rehabilitated system depicted in Figure 8.7, 50 percent of this land is considered to be leased to the local people and the remainder to be owned and managed by the government (Forest Department). This corresponds to the actual division of this land at present between land classified as 'Forest Department' land, on the one hand, and as 'civil forest' and 'panchayat forest,' on the other.

5. Livestock enumeration was done with sixty randomly selected families in fifteen villages in June 1980 (G.B. Pant University, 1980). The cow unit equivalent of various animals has been taken as follows: cow = 1 unit; bullock = 1; buffalo = 1.5; cow calf = 0.5; buffalo calf = 0.75; goat = 0.15. On an average each family owns 0.92 cow, 0.57 buffalo, 0.92 bullock, 1.1 cow calf, 0.37 buffalo calf, and 0.15 goat. In the rehabilitated system the same number of animals are envisaged.
6. Present grain production was determined by a random survey of sixty families in fifteen villages (G.B. Pant University, 1980). Total production per family is 9.2 q/yr. Assuming that 20 percent of the crop is saved for seed or spoiled, and including usual milling losses, the net edible grain per family is 5.5 quintals. Each family also purchases 4.4 quintals of grain, mostly from government ration shops. This gives a total family consumption of 9.9 q/yr (493 g per adult unit per day).

   In the rehabilitated system it is expected that a grain production of 16.6q/yr could be achieved on the same area of cultivated land. This is because of the additional compost the land would receive (see notes 15 and 16 below). The net edible grain from this much production would cover the family's need for about 9.9 q/yr.
7. Straw production is assumed to be twice the grain production on an average for all crops. Dry grass production from private grass plots (included under 'cultivated land') is assumed to be 5 q/yr at present. This would be harvested green in the rehabilitated system.
8. The figures for green grass and weeds from cultivated fields, terrace risers, and so on, are assumed to be 3 and 4 q/yr on air-dry basis for the present and the rehabilitated system, respectively: add 5 q/yr to this (see note 7 above) for the rehabilitated system.
9. Grass output from uncultivated land in the present system is taken as 12.7q/ha/yr on an air-dry basis. This value was determined by actual cutting and weighing from areas protected by wire cages and cut frequently to simulate defoliation by grazing animals (G.B. Pant University, 1980). This assumption ignores the fact that some grass from uncultivated land is harvested by hand and that some tree leaves are lopped from the few trees that are still found on this land. In the rehabilitated system 1.4 ha out of 2.8 ha is leased to farmers, and 1.4 ha is retained by the government for intensive, commercial forestry. Out of the farmer's 1.4 ha, 1 ha is assumed to be planted with fuelwood trees yielding 35 q/ha/yr of fuelwood, 20 q/ha/yr of green leaves (air-dry equivalent basis) and 20 q/ha/yr of dry leaves for bedding. The remaining 0.4 ha is planted with leguminous fodder trees which yield 120 q/ha/yr of green fodder leaves and 15 q/ha/yr grass (both on an air-dry equivalent basis). These yield data were taken from the Tinau Watershed Management Project (1980).
10. Bedding material, here considered to be fallen leaves/needles, is estimated to be 5 q/yr on an air-dry basis for the present system and 25 q, for the rehabilitated system.

11. Fuelwood consumption at present was found by questionnaire to be 32 q/family/yr in Dwarahat Block. In the rehabilitated system, 35 quintals could be produced from the one-ha fuel-tree plantation and another 5 from the 0.4 ha fodder-tree plantation.
12. Timber consumption may be of the order of 1 q/family/yr at present. In the family survey conducted in Dwarahat Block (G.B. Pant University, 1980) only a few families attempted to estimate their average annual timber consumption. This figure is thus only an approximation. In the rehabilitated system the farmer would not produce timber. They would have to purchase it.
13. Milk yields of animals have not yet been determined by actual weighing (G.B. Pant University, 1980). The basis of these estimates are as follows:

	Present system	*Rehabilitated system*
Milk yield/cow/lactation, litres	200	600
Milk yield/buffalo/lactation, litres	600	1,200
Calving interval of cow, yr	2	1.5
Calving interval of buffalo, yr	2	1.5
Consumption of milk by calf up to weaning, litres	100	200

These improvements in productivity would result from better feeding alone. The data on feed consumption, based upon the figures already given in previous notes, are:

	Present system	*Rehabilitated system*
Cow units	3.6	3.6
Green fodder, kg/head/day (air-dry basis)	2.7	6.3
Straw and dry grass, kg/head/day	1.8	2.5
Total fodder availability	4.5	8.8
Estimated requirement*	5.6	8.4

*Assuming a body weight of 200 kg for an adult unit at present, and 300 for the rehabilitated system, and a capacity of 2.5 kg dry-matter/adult unit/day/100 kg body weight.

14. The present system is assumed not to produce any saleable livestock on an average, except for the odd goat. The average complement of adult livestock on a farm is approximately one cow or buffalo, and one bullock. On an average, therefore, one adult can be replaced after 1.3 years with the calving rates given in note 13. A more realistic figure would probably be 2 years, considering mortality. At this rate every animal has

a working life of 6 years. In the rehabilitated system calving rates are increased so that the adult could be replaced every 4.5 years. This offers scope to sell animals, either the culled adult or the replacement calf. If a working life of 6 years is assumed for the rehabilitated system also, one animal could be sold every 6 years.

15. Compost is made up of dung and bedding (uneaten fodder also). In the present system the diet can be assumed to be 50 percent digestible. Dung production would thus be 29.6 q/yr on an air-dry equivalent basis. One-third of this is produced during grazing. Bedding adds 5 quintals. Total compost, therefore, is 34.6 q on an undecomposed basis. In the rehabilitated system the digestibility of the diet can be assumed to be similar, i.e. 50 percent. Total compost thus becomes 80 q/yr (all dung goes into the compost heap).
16. The nitrogen added to the cultivated land through compost has been calculated on the basis of the following assumptions:
    (a) Straw, dry grass and bedding material contain 0.5 percent nitrogen on an air-dry basis. Green grass and weeds contain 1.5 percent and tree leaves contain 2 percent.
    (b) Three kg of nitrogen are retained by livestock of the farm in the present system and 10 kg in the rehabilitated system.
    (c) One-third of all excreta at present is dropped on uncultivated land. In the rehabilitated system all dung and urine go into the compost heap.
    (d) One half of the nitrogen in fresh dung and bedding is lost by leaching and volatilization from the compost heap.

POPULATION GROWTH AND LAND-USE PLANNING IN NEPAL

The approach adopted in this section contrasts with the farm-based analysis and proposed solution of land-management problems in the Kumaun Himalaya of Uttar Pradesh. It considers Nepal as a whole, with emphasis on the Middle Mountains and the Terai. These are the most densely populated regions where relatively good accessibility, at least in some areas, provides a better opportunity for much larger-scale technological intervention. Hrabovszky and Miyan (1987) undertook their study of land-use and food-production problems from the standpoint that there are causal linkages in both directions between population growth and land use. They discuss the effects of the former on the latter and, more specifically, on adaptation patterns in local land use and in rural-to-rural and rural-to-urban migration. The particular problems of Nepal are analysed using data originating from three perspective studies carried out by APROSC (Agricultural Projects Services Centre) on request from the National Planning Commission of Nepal. These studies analyse most population and land-use problems within a framework which disaggregates Nepal into fifteen subregions, that is, the three main ecological zones of High Himalaya, Middle Mountains, and Terai, themselves divided into the five Development Regions of the country. These studies cover the recent past, from 1971/72 to 1984/85 and extend into the future, to 2005.

Some historical settlement patterns

There is growing evidence that pressure of population on the land has become the most intractable problem facing three types of situation in Developing Countries where demand for food exceeds the local ability to produce food from the available land and with the available technology. The first of these situations is the low-rainfall savannah lands of the world, which have become known as the 'dramatic drought areas.' The second type of situation where population pressure often exceeds local food production capacity is represented by the rice-bowls of Asia. Here the evolution of intensive rice cultivation under some form of water control has permitted very high population densities which now, in many locations, have exceeded the maximum possible productivity of the land, given the technologies presently being used. This situation is best exemplified by two countries of the Indian subcontinent, Bangladesh and India, where there is heavy reliance on monsoon-fed rice culture and which contain about 60 percent of the hungry of the world.

The third situation is the mountain areas. As with the savannahs, mentioned above, middle-mountain settlement also has been an early tendency in human history, as it provided a relatively disease-free environment, a climate permitting an active life, usually highly fertile soils, and in general a reliable, often high, rainfall. They were also relatively safe places against intrusion. In the case of Nepal, history relates that many of the past population explosions have been the result of immigration into the mountains from lowland areas. This movement has since been reversed as the population growth of the Nepalese mountains has not been matched by the subsistence farmers' ability to produce food, and for about the past hundred years there has been an accelerating migration from the hills to the plains. Hrabovszky and Miyan (1987) have referred to this as the 'Great Turnabout.' The reason for this has been a rapid growth in population within the confines of limited availability of potentially cultivable land. This pattern of migration is certainly not restricted to Nepal, nor to the Himalaya; it is displayed in many mountain regions, including the Andes, and has been a popular political and research topic in mountainous areas of Europe.

The drastic changes in the population – land equilibruium

As long as the Nepalese population grew at or below 1 percent per annum and as long as there were reasonable-quality land reserves available for conversion from forest or grassland use into arable land, Nepal could both feed its population and export small surpluses of agricultural products to the neighbouring countries.

Under the impact of improved health care, annual mortality figures were reduced substantially from 38 to 16.6 per thousand of population between 1951 and 1985. At the same time fertility has remained high and thus population growth is now running at about 2.7 percent per annum, and it is expected to accelerate further to about 3 percent per annum by 2005. Nepal's population grew from 8.3 million in 1951 to 16.7 million in 1985 and the projected population for 2005 is about 30 million.

Given the slow economic transformation process in Nepal, the overwhelming majority of the increases in population have to be accommodated within the agricultural sector. As has been emphasized already (Chapter 1) for the country as a whole, in 1985, 83 percent of the households were dependent upon agriculture.

There are signs that fertility has actually increased as modern communications and the welfare goals of the government have led to a lowering of local pressures on people in poorly endowed areas to control fertility.

Two fundamental land-resource use issues

Nepal is facing two fundamental land-resource issues. One of them is that the final limit of land suitability for cultivation is being approached rapidly, and the other is the speed at which land can be brought into cultivation, or its intensity of use can be increased, in order to match the growth in demand for the products of land.

Taking only the criterion of suitability of land for cultivation, Nepal still has about 730,000 ha, in addition to the 2,410,000 ha presently under cultivation, that could be converted to arable land. This would mean conversion from forest land, nearly all of which is in the Terai. There are some important considerations, however, why not all the above potential should be brought under cultivation. First, in specific locations, demand for fuelwood exceeds the supply and some lands which ultimately would be suitable for cultivation (in the Middle Mountains, often only after terracing) need to be left under forest cover. The second consideration relates to the functions of the forest in support of cultivation through its production of fodder and also forest litter for the supply of organic matter to the arable lands. These issues have also been discussed in the Kumaun Himalayan context. In the Nepalese case Hrabovszky and Miyan (1987) stipulate that, in areas with limited access to inorganic fertilizers, each hectare of cultivated land needs to be supported by 1–4 ha of forest land if soil fertility, and thus agricultural output, is to be maintained.

In view of the above constraints and of the need to take some unterraced and terraced lands *out* of cultivation, and also allowing for the increased area of settlements and roads, the Perspective Land-Use Plan estimated that only about 160,000 ha could be added to the total cultivated land, equal to only 7 percent of the presently cultivated area. This clearly points to land-use intensification as the main means for raising agricultural output. Table 8.3 shows that between 1965 and 1985 the estimated area under cultivation

Table 8.3 *Increase in area under cultivation in Nepal between 1965–66 and 1985–86 in thousands of hectares (after Hrabovszky and Miyan, 1987).*

Years	*Cultivated land*	*Cropped area*	*Cropping intensity*
1965–66	1,840	1,995	108
1970–71	2,030	2,231	110
1975–76	2,161	2,410	112
1980–81*	2,272	2,459	108
1985–86	2,410	4,002	166

*drought year

increased by 31 percent and the annually cropped area by 100 percent. This indicates that the cropping intensity (C.I. = annually cropped area/cultivated area × 100) had risen from 108 percent in 1965 to 166 percent by 1985, a major feat accomplished by the Nepalese farmers. At the same time, the price for this increase in cropping intensity has been a stagnant or declining yield.

While the Land Resources Mapping Project,[1] together with the new cadastral survey, have provided a major improvement in land-use information in Nepal, time-series information is still weak. Orderly release of Class I and Class II lands from forest to cropping use is still hampered by some of the existing land-use policies.

Adaptive patterns in response to growing population pressures

Experience in extending the cultivated land and increasing cropping intensities, and of the limits on extension of cultivated land, have been discussed above; here the constraints and opportunities for intensification and migration are analysed.

The traditional method for maintaining the fertility of the land is running into resource constraints, not only because of the declining forest/cultivated land ratio, but also because the supply of farmyard manure is rather inelastic. There are some opportunities to increase the share of manure which is available for cropland, for instance, by zero grazing, stall-feeding systems, but there is a consensus that Nepal should avoid increasing the number of animals because of the shortage of fodder and the resultant damage done to forests and grasslands.

Thus the ultimate resource for increasing production is technological change, which includes some imports of manufactured products, though this is not without its own problems. Other technological advances include improved, more disease- and pest-resistant plant varieties, more use of labour, and, above all, more irrigation where possible. As yet the impact of such interventions has been small, as it is estimated that use of inorganic fertilizer represents only 4 percent of total plant nutrient replacements and that improved crop varieties cover at present only 10 percent of the cropped area.

When land resources are constrained and technological change does not provide sufficient relief from the pressure of population on food resources, most societies adapt the pattern of distribution of the limited resources, and/or establish rules about sharing labour and thus the total product of society. Nepal shows some of these signs, but also a growing landlessness among the agricultural population.

Given the limits on all the above means of adaptation, when possible, the reaction of farming populations to a declining per capita land base is to migrate. In Nepal, just as in many other highland regions, this means 'going down the mountain' into the surrounding plains areas, both in the form of rural-to-rural and rural-to-urban migration. In Nepal there has been major migration to the main inter-mountain valleys, and especially to the Terai plains. While in the Eastern Terai there is little remaining opportunity to settle new land, in the Western Terai regions malaria control has permitted a movement into hitherto predominantly forested areas. In the Terai, the development of irrigation will also permit much higher population densities in the future, and the current man/land ratio is already higher than that of much of the Middle Mountains.

Rural-to-urban migration is important but, because of the small non-agricultural sector in Nepal's economy, it offers as yet little relief. Given the present proportion of 83 percent of total population dependent upon agriculture, even a large differential in growth rates between the agricultural and non-agricultural sectors, of, for instance, 5 percent for non-agricultural and 2.2 percent for agricultural, for a total population growth of 2.7 percent, this would raise the share of non-agricultural employment in twenty years from 17 to 25.8 percent of the total. This implies that of the total additional population of 12.3 million, 7.7 million would need to be supported by agriculture and there would be an addition of 5.3 million persons to the existing 7.6 million labour force in agriculture.

Both rural-to-rural and rural-to-urban migration in Nepal shows many different forms including seasonal, temporary, single person versus family, and organized versus spontaneous movements. In addition, Nepal has to carry the burden of Indian immigration both into the Terai and into urban areas. The latter is a new development; in comparison, in the longer-term past, there has been more temporary migration by Nepalese into India in search of employment than vice versa, notwithstanding the very large movements and permanent settlement in Sikkim, the Darjeeling Himalaya, and parts of Assam and Arunachal Pradesh over the past hundred years. An interesting parallel can also be drawn between the Swiss and Tyrolean highlanders and the Gurkhas finding outside employment as soldiers.

Looking into the future

As the limits on extensive land use have been reached, or even exceeded, in many parts of Nepal, looking into the future can provide a view of the even sharper problems and conflicts which may arise as the pressure of population on the land continues to mount. Let us first consider the development objectives of His Majesty's Government of Nepal:

1. increased food production to provide a satisfactory diet for Nepal's population;
2. income (both cash and auto-consumption) to rise per capita so as to reduce the proportion of the population below the poverty line;
3. improved regional balance in development and incomes;
4. conservation of natural resources, above all, land and forests;
5. contribution to the overall development of the economy, through income generation, export earnings, and release of labour to other sectors.

In light of these objectives and, given projected population growth and projected developments in other parts of the economy, the three APROSC perspective studies have identified a number of main lines of policies and programmes which could bring Nepal closer to those goals. The discussion below, however, presents only those which are relevant to land use.

The need for intensification of land use

Of overwhelming importance is the need to intensify land use, not only in crop agriculture, but also in the other main land-use forms, namely grasslands and forest use. Within crop agriculture a two-pronged approach may be necessary, differentiating between areas of high and low productive potential and with good or poor access. In areas of poor access primary reliance will have to be placed on local inputs, but even here the efforts for intensification should be on responsive, non-degrading types of land. While the main goal will be food for local use, some limited cash crops of high value for weight, such as spices and tea, would need to be part of the production pattern.

In areas of good access, especially those with good irrigation facilities, high-intensity land use must be pursued, often using cash-input crops, to provide high cropping intensities and high yields. This will involve major technological change in terms of production methods and in terms of supporting services for both inputs and extension advice, as well as credit and marketing. It is these areas that will have to produce a marketable surplus to feed the urban populations, to meet the raw material demands of the agricultural-processing industries, and to provide exports. The bulk of this activity is likely to be in the Terai, as is the case today. With rising urbanization and incomes, however, the role of the Middle Mountains as potential producers of fruit and vegetables could become more fully realized. Such an effort could help to release pressure on marginal areas which then can be used for less degrading types of land uses, such as, permanent tree-crops, cut-and-carry grass production, or intensive forestry.

Grasslands are used at present mainly under uncontrolled grazing systems. While there is a great scarcity of information on the use of these lands, and it is necessary to differentiate grazing adjacent to the local village, more distant forest grazing, as well as 'alpine' grazing (that is, grazing above the upper timberline), grassland overgrazing is widespread even though the actual control of the land may be rigid. The end result, common throughout the Himalaya and also illustrated in the preceding section on the Kumaun region, is low productivity by animals whose productivity is also kept low by disease and poor feeding. To improve grassland use there will need to be both more cut-and-carry stall-feeding systems, and a much improved stock. This will require improvement in the feed base through better utilization of agricultural by-products, such as straw treatment for increased digestibility and the use of short-term leguminous fodder crops to help with soil fertility maintenance as well. Given the heavy demands for food, however, the scope for a large-scale increase in the latter may be limited. Indirect means for the better utilization of the limited feed resources will also have to come from more productive animals, a larger share of productive animals in the herds, and increased feed use efficiency from reduced morbidity and mortality through improved health support. At the same time, these measures will have to be counterbalanced by policies aimed at reducing the growth in animal numbers to the absolute minimum. This takes us back to the problem of balance between forest area available to provide livestock fodder and bedding. The growing role of the buffaloes in the Terai offers some opportunities here, but a meaningful and acceptable control system for cattle populations in the Middle Mountains has yet to be developed.

Major policy and programme proposals have been formulated in recent years for the forestry sector. The central theme is to preserve forests which have protection functions, utilize optimally commercial forest areas and help community forestry to gain maximum sustainable output from forest area reflecting the diversity of rural people's needs. Both in commercial and in community forestry this means turning increasingly to plantation forestry and forestry management practices which reflect better the output mix demanded, namely a much larger share for fuel, fodder, and litter as compared to timber output. Technologies to reach these goals are known but need improvement, and successful pilot programmes point to the new institutional forms which are needed (but see the discussion based upon Mahat *et al.*, 1986a and b, 1987a and b, pp. 75–8).

In summary, intensification has to be achieved in all three major land-use types – cropped land, grazing land, and forests. Land must be allocated to uses which are not degrading and which, at the same time, represent the best and most productive use of that land. Within each of the main categories there must be a drive to use the most productive and least degradation-prone lands most intensively, so as to reduce the pressure on marginal lands, which then can be put to appropriate non-degrading uses.

Linkages between the major land-using sectors

The above arguments point to the necessity to plan and to implement land-use facilities in a manner which takes cognizance of the strong inter-relationships that exist between the three main land-use types which make up the overall agricultural land-use system. Crop agriculture depends largely on animals for its motive power and for its plant nutrients, thus on grasslands and on forest. In turn, livestock receives a large part of its feed base from crops. It also influences forest regeneration, negatively, through grazing and lopping in forest areas. Forests are the final 'givers' in this chain of inter-linkages. They provide new land for cultivation and grazing, fodder, litter, and their two wood products, fuelwood and timber. It seems that the combined production functions for these various products are as yet not well enough known to make optimal production decisions in the face of the product-mix demanded. Also of great importance is the nurturing and management of trees on private land. One of the recent trends in some areas of the Middle Mountains (Gilmour, personal communication, October 1987) is the significant increase in trees on private land over the past five to ten years. This is believed to be an auto-response to the growing pressures on communal forests (see also Gilmour, 1988).

The other over-arching consideration for improved land-use systems is the need to utilize technologies in each of the land-use types which can provide substantial protection against land degradation. The quantitative data provided by the Land Resources Mapping Project survey has shown that while there is widespread damage, its order of magnitude is much less than what is popularly stated. This, in turn, concurs with our earlier discussion of the exaggerated claims concerning landsliding, soil erosion, and downstream impacts (Chapters 4–6). There are technologies (not necessarily modern or western) for land use in each of the main types of uses that can provide substantial protection against land degradation. Their application, however, under conditions of growing population pressure requires concerted effort by all, including government and farmers, herders and foresters. It also requires that the interests of future generations be harmonized with the interests of the present one.

Timing of land-use changes

Many of the major land-use changes which will be necessary are closely tied to decisions on their timing. Little doubt exists, for example, that all good, Class I and Class II lands in the country should be under cultivation, but it is still hotly debated how fast that transfer should occur, who should settle the land, and under what conditions. In the view of some, it should happen as fast as meaningful settlement programmes can manage to implement it; others maintain that all other agricultural production options need to be exhausted before the best commercial forests in the Terai are given up for cultivation. It is also questionable whether there is enough police power in the country to prevent unorganized settlement of good land. The past indicates that there is only limited control possible on the side of government, and the so-called 'illegal encroachment' of forest lands will continue and result in an unplanned and often sub-optimal release of land from forestry.

The importance of irrigation

Irrigation is a key element in raising agricultural output and, given the natural resource endowments, the bulk of its development will be in the Terai. This creates further distortions in regional income distribution, unless counter-balanced by migration. Groundwater development is likely to be crucial, but it is predicated on the wide availability of electricity in rural areas, implying major hydroelectric development.

Estimates indicate that about 45 percent of the final total cultivated land in Nepal could be irrigated. The present area irrigated, at 587,800 ha, represents only one-third of the potential. Under Nepalese conditions irrigation adds to yields directly, enables much higher use of other inputs, such as inorganic fertilizers and high-yielding varieties of crops to be used for substantial increases in production, and above all it permits at high yields much higher cropping intensities to be reached, thus increasing total cropped area.

An overview of some major quantitative changes

There will be some substantial changes in the regional distribution of population in Nepal, with the Terai gaining in its share of the total (Table 8.4). When compared with land-use changes, the resulting figures on land per capita indicate a very tight situation with cultivated area per capita projected to decline from 0.17 ha in 1985 to 0.09 ha by 2005. Even though cropping intensities would rise, the annually cropped area per person would decline from 0.28 to 0.16 during the period. The components underlying output growth show that yield increases would contribute over 90 percent of the increases in output. Table 8.6 provides the orders of magnitudes for predicted increases in cropping intensities and in yields (for cereals). The balances between production and requirements of food by ecological regions are shown in Table 8.7. They provide a warning that some of the Middle Mountain and High Himalayan subregions will have to import large volumes of their food, and while Terai surpluses could cover these, the actual transfers may be extremely difficult. For livestock the crucial issue is whether enough feed will be available for an animal population that will grow very little in numbers beyond the 1985 total, but would have higher productivity and therefore higher feed requirements. Table 8.8 shows that a close balance could be achieved at the national level, but there will be major shortages in the mountain regions. Production demand and balances for fuelwood and timber are presented in Table 8.9. These show major surpluses of timber and frightening shortages of fuelwood as the figures emerge from calculations using present official shares of timber and fuelwood under existing forestry practices. They point to the serious need to re-orient forest production to cover local requirements.

To be able to achieve these *attractive* results, the share of current inputs in the total value output of agriculture would have to rise from about 12 to 14 percent in 1985 to 20 percent by 2005, and parallel to it, investments in

Table 8.4 *Population projections by ecological zones in Nepal, 1985–2005 (after Hrabovszky and Miyan, 1987).*

	1985	*1990*	*1995*	*2000*	*2005*	*growth rate %*
Mountain	1.4	1.5	1.7	2.0	2.4	2.73
Hills	7.8	8.7	10.0	11.6	13.5	2.78
Terai	7.5	8.9	10.4	12.1	14.1	3.21
Nepal	16.7	19.1	22.1	25.7	30.0	2.97

Table 8.5 *Projected changes in per capita availability of cultivated and cropped area by ecological zones in Nepal, 1985–2005 (after Hrabovszky and Miyan, 1987).*

Particulars	*1985*				*2005*			
	Mountain	*Hills*	*Terai*	*Nepal*	*Mountain*	*Hills*	*Terai*	*Nepal*
Population (millions)	1.4	7.8	7.5	16.7	2.4	13.5	14.1	30.00
Cultivated area (millions/ha)	0.21	0.90	1.3	2.44	0.19	0.82	1.61	2.62
Cropped area (millions/ha)	0.28	1.53	2.12	3.93	0.27	1.44	3.06	4.77
Cultivated area (ha/cap)	0.15	0.12	0.17	0.14	0.08	0.06	0.11	0.09
Cropped area (ha/cap)	0.25	0.20	0.28	0.24	0.11	0.11	0.22	0.16

Table 8.6 *Predicted increases in land-use intensity and yields per hectare of cropped area for Nepal, 1985–2005 (after Hrabovszky and Miyan, 1987).*

	1985		*2005*	
	Cropping Intensity (%)	*Yield (mt/ha of cropped area)*	*Cropping Intensity (%)*	*Yield (mt/ha of cropped area)*
Mountain	135	1.07	140	1.62
Hills	170	1.30	175	2.33
Terai	163	2.00	190	3.02
Nepal	163	1.62	182	2.86

Notes: Yields represent weighted average of main cereals (paddy, maize, wheat, millet, barley) and potato in metric tonnes (mt). Yield of potato has been adjusted into cereal terms by using conversion factor: 3.5 kg of potato = 1 kg of cereal. Cropping Intensity = area cropped × number of crops per year.

agriculture would have to grow from 10.6 percent of agricultural output to 16.8 percent. The value of agricultural output would represent only about 35 percent of total output in the economy, as compared to 52 percent in 1985.

But Nepal also needs major changes in its institutions supporting agriculture. These institutions are young and are often undergoing major

Table 8.7 *Projected balances between food production and food requirements ('000 metric tonnes) by ecological zones in Nepal, 1985–2005 (after Hrabovzsky and Miyan, 1987).*

	1985					
	Total ('000 mt)			*Per capita (kg)*		
	Production	*Requirement*	*Balance*	*Production*	*Requirement*	*Balance*
Mountain	259	293	-34	185	209	-24
Hill	1321	1590	-269	169	204	-35
Terai	1968	1619	349	262	216	46
Nepal	3548	3502	46	212	210	2

	2005					
	Total ('000 mt)			*Per capita (kg)*		
	Production	*Requirement*	*Balance*	*Production*	*Requirement*	*Balance*
Mountain	294	580	-286	123	242	-119
Hill	2074	3165	-1091	154	234	-80
Terai	5267	2925	2342	374	207	167
Nepal	7454	6670	784	249	222	27

Note: Potato has been converted into cereal terms by using conversion factor: 3.5 kg of potato = 1 kg of cereal.

metamorphoses in the experimentation for finding the most suitable solutions for Nepal's needs. Agricultural research, education, and training have all made huge forward strides in the past decades, but much further growth and improvement is needed. But if these improvements are to be meaningful they must be balanced with a much better understanding of the local environments and land-use systems.

Perhaps the single most important insight observed from the three APROSC studies (Hrabovszky and Miyan, 1987) is that the dominant factor in land-use issues in Nepal (and elsewhere in the Himalaya) is the extreme variability, both of natural conditions and of human adaptations to them. This calls for well-identified and *locally* tested prescriptions for highly specific local problems. By now it should go without reiteration that this presupposes the identification of local felt needs and direct input of local people into all levels of policy implementations as a condition for success.

DISCUSSION AND CONCLUSIONS

The two proposed approaches to solving the land degradation–population pressure problem, characteristic of much of the Himalayan region, are almost as wide apart as the two perspectives – too many people versus not enough food. They have one fundamental element in common – in either case, any chance of success will depend on major changes to the existing institutional

Table 8.8 *Feed requirements, fodder production and balances ('000 metric tonnes), by ecological zones, and major feed sources for Nepal, 1985–2005 (after Hrabovszky and Miyan, 1987).*

Sources	1985				2005			
	Mountain	*Hills*	*Terai*	*Nepal*	*Mountain*	*Hills*	*Terai*	*Nepal*
Forest	190	324	221	735	238	405	276	919
Grazing	318	677	76	1071	350	374	83	1171
Crop by-products	150	1038	1707	2895	214	1433	2817	4464
Grains and kitchen residues	2	17	5	24	4	35	9	49
Risers and bunds	75	368	102	545	55	235	96	386
Homestead fodder trees	12	65	57	134	24	130	114	268
Fallow grazing	16	77	85	178	15	62	80	156
Private forestry	—	—	—	—	16	37	13	66
Cultivated grasses	—	—	—	—	53	146	42	240
Total TDFR[1] supply	763	2566	2253	5582	969	2857	3530	7719
Total LSU '000[2]	709	3167	1964	5840	779	3562	2155	6496
Total TDFR[1] requirement	736	3286	2038	6060	893	4081	2469	7443
Feed balance	27	–720	215	–475	76	–1224	1061	276
Feed balance, % of requirement	3.7	–21.9	10.5	–7.7	8.5	–30	43.0	3.7
Total production (values in Rs million[3]				4962				8181

[1]TDFR = Total Domestic Feed Requirement.

[2]LSU = one Livestock Standard Unit, which is defined in terms of metabolizable energy sufficient for an average cow, producing the average amount of milk, and has an average calving percentage.

[3]Also includes products for poultry and piggery.

Table 8.9 *Projected fuelwood and timber production demand, and balances, by ecological zones for Nepal, (1985–2005) (after Hrabovszky and Miyan, 1987).*

Regions	*1985*				*2005*			
	Production	*Requirement*	*Balance*	*Production as % of requirement*	*Production*	*Requirement*	*Balance*	*Production as % of requirement*
Fuelwood								
Mountain	227	672	–445	–34	846	1137	–291	–74
Hill	511	3614	–3103	–14	2378	6201	–3823	–38
Terai	630	2617	–1987	–24	1208	4845	–3637	–25
Nepal	1368	6903	–5535	–20	4432	12183	–7751	–36
Timber								
Mountain	1322	228	1094	580	2627	385	2242	682
Hill	2956	1269	1687	233	5943	2204	3739	270
Terai	1957	1226	731	160	2977	2298	679	130
Nepal	6235	2723	3512	229	11547	4887	6660	236

structures. This carries with it the concern that, to date, there is little evidence that such major changes are likely to occur rapidly. We will return to this pivotal aspect of the Himalayan Problem later.

Both case studies and their accompanying prescriptions depend, to a certain extent, on the degree of reliability of their data bases. The Kumaun case is less vulnerable in this respect because it is essentially tied to a detailed university survey of a limited number of villages in a single administrative unit: Dwarahat Block of Almora district. In view of our earlier criticisms concerning data reliability and degree of representativeness, however, this point must be raised. In this context the claim for a stable population for the 1971–81 decade is fascinating. Is Dwarahat Block entirely unrepresentative, or has a large part of the problem already been solved? If so, how? Is there a massive outflow of emigrants? In any event, if the prediction of continued population stability, as used in the model for land recuperation, is borne out, what will happen to a progressively ageing subsistence–partial cash-crop community, dependent upon extremely high inputs of physical labour?

As already implied, the Nepal case study and prescription is the more vulnerable of the two because it relies upon a very much more extensive data base, that of the entire country. There are also some particular problems which are not addressed by the prescription. A large measure of the success of the long-term food plan depends upon highly efficient introduction of sophisticated technology – large-scale irrigation aided by hydroelectric power; inputs of chemical fertilizers, herbicides and pesticides; and high-yielding varieties of food crops. Who pays, and who benefits from such massive interventions? Messerschmidt (personal communication, March 1987), discussing the question of introducing higher-quality domestic livestock, as one example, indicated that so far experiments with introduced exotic or hybrid cow varieties in Nepal have been disastrous. During the recent RCUP project (United States Resource Conservation and Utilization Project), for instance, there was an attempt to introduce highly productive milk cows for breeding purposes. They all died because their requirements for survival could only be met by stall feeding under highly controlled conditions; when left to roam the village like the other more resilient local cattle, they perished! Of course, it can be argued that this is an institutional problem, and once institutional adjustment has been achieved, it will be solved. But it must also be considered that extensive increases in irrigated land will provide equally more extensive habitat for malaria-bearing mosquitoes and other disease vectors. And the spread of malaria over the past few years has already been noted with some concern.

The whole thrust of the Nepal prescription is to effect a more intensive and ultimately high-technology land-based solution. Ives, Messerli, and Thompson (in Ives and Ives, 1987) focus on this issue by asking if there are not other approaches that need to be considered – especially light industry, appropriate and clean industry, preferably, and other entrepreneurial activities that would make more off-the-farm work available. The same argument would apply equally well to the Kumaun prescription.

To proceed into an exhaustive treatment of the socio-economic and political situation in the Himalayan region is not within our current means. It would be the next logical and necessary step, so that a number of central issues must be at least introduced. Again, however, we are beset with the problem of the unbalanced availability of data and description that leaves us with an over-representation of Nepal. Here a combination of the Pitt exploration of poverty, women, and children (Chapter 7), and the Hrabovszky and Miyan predictions of agricultural trends (this chapter) can show a particularly gloomy scenario, ameliorated somewhat by a number of bright spots and 'virtuous circles,' in partial contradistinction to the seeming prevalence of vicious circles. We must suppose, nevertheless, that the situation of the women and children, and poor people in general, is not unique to Nepal, and is probably characteristic across the wider Himalayan region. It is also of more than passing interest to consider the concluding condemnation of Blaikie, Cameron, and Seddon (1980). This is based upon an extensive study of the West Central Planning Region (Dhaulagiri, Gandaki, and Lumbini zones) of Nepal, with extrapolations to the entire country. Much of their writing may be contested, yet considering the large data base that they have accumulated in comparison with the general scarcity of reliable data, their findings deserve careful consideration.

Blaikie, Cameron, and Seddon's view of Nepal is that it is a small land-locked country facing a crisis situation partly on account of its geography, its history as a near-colonial marginal frontier region, and its neo-colonial peripheral relationship to post-1947 India. The final sentences of their book: *Nepal in Crisis: Growth and Stagnation at the Periphery* (p. 284), are thought-provoking:

> Expansion of 'marginal' employment opportunities in the administration for domestic producers may have reached a peak; work for the Nepalese in India has probably declined ... and it would be safe to say that prospects here will certainly not improve. As the hill economy becomes increasingly precarious and the terai's potential disappears, the dependency relationship between the Region and India will become clearer. In addition, we see no reason to believe that the peasantry of Nepal will discover a collective political expression of its needs which reaches beyond mere populist rhetoric in time to save millions of people from impoverishment, malnutrition, fruitless migration, and early death.

This view may be too bleak and should be contrasted with the Hrabovszky and Miyan optimism that the population growth–food production gap can be largely closed if the political will and organizational ability can be harnessed. Nevertheless, even if the future lies not only in between the two, but significantly to the optimistic right of centre, there would seem to be good reason to contend that Nepal, and the entire region, *is* moving into a situation of supercrisis. This contention is based upon a political angle that sees the peasantry – that is, the subsistence farming sector that accounts for almost 90 percent of the entire population – as victims of exploitation by central elites which are themselves victim of the periphery versus the centre forces *vis-à-vis* India and the world market economy in general. This is supported by Pitt's assessment of the extremely stratified nature of Nepalese society whereby 46.5 percent of total income is controlled by the top 10 percent of households (see p. 151). The immediate question is: can the agricultural development strategy illuminated by Hrabovszky and Miyan (see pp. 194–204) function with sufficient speed and efficiency to avert a break-down in subsistence agriculture, or at least provide a longer breathing space to allow additional measures to be identified and applied? Certainly, we have some points of optimism to look up to, and the outstanding early successes of the Aga Khan Rural Support Programme in Pakistan's Northern Territory (Malik, personal communication, 1986) would indicate that sensitively assisted village level and more traditional self-help measures can make a significant difference. This view is also supported by the remarkable successes of the Chipko Movement (S. Bahuguna, personal communication, 1986; Shiva and Bandyopadhyay, 1986b). Also, in terms of development of understanding of the linkages between subsistence agriculture, animal husbandry, and forestry (Jackson, 1983, unpublished; this chapter, pp. 176–91), it would appear that a promising start can be made to reduce the degradation of uncultivated land and eliminate the pressure on common property resources. To this must be added Thompson's (personal communication, February 1987) insistence on searching out the 'bright spots' and learning how to utilize and expand them for the wider good – the Sherpa and Thakhali women entrepreneurs, the trade in Tibetan rugs from Kathmandu, Darjeeling, Dharmsala, and other centres, that have high enough value added to overfly India profitably and find effective markets in industrialized countries. Thompson concedes that the neo-Marxists do have a point but believes that history may show them as latter-day Cobbetts (Cobbett rode the English countryside and chronicled the appalling declines and miseries but completely missed the non-land-based industrial revolution where the future of Britain actually lay).

These optimistic comments, nevertheless, must be set against the depressing record of major development 'break-throughs,' such as the Green Revolution, only within recent years hailed world wide as a panacea for India's chronic food shortage (and for that of other developing countries). Undoubtedly the Green Revolution transformed India from a food-importing to a food-exporting country, but only at the high cost of a drastic increase in landlessness, a system whereby the relatively well-to-do farmer prospered and the marginal peasant suffered (Redclift, 1984).

The analysis of social forestry misdirection in southern India, by Shiva, Sharatchandra, and Bandyopadhyay (1981), adds weight to this record in its provision of a detailed case study of how easily 'outside' intervention can serve to widen the gap between the poor and the not-so-poor. The danger of this situation, of course, is that access to common property resources is restricted in various ways, thus placing further pressure on remaining commonland forests and pastures. The rapid spread of chir pine forests throughout the Garhwal and Kumaun Himalaya bears witness to this and is part of the justification to Bahuguna's protest that the World Bank poses the greatest single threat to the environment of the Himalaya (see p. 73).

These points will be taken up from a somewhat different perspective in Chapter 9. It should be apparent, however, that, to the problems of uncertainty and complexity, must be added those related to institutional failure or lack of development. At this juncture we only wish to conclude that, while the Himalayan region presents a serious physical challenge both to 'development' agencies and to subsistence agriculturalists whose numbers seem to be rapidly increasing, the main issues are social, economic, and political. To this we must add the need to identify causes and effects better, to reduce the level of uncertainty where possible, and to translate findings into political will and action.

NOTES

[1]The Land Resources Mapping Project is a joint undertaking between HMG Nepal and CIDA and is a very detailed 1:50,000-scale resource capability land-use assessment. The Final Report is not yet available (c.f. also, Shah and Schreier, 1986).

9 CRISIS, PSEUDO-CRISIS, OR SUPERCRISIS?

INTRODUCTION

Any attempt to sum up the preceding chapters would necessarily include the following. The large literature that depicts the imminence of environmental catastrophe in the Himalayan region has tended to confuse cause and effect, has largely missed the essential historical depth, and has assumed the existence of dramatic upstream–downstream inter-relationships without requiring rigorous factual substantiation. The subsistence mountain farmer has frequently been perceived as a large part of the problem, rarely as part of the solution. The 'development' agency responses have tended to be a search for widely applicable panaceas, while the extreme complexity, and especially the uncertainty, that pervade our region at all levels, have not been taken into account.

We must emphasize again that this uncertainty is not merely *technical*; that is, it is not just the absence of certainty. Rather, it is *structural* in the sense that, without their realizing it, certain actors in the Himalayan debate have succeeded in imposing their desired uncertainties within it. It is these *unwarranted uncertainties* (the latter-day myths) that have evoked the Theory of Himalayan Environmental Degradation that we wish to dismantle. Our aim in attempting this is twofold: first, to confront the full extent of the uncertainty; second, to get to grips with it, so to speak, both by reducing it, where this is possible, and by learning to live with and make the most of it, where this is not possible (Thompson, personal communication, February 1987).

Our success in achieving this ambitious goal is probably incomplete. Nevertheless, it is necessary, as part of the attempt, to face head-on the questions posed by the title of this chapter. The preceding chapters, while demanding a much deeper historical perspective, which forces a reassessment and reordering of the components of potential crisis, should leave the reader with the conviction that crisis indeed there is. Thus we cannot avoid asking, what is the scale and timetable of the crisis? What is its nature?

We have inferred repeatedly that the crisis is something to do with the growing pressure on resources, given the available technology and the pervading institutional and political situation. Or, as Blaikie and Brookfield

(1987) term it, it is 'the pressure of production on resources.' The predominance of a rapidly expanding subsistence population in the face of a depleting natural resource base, extreme political uncertainty, if not widespread unrest, and actual warfare in limited areas, together with a lack of understanding of the uncertainty, are all interlinked. The timetable would appear to be urgent in the extreme as large sections of the population exist below, or close to, minimum living, or survival, standards, and the needed institutional framework is largely lacking. It is also necessary that we attempt to define what we mean by *supercrisis* and illustrate how we think it relates to the Himalayan situation.

As we have implied earlier, we define *supercrisis* rather loosely, as relating to a region or significantly large area of the world (in the present instance, the Himalayan region, *sensu lato*), rather than to a single country, and of such magnitude as to bring millions of human beings into a life-threatening situation as a consequence of a progressive, or abrupt, deterioration of the overall life-support system. In our case this refers to the Himalaya, Karakorum, Hindu-Kush, and Hengduan mountain systems and their subjacent valleys, lowlands, and plains. This region provides the habitat for upward of 400 million people.

We use the example of sub-Sahara, Ethiopia, and East Africa as the epitome of *supercrisis* – a world-scale horror whereby many millions of people and their livestock are reduced to extreme poverty, starvation, forced migration, refugee status, and death. This human process is accompanied by ruination and desertification of vast areas of land, often submarginal land. The cause of the African supercrisis is multiple: in places, such as Ethiopia, many decades (indeed, many centuries - Hurni and Messerli, 1981; Hurni, 1983) of almost imperceptible over-use and abuse of land, leading to incremental soil losses; over most of the region, colonial disruptions of traditional societies and economies, and traditional land-management structures; these processes in turn are exacerbated by post-colonial centrist governmental abuse, developmental emphasis on commercial agriculture; guerrilla, civil, and international warfare, from struggles of liberation from colonial subjugation to post-colonial dictatorships. Rapid population growth resulting from reduction in the mortality rate without significant change in the fertility rate seems to be an essential component. Further pressures arise from misdirected development aid, and in some countries very large expenditures on armaments. The whole tinderbox is then torched by one of the more pronounced and prolonged, but nevertheless naturally recurring, droughts. The consequences of such a supercrisis have been too recently and too graphically illustrated in the news media to warrant detailed reference here. Nevertheless, we must emphasize again that the costs of a fire-brigade relief action, both in terms of the consumption of resources, further socio-economic disruptions, and human misery, and, in the longer term, serious land degradation, are enormous.

Since, as we have illustrated in Chapter 1, the Himalayan region has been depicted for three decades as facing imminent environmental and socio-economic collapse, with the year AD 2000 as a kind of rapidly approaching doomsday, it would appear that we must face the stark moral necessity of assessing this situation. AD 2000 is little more than a decade away. Will Nepal indeed be washed down the Ganges to become new islands in the Bay of Bengal? If so, will this be necessarily bad? Will this be the moment for the Dutch polder engineers – but we are back to the myths! Will it be the result of human interference in the mountains or of the inexorable processes of nature, or a combination? Can anything be done about it? Or should we throw up our hands and despair? Alternatively, as may be assumed from our attempts to dismantle the myths, and to denigrate the intellectually satisfying Theory of Himalayan Environmental Degradation, are we, in effect, dealing with a pseudo-crisis, something that has evolved in the minds of many people and institutions because of the mismanagement of the sea of uncertainty that laps against the highest ramparts of the world's most beautiful and massive mountains? Or are we *merely* witnessing a common-or-garden crisis, or better, a multiplicity of such crises, the likes of which the world is awash with, always has been, always will be?

It is our contention that, only one step down from the arms race and the threat of nuclear annihilation facing the entire world, lies a series of issues, one of which is the question of crisis, pseudo-crisis, or supercrisis in the Himalayan region. We believe that the extent of the uncertainty surrounding this question is so pervasive that it would be unwise to under-react by taking a *laissez-faire* attitude. In contrast, and again because of the uncertainty, immediate development of an emergency response as a top international and regional priority may be an over-reaction. From this it follows that a *rational* rather than an *emotional* response is required: it is vitally necessary to determine just what the problems really are; that is, a special kind of crisis response is needed, but in terms of seeking plural problem definitions and the enumeration of plural solutions. Multiple approaches and operational and institutional flexibility are seen as prerequisites leading to the management of uncertainty and complexity. But this will not just happen. Thus we recommend that some form of special United Nations and regional, international, initiative be called for so that a number of carefully defined tasks can be identified, and the necessary resources concentrated to lead into sustained action.

We believe that such an undertaking must be regarded from the beginning as something that is extremely ambitious in view of the existing political tensions, but that is also most urgently needed. In addition, a priority objective must be to improve the living conditions of the poor by providing them with enhanced access to natural resources and by incorporating them fully into the development process itself. This will require a major adjustment in access to and control of land, perhaps a total restructuring of land tenure and taxation systems, and ruling elites throughout the world have shown reluctance to face up to such reforms. We believe, however, that rapid and sustained progress along these lines will be essential if there is to be any hope of reversing the overall regional trend toward debilitating resource degradation, and rapid enlargement of the current level of socio-political unrest.

In this chapter we will attempt to justify our adherence to these rather draconian statements and the even more dire inferences that lay behind them. From there we will be able to move more directly into a discussion of a preliminary research strategy, adequate perhaps for future modification and refinement.

WHAT IS THE NATURE AND EXTENT OF CRISIS IN NEPAL?

As we have already indicated, we believe that the environmental deterioration in Nepal has been over-dramatized, that correlations have been represented as cause and effect that have often been confused, and that a perceived supercrisis has evolved in the popular press, the conservationist literature, and even in the mainstream scientific writings. We strongly support the contention of Thompson and Warburton (1985a and b) that the very uncertainty that has emerged from this discharge of emotion and latter-day myth has become part of the problem. Thus, our primary aim has been to challenge what we perceive as a series of ingenious but unsupportable linkages connecting massive landscape changes by the subsistence farmers with increased flooding and siltation on the Ganges Plain.

We think that while we have succeeded in this task, the real task lies ahead – namely, to convince the vested interests that a broader, more holistic, more critical approach is needed, and that a better historical perspective is essential. On this premise we have concluded that the causes of the crisis are not environmental but relate to the social, economic, institutional, and political situation in which the Himalayan region finds itself. While the ultimate effects may be similar – that is, an environmental catastrophe – clarity on this point should result in more rational and more effective responses.

This does not mean that the environmental trends and constraints can be put aside. If two billion dollars are to be spent, for instance, on construction of a cascade of hydroelectric facilities along the River Arun in eastern Nepal without anything having been learned from the destruction of the Namche Small Hydel Project in adjacent Khumbu Himal, then Nepal and the entire region may be facing a critical wastage of resources. The large hydroelectric project solution that is in play throughout the region needs critical re-examination. For World Bank spokesmen to claim, for instance, that the Karnali high dam project has received positive feasibility assessments by three separate engineering studies is not an effective justification: these are narrowly focused engineering assessments, not holistic studies, and it must be a very rare case for engineering consultants to proclaim that such vast projects (in this case multi-billion dollar) are not feasible. The, at least temporary, blockage of the Tehri Dam project in Uttar Pradesh by the efforts of the Chipko Movement, is a ray of hope. Similarly, as Mahat *et al.* (1986a, 1986b, 1987a, 1987b) have demonstrated, continued deterioration of the forest resources of Sindhu Palchok and Kabhre Palanchok districts will threaten the very basis of subsistence agriculture as it exists today. Even if we have been successful in demonstrating that loss of Middle Mountain forest area is not a simple post-1950 process, loss has undoubtedly occurred, but over a much longer time-span, and population has increased dramatically. But the long-term causes of this loss of forest resources have been primarily the direct result of government policies rather than high population densities (Griffin *et al.*, 1988).

Population pressure on resources, as debated extensively by Blaikie and Brookfield (1987), is itself a conundrum. Land degradation can proceed as a result of *reduced* population just as it can arise from increasing population. Our own conviction is that, hypothetically, a reduction in the total population of the Middle Mountains, for instance, by large-scale abandonment, or by state-enforced out-migration, would induce far greater slope destabilization and soil erosion than a continuation of the present situation of population growth: an active viable agriculture augments slope stability in most instances. The lack of success of Nepal's strongly supported family planning programme must be seen, despite an unreliable data base, as a result of the villagers' perception that more children add to their capital resources. From this it follows that the real questions are: what are children? what do they provide? Then we must ask, can we seek other ways to accomplish the same ends and encourage changes in behaviour accordingly? If the need for children is a barrier to effective population control, then it is necessary that we understand the character of the original barrier. In this context, recent years have witnessed the publication of a number of detailed treatises on the social anthropology of individual villages in the Nepalese mountains. These lead to the general conclusion that one of the major factors behind large families is that it is a risk-avoiding strategy (Goldstein, 1977, Goldstein *et al.*, 1983; Manzardo *et al.*, 1975; Messerschmidt, 1976b, 1982, 1987; Fisher, 1986; Fricke, 1986). Studies of this kind are especially interesting in terms of our discussion because they bring out a number of important points derived from a strong, if strictly local, data base. Mountain villagers are responding to a clearly perceived dangerous increase in production pressure on available resources. Part of this response is to have as many children as possible. Fricke's description of this apparent dilemma is quite graphic:

> The organization of Timling's economy-household as the primary economic unit, kinship as the nexus of exchange, diversified pursuits as the key to survival and avoiding risks – reinforces the desire to have as many children as possible [which in turn] become important resources in themselves for expanding and diversifying the domestic economy.
>
> (Fricke, 1986:190)

Whether conclusions such as these, apparently valid for a number of individual villages, can be extended to embrace all of Nepal or the wider Himalayan region, defies a firm answer at present. Thus we must inevitably turn to the national census-based data and larger-scale studies, such as those of the Asian Development Bank, Nepal Agriculture Sector Strategy Study (HMG Nepal and ADB, 1982), of Shah and Schreier (1985), and of Hrabovszky and Miyan (1987, and see Chapter 8) for wider insights.

The Asian Development Bank study, hereinafter referred to as ADB (1982), indicated that between 1975 and 1980 the cultivated area of Nepal expanded by 34 percent. This was accompanied by a decline in total area under forest, mainly in the Terai, of 15 percent (while we are not necessarily accepting these data as accurate the trends are most likely reasonable), yet total agricultural productivity has actually declined slightly between 1960 and 1980 (this more or less corroborates the findings of Hrabovszky and Miyan, 1987). Thus substantial investments, both domestic and external, have failed to increase productivity while population has grown from 9.4 million in 1961 to 16 million in 1981. As Hrabovszky and Miyan have shown (pp. 191–206, above), the increase in area under cultivation derives primarily from a combination of the conversion of Terai forest land to agriculture and the progressive spread of intercropping and double- and triple-cropping with the expansion of irrigation, and intensification in the more favoured parts of the Middle Mountains. Further progress in this direction, however, appears to be inhibited, or at least restricted, by institutional break-down (see pp. 261–67). But it is also significant that in 1966 yields, especially of rice, in Nepal were amongst the highest in South Asia whereas by 1982 they had dropped to be amongst the lowest (ADB, 1982 (2): 34). This poor showing, despite the considerable investments, is assumed to result from use of more marginal lands for cultivation together with a decline in soil-nutrient status.

There are other underlying factors. First, in those areas where it is feasible to use chemical fertilizers, mainly the more accessible parts of the Terai, applications appear to be limited to only about one-fifth to one-quarter of the amounts used in other South Asian countries. Again the real problems appear to be organizational and institutional, although we are omitting for the moment the negative aspects of the introduction of artificial products. Thus, there is little co-ordination in target fixing and monitoring, in achieving correct timing of application, and in linking availability of irrigation water to cropping cycle requirements.

The potential for increased irrigation and the failure to realize it rapidly enough have been used to illustrate the problem of institutional break-down. Because most of the existing cultivation in Nepal is maintained under rainfed conditions, the usefulness of supplemental water to produce higher yields and increased cropping intensities is well recognized. An assured supply of irrigation water and the provision of drainage, where necessary, could conceivably permit the planting of a second, or even a third, crop, thus increasing cropping intensities to well over 200 percent independent of rainfall. Because of the ensuing reduced risk of crop losses due to drought, farmers can safely enhance the level and quality of inputs, such as seed, chemical fertilizers, pesticides, and labour, with resultant higher yields, production, and net income. The Terai offers by far the greatest opportunity for such development, although the Middle Mountains should not be ignored. This, once again, is the Hrabovszky and Miyan solution, which has its own limitations.

It has been pointed out by Messerschmidt (personal communication, February 1987) that expanded irrigation increases the breeding grounds for malaria-carrying mosquitoes, and applications of fertilizer, pesticides, and herbicides, in addition to their direct negative aspects, also tend to enable the relatively rich farmer to prosper and the larger number of poorer farmers to suffer additional privation. Of course, a more holistic approach is needed; there are no simple solutions. Our argument nevertheless can be supported best by a discussion of the situation in the Terai, borrowing extensively from the Asian Development Bank study (ADB, 1982 (2): 74–92).

The introduction of most forms of irrigation is a relatively high-cost exercise. In the general context of Nepal, however, and of much of the wider Himalayan region, declining land/man ratios, increasing population and food imports, and plentiful water resources, it would appear that irrigation can contribute to enhanced agricultural production *if* it is fully integrated as part of an agricultural development package.

While data availability and reliability places constraints on any discussion of the potential for agricultural enhancement by irrigation, nevertheless, it would appear that there is considerable room for improvement. Estimates, regardless of accuracy, would indicate that in 1980 irrigation on a year-round basis was only practised on about one-third of the potential command area in the mountains and on about one-fifth of the potential area in the Terai. In the Terai the area under year-round irrigation is estimated at 65,000 ha, of which 40,000 ha have been developed by the government. An additional 194,000 ha for irrigation are under design or construction by the government, leaving a potential of 1 million ha for future development.

Given the considerable potential and the high cost of infrastructural development it is obvious that massive expansion of areas under irrigation will depend upon effective government and foreign aid organization. The current situation, in view of the critical need for such progress, is depressing. Irrigation developments in Nepal are under the responsibilities of three Ministries: Water Resources, Agriculture, and Panchayat and Local Development. However, the division of responsibility is loosely defined and co-ordination between the three Ministries is not well established. Thus there is frequent duplication of effort and a serious loss of effectiveness. A direct quotation from the ADB (1982) study will serve to illustrate the fragile nature of the prevailing situation.

> A number of problems are evident in DIHM's operations [Department of Irrigation, Hydrology, and Meteorology]. An apparent shortage of overall manpower may be less related to numbers than to questionable deployment of available staff among head and regional or project field offices. Also, specific staff are not always given assignments for which they are best suited. There are unquestionably areas of expertise which are under-represented among DIHM's staff vis-à-vis the role required of DIHM.

Among these, lack of ground-water development and construction-related specialties are noted. There is also in DIHM an apparent lack of basic management skills, particularly at field levels. Coordination of activities and paper flow within DIHM, particularly between head and field, or regional, offices is slow and problematic. External coordination, particularly with Ministry of Agriculture agencies which supply agricultural support services essential to the success of DIHM projects is inadequate. Probably because of the difficulties which arise in such coordination, too few of DIHM's projects incorporate provision for assured supporting services, and for tertiary and/or farm-level irrigation and drainage facilities necessary to achieve success. Many of DIHM's projects have badly deteriorated because of inadequate or faulty operation and maintenance due to both manpower and budget limitations.

(ADB, 1982 (2):87)

There follows a depressing account of incomplete projects, poor maintenance, large short-falls of budgeted targets, unsound planning. Even the once-constructed irrigation systems that had been deficient from their initiation have experienced accelerated deterioration because of inadequate operation and maintenance. This has led to lack of water distribution at the farm level with a serious loss in confidence by the intended 'beneficiaries' and consequently low collection rates of irrigation service fees. The message introduced here is that if the large-scale technological 'fix' is to be applied, either on its own, or as part of a broader 'package,' *at least it should be undertaken efficiently*.

This tale of discouragement must be set against the Hrabovszky and Miyan (see pp. 191–206) projections for increased cropping intensity and production levels needed to match and offset projected population growth. The discussion of irrigation potential relates primarily to Terai commercial undertakings; but it must be remembered that the largest proportion of the population continues to live in the Middle Mountain belt and to depend upon subsistence agriculture. It is relevant therefore to look briefly at the mountain subsistence farmer's situation as perceived by the Asian Development Bank study.

The National Planning Commission defined poverty as below N. Rs 2/capita/day income in 1977 and stipulated that 41 percent of rural households were living below this level with the highest incidence occurring in the mountains (71.5 percent). These figures are corroborated by information on the land-tenure system. In 1971, 55 percent of the farmers owned less than 12 percent of the land with an average farm size of 0.21 ha. At the other end of the scale, 6 percent owned 44 percent of the land with an average holding of 6.8 ha. This group were mainly absentee landlords who organized sharecropping under which the sharecroppers have neither the motivation nor the resources to increase production. The critical nature of this situation has long been recognized and the 1964 Lands Act was an attempt to rectify it.

Ceilings on ownership of land and the distribution of excess land to tenants and landless farmers, award of tenancy rights to those who actually tilled specific holdings, security of tenancy, fixation of rents, scaling down of peasant debts, and so on, were features of this Act. However, the benefits were extremely limited in terms of the overall agrarian system and there were no tangible effects on the uneven distribution of land holdings. Successful evasion of the legal provisions regarding land-holding ceilings, for instance, resulted in redistribution being limited to 23,000 ha, or less than 1 percent of the total cultivated land. There *was* an improvement in tenancy conditions, however; about 1.8 million tenants were identified and issued temporary identification slips. Formal certificates of tenancy were issued to 300,000 tenants. Nevertheless an estimated 40 percent of the total were left out of the process and the drive for further identification rapidly dwindled away because of lack of firm political commitment.

Similar situations prevail in the areas of indebtedness, access to government-supported loans versus local usury, and rental. For instance, despite government attempts to make loans available through a variety of institutional arrangements, co-operatives and state banks, at interest rates of 6–15 percent, small farmers tend to rely on private loans at the village level at interest rates as high as 150 percent. In the same vein, despite government intervention to restrict excessive rentals, the traditional rate of 50 percent of the annual crop yield, and higher, is still widely extracted, if only because many farmers privately volunteer to carry the traditional burden simply because their forefathers had done so; this is a matter of generations of inherited tradition and pride, despite fairly widespread knowledge of the recent government attempts at reform.

The ADB (1982) study characterized the Nepal agricultural sector as follows:

1. a high man–land ratio
2. great disparity in land ownership
3. high, debilitating rentals
4. large number of poorly fed livestock of low productivity and high level of disease
5. declining forage base
6. inadequate dissemination of new techniques
7. ineffective extension services
8. lack of timely availability of inputs
9. weak institutional support for small farmers
10. deteriorating environment
11. declining soil fertility and reduced yields
12. reduced availability of the full range of forest products.

To these must be added the increasing control of agricultural raw materials from across the open border with India, depriving Nepal of potential industrial growth, as well as an extensive loss of revenue due to smuggling and illegal transfer of products. This is further exacerbated by direct control by Indian entrepreneurs of small businesses, including retail enterprises, in the villages and towns of the Terai. To this general list one might add: lack of input of institutional support structures, and especially a lack of demonstrable encouragement to and sincere appreciation and understanding of farmer-based solutions (Messerschmidt, personal communication, February 1987). Also, as pointed out by Shrestha (1985), many of the mountain families are virtually held in place by the remittances of the Gurkha soldiers in the British and Indian armies. Both direct remittances and the benefits of pensions taken back to the home villages by retirees may merely serve to hold a deteriorating system in place rather than provide the capital for re-investement in farm productivity. Shrestha (1985) argues that, while the internal institutional policies were responsible for initiating under-development and external migration, British India's Gurkha recruitment policy contributed to their perpetuation. This leads to the stipulation that today Nepal's economy, in part at least, is trapped in a negative feedback cycle in which under-development fuels out-migration which, in turn, propagates under-development. Regardless of the nature of the socio-economic effects of this additional vicious circle, with its highly localized impacts in terms of the selection of army recruits, out-migration from the hills is perceived as a major economic survival factor for many migrant households and local economies.

We have discussed earlier the unfortunate and destructive tendency of using the 'ignorant hill farmer' as a convenient scapegoat (Chapter 1). What is perhaps even more critical is the potential loss of a great wealth of traditional environmental knowledge that goes with this. Whiteman (1985) has demonstrated this problem most effectively and we suspect that it is a world-wide pervasive phenomenon that characterizes under-development. While international agency attitudes are changing in this respect, it is by no means universal to perceive the traditional subsistence farmer's actions as fundamentally rational and based upon generations of careful experimentation. There is still a long way to go before we can approach a solution to the difficulty of matching what is most valuable and vital from traditional practice with what is both technologically *and* institutionally appropriate from the 'modern' sector in terms of foreign aid.

Development of water resources as a perceived solution

Another extremely important potential for resource development is water. While we have discussed irrigation, here we will add a brief commentary about Nepal's much vaunted waterpower potential, set by some estimates as equalling the entire actual waterpower generation of North America. The current Five Year Plan is following the trend of the last two by placing increasing emphasis on this apparently ready solution to balance-of-payment problems. The Arun Cascade, with its allocation of almost two billion dollars, has been mentioned briefly already. There is also the truly giant Karnali High Dam project, and many more. The obvious objectives include increasing the proportion of hydroelectric power within the overall national energy consumption budget, and thereby reducing the dependency on fuelwood, and selling to India very large amounts of energy. It is estimated that only 2 percent of the national energy requirements were being met in 1975 by hydroelectricity and much less than 1 percent of the potential had been developed (Sharma, 1983). These large, even macro, engineering projects are the types of undertakings that, throughout the world, have attracted vast sums of international money relatively easily.

We need only make a passing reference to what may be regarded as India's atrocious record of ill-conceived waterpower development in terms of the unanticipated siltation rates of reservoirs (Chapter 6). Even so, at a recent conference on the environmental problems of water resources development in the Himalayan region, the chairman of India's Central Water Commission made the revealing statement:

> with our circumstances and needs we cannot cry halt to the development of water resources *merely* for the fear of impinging on the environmental balance. A large segment of our population, which resides in the Indo-Gangetic plains, is dependent on the Himalayan resources for its survival. For any programme aimed at bettering the lot of these people, the exploitation of the resources of the Himalayas is a primary requirement. We, therefore, cannot afford the luxury of totally stopping development with a view to preserving the environment. (our emphasis)
>
> (Y. K. Murthy, 1982:67–8)

In this general context, recent hydroelectric developments in Nepal deserve attention. The following remarks, based in part on Bjønness's (1982–83–84, 1987) investigations of the Kulekhani Project, are, by comparison with the projected Karnali high dam, a second-order undertaking with a generating capacity of 600,000 kwh. Delivery of power began in March 1982.

The Kulekhani River is a tributary of the Bagmati which, lower down, flows through Kathmandu. The extent of the Kulekhani watershed above the dam is 212 km^2 which had a 1971 population of 30,000, growing to about 36,000 by 1979. Kathmandu is the primary market. The upper Kulekhani River has been diverted by tunnel into the Rapti River, giving a hydraulic head of 600 m. The study by Bjønness concentrated on the socio-economic and environmental impacts of the power project and her main findings are itemized here:

1. 1,200 people in 235 houses had to be removed from land that was subsequently submerged by the reservoir. Compensation was offered in cash or by provision of land in the Terai. More than 80 percent accepted the cash option. However, payment was not completed until two years after dispossession, during which time land values had risen sharply and much of the cash had been spent on subsistence; this led to pauperization and landlessness;
2. No consideration was given to the possible downstream effects, yet, below

the dam, villages and households were presented with a dry river channel resulting in a loss of irrigation water, and the rendering useless of many water mills;
3. Above the dam the reservoir severed many households from access to markets and water mills, or enforced a long detour;
4. No account was taken of the importance of communal facilities, nor compensation paid where they were destroyed;
5. Progressive deforestation of the adjacent upper watershed and landsliding into the reservoir is projected to reduce the design life of the project through accelerated siltation (electricity outages in Kathmandu during our visit in October 1987 were explained as the result of reduced Kulekhani capacity due to serious slope instability above the dam which necessitated the lowering of the level of water in the reservoir).

During the summer monsoon of 1986 one of us (JDI) was able to make a brief visit to the Kulekhani reservoir. Certainly the fears of excessive siltation of the reservoir by landsliding directly into the water had not so far been fully borne out, and the inconvenience (or economic annihilation?) of families stranded on the far side has been mitigated by provision of a motor boat service (yet even in this instance questions must be raised: who pays? and, is this an appropriate solution?). But also apparent was the augmentation of soil erosion and siltation from the mining of clays and silts in upstream tributary valleys that had been undertaken during construction for the purpose of sealing the bottom of the reservoir. This flagrant violation of good practice (and actual defiance of contractual agreement) could hardly have gone unnoticed by the government inspectors; it must be concluded that it was undertaken to augment the profit margin of the contractors.

The implications of this discussion for the Karnali and Arun Cascade projects are significant, especially since they have been declared technically and economically feasible with the prospects for the expenditure of 6–7 billion dollars and massive environmental and socio-economic adjustments. In particular, the problem of the regular sale of large volumes of electricity outside of Nepal and the usual disassociation between a major power supply and the incremental and minute needs of scattered mountain people, not to speak of the potential release of violent natural hazards, need attention. It is remarkable, despite all the lessons of past failures, or of only partial successes, that hydroelectricity, and other macro-engineering works are rarely viewed holistically, nor are real costs assessed against real benefits.

Institutional problems

An answer to the question: what is the nature and extent of the crisis in Nepal? clearly takes us out of the realm of the physical environment and into that of politics at various levels. Increased foreign aid, rising population numbers, falling agricultural output in an overwhelmingly subsistence rural economy, deteriorating environment, and institutional maladaptions, if not actual break-downs, would seem to combine into a set of rapidly blinking warning lights. We then must add the uncertainty issue and the confusion of competing foreign-aid agencies that renders effective government doubly difficult. Thus we must raise the spectre of Hrabovszky and Miyan's attempts to balance food production through further conversions of land and cropping intensification against rising population as falling short of requirements. Does this constitute Nepal's contribution to regional supercrisis? A degree of insight is provided by the proceedings of a recent conference held in Kathmandu on Foreign Aid and Development in Nepal (IDS, 1983).

We shall use the substantial and thought-provoking 340-page proceedings of this conference simply by introducing a number of direct quotations. Before doing so, however, we wish to emphasize that encouragement should be derived from the very fact that such a meeting could be held in Kathmandu and the proceedings published there. The participants included present and former high government officials, representatives of the major aid agencies, and Nepalese intellectuals. The severe institutional criticism that was provoked should perhaps be regarded as a sign that change and progress can occur.

'Foreign Aid and Women', by Bina Pradhan and Indira Shrestha, pp. 99–154:

> By adopting and using such concepts and terms [the western concepts of 'housewife', 'economic activity', and 'household head' with which the authors charge the foreign-aid agencies] without looking into the realities of the rural household production system in Nepal, the productive roles of women have been completely ignored and distorted which has led to women being by-passed in development. This in turn has meant that both women and the development process have suffered. (pp. 104–5)
>
> Women simply do not appear in the agricultural component of IRDPs [Integrated Rural Development Projects]. What is even more disturbing is that they seem to have been deliberately excluded at times. (p. 119)
>
> It is well worth mentioning an exchange of words that occurred on this huge trade school layout. In response to a query about why there was no provision for girls in the envisaged training scheme, the (foreign) expert said, 'Do you think that a 60 year old farmer (by implication: male farmer) would listen to a young female JTA [Junior Technical Assistant]?' It is difficult to find any other explanation for this than that the expert was unaware of the existence, even prevalence, of female farmers in Nepal. (p. 120)
>
> This obviously undermines women but perhaps less obviously, may unbalance the delicately balanced unit of the economic-system, the family farm households. Further, it seems highly probable that development itself is hindered by this incorrect focus, or non-focus, of omitting women from all development activities in agriculture. (p. 120)

'Technical Assistance and the Growth of Administrative Capability in Nepal', by Bihari K. Shrestha, pp. 219–69:

> How much does an expatriate really cost? ... one man-year is roughly

> budgeted at one million rupees (excluding agency overhead), depending on the grade, of course. On the other hand an average gazetted HMG official costs . . . 20,000–30,000 rupees per year which gives a ratio of one expatriate for 30 to 50 of our counterparts. An informed source has it that currently there are a total of 334 expatriates working in Nepal for several bilateral and multilateral projects covering almost all the development sectors in the country. This list, however, excludes those free-lance expatriates who, because of the lenient immigration policy of the country, stick around until they land a job (paid at international rates mostly) directly with a donor or indirectly through a local consulting firm. Everyone of them included, it is a mammoth population of expatriate advisors. It is probably this phenomenon that prompted a UN expert to lament as long ago as 1970 that Nepal was over-advised and under-nourished. (pp. 219–20)
>
> Administrative capability, however, remains a relative concept. Is the administration capable of performing to a degree sufficient to meet the challenges of the situation? The answer to this question in Nepal is clearly in the negative. The economy is considered to be in a shambles, environmental deterioration is rapid and rampant ... agricultural production is declining. If some authors chose to call Nepal a country in crisis they are no more too wrong. (p. 226)
>
> In sum, what one encounters in Nepal is a potentially competent bureaucracy that does not perform except in extraordinary circumstances. (p. 232)

'Foreign Aid in Nepal's Development: An Overview', by Devandra Raj Panday, pp. 270–326:

> Why, in spite of these major accomplishments in the infrastructural field and increasing investment even in the productive sectors of the economy, should there be almost a consensus that overall poverty is increasing. . . ? This is the principal issue which all the five accompanying papers have tried to address. Their assessment is that agriculture has not benefitted; the poor have been bypassed; the women have not even been understood; the relations of production and distribution of power have gotten worse and the technical assistance has not contributed to the improvement of administrative capability. (p. 282)
>
> One of the primary functions of foreign aid assistance is to buy time – time enough to mobilize and manage an internally generated momentum of growth. What may have sustained us, or even saved us from total disaster in the past, will not have played its proper role if it is eventually going to ruin us in the future. . . . To what extent does excessive dependence on foreign aid financing transform, co-opt or even obviate the need for fashioning a rational, workable and committed development strategy for Nepal? (pp. 283–4)
>
> I am, therefore, inclined to submit that it is in its role as the purveyor of changing concepts that foreign aid has been most counter-productive in Nepal. (p. 289)

> How one is to provide for 'basic needs' without diversifying away from agriculture in a country where the man–arable land ratio is about to reach or cross the threshold of disaster is not even discussed seriously. (p. 292)
>
> It is understood and agreed that aid to Nepal is for the socio-economic upliftment of its people.... With the record of foreign aid's performance having been what it is, there is no alternative to taking steps in the direction suggested above; and the proposals are far from being radical. If the status quo is maintained any further, it will reinforce the arguments of Mishra and Sharma that foreign aid has been only an instrument of collusion between the urban elite and their rural counterparts and the country's ruling class and donors.... From the point of view of the needs and problems of the Nepali people, such aid might as well be stopped altogether. The advantage would be that the contradictions can be settled internally, however painful a process that might be for some of us.
>
> (pp. 303–4)

While certainly a minority voice amongst the conference participants, the recommendation that foreign aid be terminated is interesting if only because it was actually proposed, and the proposal subsequently published. A less radical, but still highly significant issue relates to the apparent claim that it would be inappropriate for donors (foreign-aid agencies) to seek to influence government policy because that would amount to interference in the internal affairs of an independent state. This was countered by the statement that the very fact of non-interference while continuing to supply aid was in fact interference – in favour of the status quo.

Nevertheless, as indicated repeatedly throughout our presentation, there is another side, or many other perspectives, to most elements of the Himalayan debate – the very essence of uncertainty! The points of view quoted above are no exception and it is our concern that we make some effort to counter the accusation of insincerity on the part of the government, and the 'donors' – the UN and bilateral government-aid agencies and their personnel based in Kathmandu. It is possible, as one example, to counter Bina Pradhan's criticisms. The United States-funded RCUP project (US Resource Conservation and Utilization Project), under the guidance of a social scientist with assistance from a host of Nepalese male staff, worked overtime to instigate not only a quota for women to study forestry (to become Junior Technicians and Junior Technical Assistants) at both the Institute of Forestry and the Institute of Renewable Natural Resources, but also to make sure that those quotas were filled with the best possible candidates. And there was no trouble in finding qualified candidates once it was understood that a sincere effort was underway to attract, train, and employ Nepalese women. (Messerschmidt [personal communication, February 1987] explains that Bina Pradhan missed this point in her presentation, probably because she did not believe it could happen – in the sense that it has not worked elsewhere in Nepal is partial justification for her attack.) Similarly, it must be emphasized that many millions of dollars are provided by donor agencies for the training of young Nepalese in appropriate fields for the development problems of their country, both in other Asian institutions as well as in western countries. The objective of these extensive and varied training schemes (and the scholarly results of some of them can be inferred, if only in a very small way, from the authorships of papers published in *Mountain Research and Development*) is eventually to replace the expatriate workers with qualified and well-trained Nepalese. The development, or 'donor,' agencies have been at least partially successful in attaining this objective. Nevertheless, one of the obstacles facing greater success is the system within Nepal itself. Many of the recipients of training assistance face great difficulty in getting down to work when they return due to lack of resources, low pay, frustration with a complex bureaucracy. The point can be made that the 'donors' are sincere enough about training, a large part of the onus is on the Nepalese system to produce. Nevertheless, the government's Decentralization Act of 1983 is a very important and promising initiative. But despite this partial counterweight, it can be argued that there are probably too many expatriates in Nepal, and that their presence, together with the large, and often competing array of their agencies, does complicate the business of governing.

SOME CRISIS INDICATORS FROM THE WIDER REGION

Much of the foregoing discussion on Nepal is applicable, in part, to the problems that impose on the wider region, although some major differences must be borne in mind. Perhaps the single most important difference is the all-pervading presence of foreign and international aid agencies in Kathmandu and their correspondingly greater, and sometimes confusing, impacts on central government policy definition (along the lines that policy may follow aid availability rather than the reverse). China and India stand at the other extreme and, relative to Nepal, are able to exert much greater, if not total, control over developmental policy. Nevertheless, the Himalayan sectors of the Indian states find themselves in the same, or a very similar, situation of periphery with rather modest political clout and inability to control their own destinies, as is true of Nepal. In terms of the pressure of production on resources and agricultural decline, the Indian Planning Commission has pointed out that population densities in relation to arable land for the Himachal Pradesh and Uttar Pradesh Himalaya are four times that of the neighbouring plains, while productivity is very much lower; subholdings are divided into non-viable fragments with a single family working up to twenty-five individual scraps of land, and up to 60 percent of the family income being derived from the remittances of male members working on the plains. It is contended that once out-migration is initiated, often on a temporary or seasonal basis, rural productivity declines still further. It can be argued that the Himalayan states of northern India, or the mountain sections of northern Indian states (such as Uttar Pradesh and West Bengal) are subject to 'internal colonization' on the part of the federal and/or state governments. This is analogous to the neo-colonialism effected by 'outside' countries. Development in Nepal can be characterized as being under the influence of neo-colonialism at the hands of India; that of Himachal Pradesh and northern Uttar Pradesh, as under the influence of 'internal colonialism.' These processes, at least in part, are responsible for some of the political unrest and activism, as exemplified by the Chipko Movement on the one hand, and by various levels of pressure for local autonomy or independent statehood, ranging from the Kumaun Himalaya to the Darjeeling district of West Bengal.

Similar conditions can be cited for the Pakistan Northern Territory and Hindu Kush in general, but exacerbated by the influx of the world's largest single group of war-torn refugees, perhaps as much as one-third of the entire population of Afghanistan. This constitutes a very special circumstance and warrants further emphasis.

Massive migration into northern Pakistan

This topic has been dealt with extensively by Allan (1987). The flow of refugees from Afghanistan into northern Pakistan constitutes one of the largest migrations in recent times. Approximately three and a half million refugees now reside in Pakistan (another million are believed to have entered Iran). Most of the refugees in Pakistan live today in the North-West Frontier Province. It is contiguous with Afghanistan and contains Pakistan's most extensive forests and mountain pastures. Allan (1987) has demonstrated that the impacts of this vast number of refugees vary greatly in accordance with the type of environment from which the widely different groups have originated and the types into which they have settled. The far-travelled refugees, from north of the Hindu Kush mountains, have caused the most extensive environmental damage. And the maximum disturbance has occurred where refugees have been settled into forest land as distinct from sparsely vegetated arid land. Allan relates this tendency, in part, to the fact that the severest impacts are perpetrated by refugees originating from arid to semi-arid homelands where forest depletion had occurred over the past several centuries, implying a total absence of local, indigenous institutions for the proper management of forest lands. Figure 9.1 shows the principal refugee travel routes and the areas of most extensive deforestation (Allan, 1987:201). This treatment does not take into account that area of northern Pakistan called 'Tribal Territory,' as distinct from the 'settled areas'; the small forest resources have virtually disappeared in the districts between Peshawar and Quetta. The Tribal Territory is an autonomous administrative area and is closed to foreigners and non-resident Pakistanis. Thus no attempt was made by Allan to establish ground verification of what appears as almost total elimination of forests as determined from inspection of satellite images.

In addition to the environmental impacts of the refugees, however, it appears that Pakistan nationals have taken advantage of the ensuing confusion to indulge in illegal logging. Quite apart from the social, political, and humanitarian issues, this largest of recent migrations has caused extensive depletion of northern Pakistan's natural resources. Allan believes

Figure 9.1 Northern Pakistan showing the routes and impact areas of refugees from Afghanistan (after Allan, 1987: 201, Figure 1).

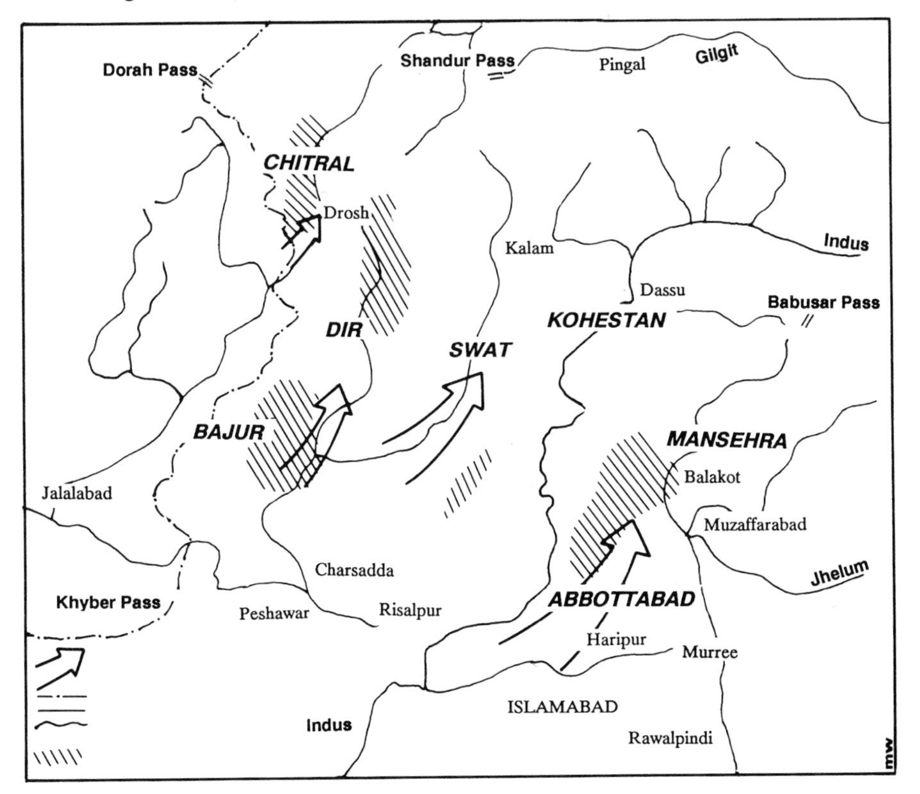

that much of this loss is probably irreversible. Certainly, the human costs are much greater, but much of the environmental damage could have been avoided had refugee camps been located further south near the Grand Trunk Road[1].

Eco-politics and indicators in the India Himalaya

The degree of similarity between the status of Nepal and that of the other sectors of the Himalaya may be much closer than may appear at first glance. The second report on 'The State of India's Environment, 1984–85,' produced by the Centre for Science and Environment (1985), contains a wealth of material comparable to much of that introduced for Nepal. One small section on mountain forest problems is worth quoting, if only to emphasize again the political nature of the problem – the apparent perceived need to exploit the poor, along with the environment.

Under the heading 'Ban the People' the 'Citizens' Report' cites a government document submitted in 1984 to the Union Ministry of

Agriculture by Dr. M. S. Chaudhary, former chief secretary of Madhya Pradesh, which recommends the curtailing of the rights of the rural people in virtually every area vital for their survival as 'in the national interest' (a prime example of would-be 'internal colonialism'). The chapter concludes (p. 98):

> To stop people using forest resources, the ridiculous recommendation has been made that wooden implements should be replaced by steel, and monetary assistance given to transport cement to the mountains, so people stop making wooden houses. The report does not calculate the energy and environmental costs of steel and cement production and the subsidies that would be required to get these high-cost and high-energy materials to the poor.
>
> The report is suggesting a transformation in the use of forests but without any understanding of the socio-economic context in which the forest resources are used in India, or any quantification of people's present and future needs, and of energy and environmental management issues. No report makes foresters' biases more evident than this one.

While this document is a report and not a policy, it is useful as an indicator that western expatriate experts do not have a monopoly on lack of understanding of the rural poor and their needs. It is worth quoting, however, the introductory statements to the chapter on 'Dams' in the Citizens' Report (1985, p. 99).

> Large dams are today India's most controversial environmental issue. Silent Valley dam has already been given up [a success for the pro-environment forces]. Groups are protesting against another half dozen.
>
> The key issue is not nature but people. Energy and water planners are stressing hydro-power and canal irrigation but have made no study of how many people will be displaced.
>
> Government officials argue that 'someone has to suffer for progress.' Usually these 'someones' are tribals, the poorest and the most powerless.
>
> The colossal Narmada Basin Development Programme, which will involve the building of 329 large dams, may end up costing Rs 25,000 crore. It will also displace a million people.
>
> Experience has shown that people, for the moment, are prepared to move but they want new land for old. Except in Maharashtra, this principal is not accepted by any state government.
>
> The cost of forests lost is also high. Large dams have drowned half a million hectares of forest – about a tenth of the area that has benefitted from canal irrigation.
>
> With 176 major and 447 medium irrigation projects under construction, most behind schedule, some experts argue: 'No new projects now; consolidate and learn to get good crops from what we already have.'
>
> Small earthen dams for water harvesting are both ecologically sound and economically profitable. Three small reservoirs have transformed the economy of a village in Chandigarh. There is no soil erosion, no

deforestation, no desertification, and no one has been displaced. The lesson: water conservation, yes; big dams, no.

While the above-quoted remarks refer to India as a whole, they are especially apposite to the Himalaya and the Ganges and Brahmaputra Plains. In the same vein, the Indian sectors of the Himalaya have felt the impacts of large-scale mineral resource extraction far more than have Nepal or Bhutan, if only because of the vastly larger industrial sector that has developed in post-1947 India. Shiva and Bandyopadhyay (1985) provide a graphic description of the devastation wrought in the Doon Valley and along the Mussoorie ridge of the Lesser Himalaya. The recklessness of uncontrolled and irresponsible development of limestone quarrying and cement-works construction was apparent to one of us (JDI) during a short visit in August 1984. Perhaps the most shattering impression was the systematic avoidance of responsibility by the large industries for damage to property and even death. In this instance truck drivers, ferrying limestone from the quarries to the cement works in Doon Valley below sign a contract for each round trip. Thus they are contractors, so that when a truck overturns on the curve of an appallingly constructed road and crashes through a house killing its occupants, the driver (usually poor) is legally responsible rather than the industry. The exploitation of the environment, the truck drivers, the workers, and the neighbouring people, as well as schools, colleges, agriculture, tourist industry and once profitable horticulture, as well as the originally pure air of Doon Valley is complete – for short-term financial gain for the few.

And yet, as indicated in some of the comments from 'Dams' an Indian environmental movement is beginning to achieve successes. In the case of Doon Valley, Shiva and Bandyopadhyay (1985) reported a benchmark success in the form of a Supreme Court of India judgment allowing quarrying to be continued in only seven of the sixty limestone quarries along the Mussoorie ridge, so that rehabilitation becomes a possibility. Furthermore, the Doon Valley citizens' environmental action groups have filed a public interest litigation in the local court against the limestone-based industries that are polluting the 'closed valley with foul dust transforming the serene Doon Valley into a "gas chamber."'

Western Sichuan and Yunnan, People's Republic of China

Somewhat comparable to the worst aspects of virtually uncontrolled commercial interests in India is the apparently uncontrolled and unco-ordinated industrial interests in northwestern Yunnan and Sichuan. Most depressing of all are the wide areas of forest on the interfluve between the Jinsha Jiang and Lancang Jiang (Mekong) that have been clear-cut. In 1985 the large area of felled logs was a scene of waste and decay because the transport and logging sectors were not able to co-ordinate and the timber was left to rot on the ground (Figure 9.2). The foregoing comment relates to state and central government needs for the timber resources of the Hengduan

Figure 9.2 A once densely forested section of the Yunnan Plateau between the Jinsha Jiang and the Mekong trench. Lack of co-ordination between loggers and transport authorities resulted in many square kilometres of logs being left on the ground to rot.

Mountains and the difficulties of co-ordination. At the local scale (rather, the scale of the innumerable local villages) heavy site-specific forest exploitation is also an important factor. This results both from the increased extraction of forest resources by rapidly increasing local populations, and from an increase in living standards of those same populations. For instance, and in regard to the latter point, local Naxi officials (Lijiang autonomous county), as well as villagers, indicated that their living standards had doubled between 1979 and 1985. This was accredited as a direct response to the agricultural reforms initiated by the central government in 1979 ('responsibility systems in rural areas'). It was stated that about a third of all houses in the area had been built since 1979. The impacts of this, albeit humanely necessary, process on the immediately accessible forests were graphically apparent (see Figure 3.2 p. 54). These impacts, however, were exacerbated by the very wasteful and primitive methods of logging (Ives, 1985). On a very minor scale one of us (JDI) observed the relentless hunting of the endangered endemic pheasants (especially *Crossoptilon c. lichiangense*) by the subsistence farmers in the Yulongxue Shan of northwestern Yunnan, and the sale of a skin and plumage for about US $1.00 when, if properly organized, the western photographer-ornithologist would pay the equivalent of a fortune in local currency to photograph such a bird. But also these tales of woe must be tempered by an encouraging growth of awareness and a determination to effect a better balanced utilization of natural resources.

Bhutan: A study in contrasts?

Several passing references have already been made, indicating that Bhutan is an exception amongst the Himalayan states and territories: its being dropped off the World Bank bottom position of Least Developed Countries; the much more healthy state of its forest resources, to the extent of being heavily in surplus; and the more stringent internal control of affairs by His Majesty's Government, seen especially in the refusal to allow tourism to run rampant.

The Kingdom of Bhutan is a very small and rather isolated state; the total population is a little over one million and its area is 46,500 km^2. Karan (1987a) has drawn a useful comparison between Bhutan and Sikkim in terms of contrasts in remaining forest cover, population totals and annual growth rates, and in environmental stress (cf. also Chapter 3, pp. 53–7). The following brief account of the country's involvement with 'development' is extracted from a report by UNU Fellow D. N. S. Dhakal (1987). He explains that it was as recently as 1961 that Bhutan launched a planned process of economic development. The first Five Year Plan (1961–66) primarily concentrated on development of infrastructures. Of all the government departmental outlay, the construction of roads topped the priority list, accounting for 66 percent of the total US $10.7 million budget. Some of the contract funds for the construction projects went to local contractors who made rapid profits and formed the core of today's affluent group.

Another major thrust of both the first and the second Five Year Plans was on education. In this way the government established a number of primary and secondary schools, and also provided scholarships for brighter children to study abroad. This supplied the government, both at home and abroad, with educated school graduates who, by the early 1970s, were able to shoulder government responsibilities. Thereafter, the white-collar sector found public popularity and created momentum to form today's educated elites.

Thus, the dawn of development broke the Kingdom's traditional homogeneous society into classes and ushered the people into a new era – an era of business opportunities, free education, free medical care, and job prospects outside the traditional sphere.

'Perpetual aid syndrome.' The total outlay of the second Plan was 90 percent larger than the first (US $20.2 million), and the third was 139 percent

larger than the second (US $47.5 million). All three Five Year Plans were developed entirely by foreign expertise; and, in fact, the total outlay of the first and the second Plans came as grants from the Government of India. As a result the people knew little about what these early development investments would mean to them in future. Furthermore, the continual aid assurance of the Government of India, the availability of the United Nations' development assistance when Bhutan joined the UN in 1971, and the kind gestures of other friendly nations (such as Switzerland, Japan, and Australia) to participate in development work escalated the confidence of the people in external aid.

Assured of the bilateral and international aid, the government continued to add development projects, rehabilitate settlements, reduce taxes, and subsidize prices on essential commodities, fertilizers, agricultural implements, and cement. This made the general public heavily dependent on the government: dependent on jobs, education, and health care. From the dawn of 'development' to the third Five Year Plan the 'Perpetual Aid Syndrome' preoccupied most Bhutanese.

The transition phase. This phase began once the newly ordained Planning Commission took responsibility for preparation of the fourth Five Year Plan (1976–81). Until then, no single responsible government body had existed to co-ordinate development plans for review and planning; most tasks had been undertaken on an *ad hoc* basis. The Planning Commission established statistical units, channelled information from audit and central account units to generate at least the basic data deemed necessary for checking on progress toward its goals. Having routed most information through its administrative units, the Planning Commission launched a US $101.6 million fourth Five Year Plan in 1976.

Although the format of the fourth Plan differed little from that of the earlier plans, emphasis this time shifted to agriculture, which was allocated 29 percent of the total budget. The government hoped to boost agricultural yields, and thereby reduce the already staggering dependence of the people on the government. Simultaneously, large investments, outside the plan outlay, were made in a 336 megawatt hydroelectric project, various industries, and major irrigation works, in order to establish the revenue-generating base for increasing the internal contribution in the fifth Plan outlay. The capital investment came from the Government of India both in grants and loans. But upon evaluating the feedback from the fourth Plan in 1982, the government noted a record food deficit of 25,000 tonnes, unbalanced regional development, and a huge overhead cost due to the burgeoning bureaucracy which consumed most of the revenue from capital investment. This convinced the government that there was a basic flaw in the system, and it was decided thereafter to lead the people slowly away from 'The Perpetual Aid Syndrome' toward 'self-reliance.'

The change to self-reliance. The self-reliance policy of the fifth Plan (1981–87) required a structural change in the bureaucracy. As the first step, the government decentralized the development administration into districts (*dzongkhags*). In the *dzongkhags, Dzongdhas* (district commissioners) were entrusted to constitute district planning committees (*Dzongkhag Yargye Tshokchungs*) to decide upon the nature and quantity of aid required by the people in each district.

Every development proposal was to be submitted through the people. The government officials would help the people to understand the feasibility and cost of specific projects. Also, it was made mandatory for the public to contribute labour, cash, or materials. Only upon meeting these conditions can the *Dzongdha* forward plan proposals to the National Planning Commission for final approval.

Another obligation requires that the district plan must conform with the general guidelines stipulated in the National Plan document. Also, once the plan has been initiated no interim alteration is permitted without the prior approval of an appropriate authority. In addition, the progress of the plan would be periodically monitored and evaluated by an expert team deputed by the National Planning Commission. The *Dzongdha* is accountable for mismanagement or slow progress.

This work format was intended to streamline responsibility and bring a general awareness to the people of the amount of money the government had been committing to the *dzongkhag*. Also, the villagers would become cost-conscious and be encouraged to use the free development-support facilities in agriculture, animal husbandry, education, and health. In addition, the *dzongkhag*'s elites would become familiar with the recurring expenditures from the development infrastructure, and would comprehend the necessity that some day the people should be left alone to manage their own affairs.

Other reforms in the fifth Plan were concerned with reorganizing the bureaucracy, cutting down on unnecessary staff, and commercialization of public enterprises. These reforms helped the government reduce overhead expenditure and increase working efficiency, and they produced significant annual revenues. It is expected that the revenue-generating sources, such as tourism, industry, power, and forestry would continue to improve in efficiency and would provide a significant internal contribution to the budget of the sixth Five Year Plan (1987–92). The government estimates total revenue at US $94.2 million during the fifth Plan period.

Problems to overcome. It seems that everything that can be accomplished by the bureaucrats has been done. But there are remaining tasks of a scientific nature if Bhutan is to pursue a sustained, holistic approach to the development of a complex mountain system. Of the many important issues, the following require immediate attention:

- To investigate scientifically whether or not a subsistence community with an average of about 1 ha of land per family can become self-sufficient within a certain time period.
- To suggest an alternative approach to tackling the household problems of

self-sufficiency if the present milieu inhibits the society from moving toward this goal.

- To determine the present status of soil erosion, and integrate a new village-wide soil-management system based on agro-climatic data.
- To assess the present status of environmental degradation due to road cuttings, commercial loggings, industrial development, hydroelectric dams, and mining, and then to establish a regular monitoring body to record the environmental changes within a certain time period.
- To create facilities for recording time-series data on rainfall, temperature, and wind and also on stream-flow and sediment transfer. This information is important for deciding whether or not a cycle of catastrophic events would effect developmental or capital projects in the specified time period.

The development process, which invariably creates disturbance in a self-contained, subsistence farming community, is difficult to reconcile with native values and goals. As more developmental programmes are initiated, they create disturbances that require more attention. Whether or not a developing country such as Bhutan can some day succeed in overcoming these problems, it is still necessary to make an honest effort. Is there not something of critical importance to be learned from the Bhutanese that is highly relevant to the entire 'Himalayan Problem?'

The Xizang Autonomous Region (Tibet) People's Republic of China

The northern flank of the Greater Himalayan range and the adjacent southern sections of the Tibetan Plateau, while warranting comparisons with northwestern Nepal and Ladakh, are in a very different category to most of our region. Characterized by extreme altitudes and a mountain desert or semi-arid climate, and sparse population density dependent traditionally on a combination of pastoral nomadism and trade, even the more densely populated southern sections of Tibet have remained remote from the mainstreams of international economy for over a thousand years. Much emotional writing has led to a strangely biased western perception such that Tibet and the Tibetans are as much caught up in 'uncertainty' as any part of the Himalaya. Tibet maintained full independence from or only nominal suzerainty to China throughout the past thousand years until 1950–59. It has even been difficult to estimate total population and to account for its population growth rate between the fifteenth century and 1950. In a profound analysis of these issues Goldstein (1981) discusses the social and demographic implications of Buddhist Lamaism and fraternal polyandry within the context of Tibet as an 'encapsulated environment.' By 'encapsulated' Goldstein means that all available arable niches must have been fully occupied several centuries ago with maximum intensification within the available technological limits, and concludes that by 1950–59 the population was of the order of three million.

We do not intend here a detailed discussion of the military and political issues that colour the present-day situation in Tibet. As indicated earlier, there is some reason for hypothesizing that Tibet had much more extensive forest cover in the more distant past than is usually assumed, and that this was progressively lost over a period of several centuries. About 100,000 Tibetans are believed to have fled with the Dalai Lama in 1959, not an insignificant number, but by no means influential from a purely demographic point of view. There are conflicting figures on loss of life during the Chinese military confrontations and the subsequent Cultural Revolution. Our Chinese and Tibetan hosts in 1980, when we travelled extensively between Lhasa and Kathmandu as guests of the Central and Tibetan autonomous governments, indicated that there were 300,000 members of the People's Liberation Army stationed in Tibet and about 125,000 Han Chinese settlers had been encouraged to take up residence through financial inducements. The western news media have claimed, from time to time, that Chinese settlers number as high as seven million (or as low as 150,000). The high figure appears absurd in view of the 'encapsulated environment,' the cost of maintaining such a large population in the face of extremely long, tortuous, and uncertain supply routes, and the openly admitted (by Chinese authorities) high incidence of infant pulmonary oedema in children born to Chinese settlers. Similarly chronic altitude sickness in civilian adults and military personnel alike constitutes a severe problem.

Introduced new technologies, including a revised forestry and agricultural programme, and light industries, are undoubtedly effecting changes. The rapid opening of Tibet to tourism, especially in recent years when it became possible for travellers to cross directly from Nepal into Chinese territory at many points without central government control and enforced entry via Beijing, is startling. The 1987 October riots in Lhasa brought a temporary halt to this openness (unmatched almost anywhere in the Himalayan region except for parts of Nepal).

Some of the results of post-1959 events in Tibet, however, must be mentioned, both positive and negative, even if only in a speculative manner. Early lack of central government integration of its new forestry and agricultural policies led to accelerated soil erosion (Sun Honglie, 1983). There appear to have been widespread losses in wildlife and this appears not yet to have been matched by adequate appreciation by the Tibetan authorities of the need for thorough-going conservationist policies. The admittedly widespread destruction of monasteries and temples after 1959 and during the Cultural Revolution is being repaired, and the major elements that did survive effectively maintained. The Potala, of course, is a notable example (Figure 9.3). Tibetan ways of life have changed dramatically, although recent years have witnessed a liberalization process that has evoked a religious resurgence. On the other hand, the desperate poverty of a majority under serfdom, maintaining a very large theocratic and aristocratic elite, together with the constraints of the 'encapsulated environment' that seems to have maintained a rate of population growth at the incredibly low figure of 0.21 percent per

Figure 9.3 The Potala, Lhasa, former palace of the Dalai Lama, now a museum and major tourist attraction.

annum (Goldstein, 1981:11) (with all its implications), have disappeared. While the existing living standard of the average Tibetan is undoubtedly low, given the difficulty of even approaching any meaningful statistics, it is probably comparable to that of the Himalayan people as a whole. There is no available *and* reliable information that would lead to a conclusion that it is necessarily lower. One of the especially unfortunate aspects of the Tibetan brand of uncertainty, however, is the persistence of heavily biased reporting by the western news media.

Militarism and regional politics

The issue of militarism and regional politics has been raised already, if obliquely, but especially in the foregoing section on Tibet. We cannot handle such an issue exhaustively, or completely, within the context of this book. But nor can we ignore it, since politico-military eruption must be regarded as one of the possible triggers that could prove to be the torch that would set the fires of supercrisis.

The international frontiers throughout the Himalayan region are extensively under challenge and confrontation. The Indo-Chinese borders, now that the convenient Tibetan buffer of the British Raj is gone, are a matter of grave differences. Equally so, the Kashmir Question remains in a state of constant tension. Reports of serious military exchanges between India and Pakistan (even above 6,000 m) in the Karakorum, and the embroilment of the Soviet Union in Afghanistan represent major disruptions on any standard.

There are, of course, severe environmental and socio-economic repercussions that derive from the politico-military situation, even as it stands today and without considering the possible consequences of much more widespread military activity. Recent reports (October 1987) from Kashmir and Pakistan's Northern Territories, while not authenticated for obvious reasons, suggest that there may be as many as 700,000 Indian troops facing about half that number of Pakistan troops in the general region of the Karakorum, or northern Kashmir. The probable environmental impacts of such a concentration (even if the numbers quoted are substantially inflated), in terms of supply, loss of forest and wildlife, and water pollution, together with the support infrastructure, is alarming. Against this we must rank the construction by India of some 40,000–60,000 km of generally badly engineered roads in the Himalaya as a direct response of the 1962 border war with China. This has already been discussed in Chapter 5 in terms of landslide inducement and changed access to hitherto remote regions. Similarly, the environmental impacts in northern Pakistan of three and a half million Afghan refugees has already been emphasized. But the situation in Afghanistan itself is far worse, of course, even though not demonstrable in detail.

Grave concern over these politico-military problems was expressed by all participants of the Mohonk Mountain Conference of April 1986 (Ives and Ives, 1987). And this prompted one of the more important resolutions of the Conference, unanimously approved:

> *Resolution 3*
> Realizing that nature recognizes no international boundaries and that many of the issues and challenges facing development and conservation cannot be dealt with adequately without co-operation between the countries of the Himalayan region, the Mohonk Mountain Conference strongly urges the governments of the Himalayan region to take steps to establish international parks in border areas (Parks for Peace) to promote peace, friendship, and co-operation in research and management, for the optimal sustainable use of the natural and human resources, and to improve the quality of life of all the peoples of the region.
>
> (Ives and Ives, 1987:185)

Figure 9.4 Proposed enlargement for Sagarmatha (Mt. Everest) National Park, Nepal, and creation of the contiguous Qomolangma Nature Reserve, China.

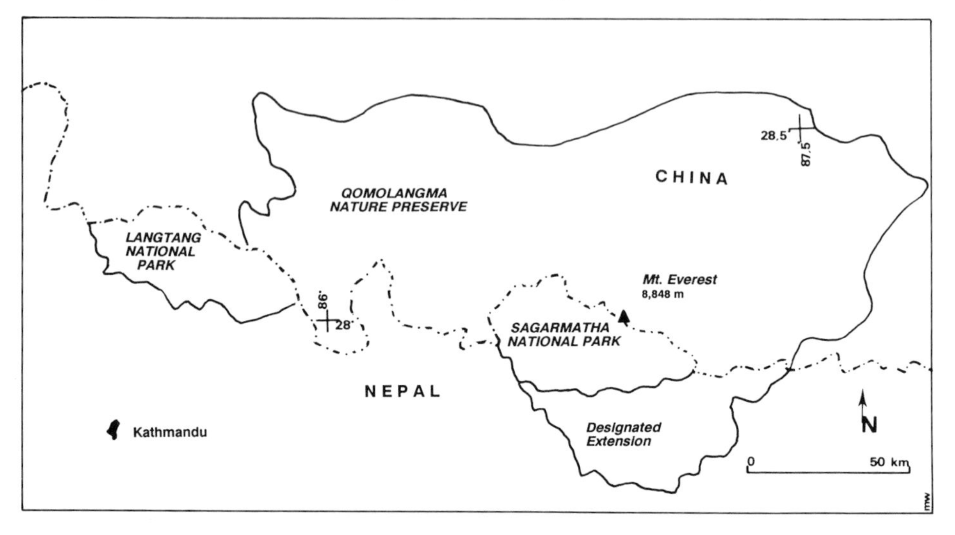

It is gratifying to learn that during 1987 China and Nepal took a first major step in this direction. Nepal has agreed in principle to more than double the area of Sagarmatha (Mt. Everest) National Park, while China and the Tibetan autonomous authorities are in the process of establishing a very large national park (Qomolangma Nature Reserve) on the northern flank of the Himalayan crest which will be contiguous with the Nepalese park, but extend much further westward to include Mt. Xishapangma (Figure 9.4) (Garrett, 1987). This again is another 'virtuous circle' of the Himalayan Problem.

CONCLUSIONS

In this chapter we have raised the question of crisis, pseudo-crisis, or supercrisis in the Himalayan region and have introduced the main group of indicators that we consider relevant to answering the major question so posed. While continuing to side-step the fundamental question about the cause of poverty and the relationship between poverty and rapid population growth, it nevertheless appears to us that a population growth rate of close to 3 percent per annum, with declining agricultural productivity and increasing pressures on land, water, and forest constitutes a very serious threat to regional stability. When reckless development of resources and, for whatever reason, exploitation of the poor, and lack of understanding, or unwillingness to understand, the role of women and the subsistence sector as a whole, is added, together with continuing decline despite large financial inputs, then we feel justified in claiming a state of rapidly developing emergency. The political friction along the mountain frontiers, in the recent past especially between

India and Pakistan and India and China, at present between India and Pakistan, and Pakistan and Afghanistan, together with the neo-colonial status of Nepal and the unenviable downstream position of chronically impoverished Bangladesh, provides further stresses. Within India itself the continuing demand for home rule or independence by the Sikhs, if anything, serves to divert foreign attention from the plethora of smaller, but potentially very serious ethnic, religious, and nationalist conflicts that affect much of the mountain perimeter. These include the demands by the majority of the hill people of the Darjeeling area (originally Nepalese migrants) for an independent state, similar but less clearly articulated demands in the Kumaun districts, in Himachal, and in Arunachal and Assam. There is also the plains peoples' perception that the alternation of floods and dry-season low water that besets them is due to the reckless landscape changes brought about by subsistence farmers in the mountains.

In comparison with the actual African supercrisis, however, the Himalayan region does not and is not likely to face the supercrisis-triggering mechanism of a region-wide drought. It may be argued, therefore, that without an effective triggering mechanism in the wings, a supercrisis is not imminent and the scene is set rather for a deepening of a series of chronic problems. This somewhat optimistic statement, however, could be rudely negated by any significant escalation of the extremely fragile politico-military situation. Yet we can also see progress: the growing strength and successes of movements such as Chipko and the Citizens' Report in India, the growth in understanding of the need to improve the welfare of the poorest people and to integrate them into policy initiation, development, and monitoring – and this by local government, UN agency, and bilateral aid agency alike.

Despite this, we believe that real progress will depend upon the region's ability to initiate macro-scale watershed management, reduction or neutralization of border conflicts, massive realignment of access to natural resources, including fundamental changes in land tenure and taxation systems, and effective peoples' participation in each of these processes. Thus, we feel that it is not necessary for us to stipulate a preference between imminent supercrisis or the gradual development of numerous sub-regional crises, in order to justify a recommendation for an immediate, urgent, and supra-national response. In this sense it is not necessary to argue whether the perceived environmental degradation is cause or effect; if the socio-political elements of the problem are not soon taken in hand we believe that environmental collapse could rapidly ensue. We will then have supercrisis enough for all.

NOTE

[1]The situation is in a further state of flux with the 1988 withdrawal of Soviet military forces from Afghanistan and the beginnings of refugee resettlement.

10 RESEARCH STRATEGY FOR THE HIMALAYAN REGION

INTRODUCTION

We have concluded that our region is characterized by extreme uncertainty and great complexity. We have tried to dissipate some of the uncertainty and this has led us to two postulates. First, the Theory of Himalayan Environmental Degradation is not a valid entity and must be broken down into its component parts and each part must be evaluated on its own merits. Thus the major linkage, population growth and deforestation in the mountains leading to massive damage on the plains, is not accepted. We favour the more cautious approach based upon acknowledgment that the long-term geophysical processes are more than adequate to account for the on-going formation of the plains as the continuously rising mountains are progressively eroded. We would argue, for instance, that *appropriate* forest establishment in the mountains is vital, but only for the well-being of the mountain environments and the mountain peoples dependent upon those environments. If reforestation in the mountains is conceived as a palliative for the problems of the plains, it is likely that vast resources will be expended to reap only disappointment. Moreover, this conception may divert attention away from the necessity of water resources management and adaptations to the natural environment of the plains.

In terms of the need for a research strategy, however, we must confess that we cannot totally disregard the importance of learning more about the natural process linkages between mountains and plains. In part, this is because we are not satisfied with being in the position of concluding that human intervention in the mountains has little or no impact on the plains because the inherent problem of shortage of data pervaded our own counter-arguments. In assessing future prospects, therefore, it is better to stipulate that, while we are comfortable with our refutation, the linkage between events in the mountains and events on the plains is unproven rather than proven false.

Thus we will outline a standard approach that should serve to improve our understanding of the highland–lowland physical linkages. This, at the same time, will provide two important practical advantages: (1) as our understanding of Himalayan natural science grows (but without comparable and integrated growth in human science this will be inadequate), 'development' agencies and governments should be progressively better positioned to effect successful interventions in terms of resource utilization; and (2) the same learning process will provide the necessary benchmarks against which the magnitude of any future possible impacts on the plains deriving from land-use changes in the mountains can be assessed.

The second postulate is that, despite our setting aside the physical highland–lowland elements of supercrisis, there are enough indicators of potential disaster for us to recommend that the national governments of the region, and the international community, should begin to react immediately to the *probability* that the socio-economic situation in the Himalayan region will run out of control. It must be recognized that this could give way to a level of human suffering not yet witnessed in this harsh world of the late twentieth century. Nevertheless, this should not veil our conviction that there *are* grounds for optimism. While we cannot understand whether population pressure on an inadequate natural resource base is driving a downward spiral toward increased poverty and starvation, or whether poverty is the basic causal factor itself, we are inclined to regard population growth, at least in part, as the symptom. The relationship must be more fully explored and a major adjustment in thinking and planning implemented. This will be very difficult, not the least from a political point of view – both from within national societies and between nation states. It is because of the latter conviction that we urge a co-operative and unprecedentedly large-scale undertaking between governments with some appropriate initiative being provided by the United Nations.

Two issues are apparent already, however: the level of uncertainty enshrouding the Himalayan Problem must be both lowered *and* utilized. Both approaches need to be faced and we believe that our own modest efforts have merely laid a foundation for this. This chapter, therefore, is essentially a first tentative suggestion of a series of considerations that might be useful for the delimitation of a major change in the agenda of research planning and 'development' implementation. As might be anticipated, there are several possible approaches to the definition of a research strategy, and a healthy strategy will be one that commands many different initiatives. Because we ourselves endorse the need to avoid rigidity we will introduce a number of perspectives here. The next section is largely derived from one of the Mohonk Conference panel discussions, led by Michael Thompson (Ives *et al.*, 1987). We have identified as many as possible of the specific ideas with the names of the individual conference participants. While this may be somewhat unconventional in a book of this kind, we believe that our approach will help in indicating the spirit of enthusiasm and the intellectual stimulation that the Conference brought forth. This could be considered of some importance as a demonstration that international co-operation is feasible. This was true, even amongst a most diverse and complex group of participants, with the seeming potential for the maximum possible disagreement and contention, almost as a mirror of the Himalayan region itself. The unanimously approved Resolutions are included at the end of this chapter.

UNCERTAINTY REVISITED: WHAT ARE THE CRITICAL GAPS IN KNOWLEDGE?

The task of the panel was approached in a deliberately general and non-disciplinary way. Our view is that it is not necessary to define the research strategy in terms of its specific contents, as if it were a box full of apples. Rather, research is a process – an ever-rolling stream – and our task is to discern the change in its direction that appears to be desirable (perhaps inevitable) in light of the new thinking that we have tried to promote. That there has been a significant change in direction we have no doubt but we are, at present, less certain as to exactly what that change has been.

The general feeling is that there is much progress to be made at present through a continuation of the process of systematic questioning of unquestioned assumptions. Since this, of course, was a major component in the strategy for the Conference itself, we are pleased to see that it did indeed catalyse many of our discussions; never have we seen so many cherished assumptions evaporated into so many despicable myths! We feel that all this must have some central strategic relevance for thinking about the future direction of research. But the trouble is that such turmoil – such wholesale demolition – creates an awful lot of dust. In consequence we feel that, though we can already begin to discern some of the more prominent features of the new direction for research, things will become much, much clearer once the dust has settled. A small working group charged with this clarificatory task, and coming together in months rather than years, is what is needed. That is our first recommendation.

In the meantime, it is possible to set out (in a rather disjointed and tentative way) what we feel are some of the discernible features of this new direction. Some of the points we list are abstract guiding principles; others are quite specific proposals for future research inputs. We have taken the liberties of fleshing out some of these points, of making some connections between them, and of filling in some of the more obvious gaps in the hope that this report can then serve as some sort of initial orientation for the proposed working group and for more widespread questioning and response.

1. Anticipation Versus Resilience

There are two very different ways of coping with this apparent antithesis. We can try to anticipate future events and then make specific preparations for their arrival; in contrast we can resist this anticipatory urge and, instead, increase our generalized resources so that we will be well placed to absorb or exploit whatever the future brings. The first mode tends to lead us into planning, hierarchical patterns of organization, and centralized information; the second tends to give us markets, myriad autonomous agents, and diffused information. In consequence, much research ends up by being 'captured' by the institutionally generated assumption that anticipation is the right (indeed, the only) mode. The danger here, however, is that not everything in this world can be anticipated, and this means that there will always have to be some trade-off between the two modes: anticipation and resilience.

Resilience is the generalized ability of a system to cope with the unexpected. When the bridges along a 40-km stretch of the Dudh Kosi, in Khumbu Himal, were swept away by the lake break-out from the Langmoche Glacier, they were re-built by the local people in a few weeks, despite some shortcomings. By contrast, the Austrian-funded hydroelectric project at Thame (which was also destroyed) was not replaced. That is an example of the lack of resilience. It is important to realize, however, that resilience, for all its grass roots and diffuse nature, is never something that is just *there*. It can be fostered (by institutionally appropriate development) and it can be discouraged (by institutionally inappropriate development).

All of us acknowledged familiarity with a particular breed of forestry expert that is much given to pronouncing that the only way to ensure establishment of a good forest in the Himalaya is to station armed guards all around it. Such a solution, which defines the local farmer as part of the problem, is a perfect example of institutionally inappropriate development. And it is revealed as such by the many forests that are now flourishing without any armed guards, or even without wire fences around them. These forests grow because the local people want them to grow, and because they have succeeded in securing (with the help of development projects) the generalized resources to ensure that they do grow. These institutionally appropriate developments succeed because they have been able to conceive of the local farmer as a part of the solution (Griffin *et al.*, 1988).

So *anticipation* (the 'outside' designation of a village forest, for instance) and *resilience* (the capacity of the villagers to recognize and manage a renewable resource, for instance) are alternative modes, each of which will be more or less appropriate in specific physical and social contexts. The trouble is that anticipators (institutional agencies, and national ministeries, for instance) fund research while local farmers tend not to. There is always the risk, therefore, that research will become so tightly harnessed to anticipation that it will lose all sight of this crucial and multi-levelled trade-off between itself and resilience. In contrast, if we (as researchers and practitioners) insist that these two modes must always be weighed against one another, then the kind of *fine-graining* so essential for successful implementation will come about quite naturally as specific contexts variously induce their appropriate weights.

This is a very general principle that reappears in more specific form in many of the points that follow. In a strategic sense, therefore, we can use it as the *leitmotif* for the explication and development of the new research direction.

2. What is the Problem?

When people talk about the environmental problems of the Himalaya (even, or especially, when they talk quantitatively) they are not talking about the same thing. During the Conference we heard speakers from governments and agencies pinning 'the problem' on increasing population, and showing signs of exasperation when people from the Chipko Movement insisted that population increase was little more than a symptom of the real problem: a lack of control over local resources by local people. In much the same way, the planting of trees was seen as the solution by some whilst others were adamant that the only answer was better land-use management.

The inescapable conclusion is that there are *plural problem definitions* and *plural solution definitions* and, more importantly, that they do not go away. If we insist on just one perception of the problem then inevitably we will be excluding many of the people and, in the process, committing investment to a development path that, sooner or later, will prove to be unsustainable – physically, socially, or both. 'Single problem/single solution' approaches (which, again, are prevalent at present) inevitably foster wrong thinking. Right thinking requires us to develop and strengthen 'multiple problem/multiple solution' approaches.

3. What is a Resource?

Those who speak of 'natural resources' locate resources in the physical world; those who speak of 'raw materials' see resources as flowing from the successful interaction of those raw materials with culture – with such essentially intangible assets as human skill, knowledge, and enterprise.

Time and again we slipped into the easy assumption that for the Himalayan people to have enough food it is essential that the Himalayan land produces enough to feed them. To be precise, this basic need is satisfied once the people are in a position to *command* enough food. Increasing the productivity of the land is, of course, one of the ways of achieving sufficient command over food, but so too is the sort of service- or manufacturing-based economic growth that has occurred, for example, in Ladakh, in the Khumbu, and in the Kathmandu Valley.

Deepak Bajracharya, Dipak Gyawali, and Hementa Mishra were not convinced that the Chipko ideal of land-based self-sufficiency was appropriate for Nepal and argued that Nepal was always a country of traders and that trade was the very foundation upon which their nation had been built. The slogan 'Trade not Aid' neatly expresses this clash of resource perceptions, although it can be argued that the majority have always been members of the 'cautious cultivator' community.

Again, each of these contradictory definitions will have to be incorporated into research and legitimated in policy debate if the decision-making process is not to be captured by sectional interests. Here, in the crucial distinction between land-based and non-land-based development opportunities, we can see, perhaps more clearly than in the 'too many people/not *enough food*' distinction, the positive advantages of striving to maintain the legitimacy of contradictory definitions.

Points 2 and 3 together begin to suggest research strategies for explicitly recognizing and dealing with the *political economy* and *political culture* dimensions (both local and geopolitical) that often underlie what, at first sight, look like merely technical questions.

4. Who is the Client?

As Sunderlal Bahuguna repeatedly pointed out, the interests of the researcher's *non-paying client* (the villager, in most cases) are not always consonant with those of his *paying client* (the agency that sponsors his research). A researcher who identifies himself with the interests of just one of these clients ends up doing both of them (and his science) a disservice. The debate between Shah and Schreier (1985) and Messerschmidt (1985) in the pages of *Mountain Research and Development* is very much to do with this question of client ambivalence, and much of the discussion during the Conference (David Griffin's *Participatory Action Research*, for instance, and Sunderlal Bahugana's *Head, Heart, Hands*) can be interpreted as appropriate methodologies[1] for ensuring the visibility of both clients.

5. Problems know no Boundaries

The research apple is already sliced (by disciplines, by institutions, by departments) but problems tend to be *un*sliced. There is a crucial need to make the system connections across the slices (and this is not achieved just by including the word 'integrated' in the project title). There are deep philosophical implications to this *systems approach* and it is something that has to be worked at hard – practically *and* conceptually.

6. The Heterogeneity of Problems and Capabilities

Maps (in the widest sense of the word) are a particularly apt way of respecting and making the most of the heterogeneity that, we all agree, is such a key feature of the region. 'Trouble spot' maps, for instance, readily capture Romm's elusive idea that '90 percent of the damage may be caused by 10 percent of the land' (Jeff Romm, personal communication, December 1983).[2] Such gradations of red colouration help focus remedial action where it is most needed, but they should also be accompanied by 'opportunity maps' in which the gradations of green colouration tell us where local enterprise (tourism in the Khumbu, for instance, agricultural productivity in the Kathmandu Valley, and the Tibetan carpet industry in a number of locations) is headed for 'take-off' under its own steam (that is, without development aid).

7. Everything is Not Getting Worse Everywhere

Point 6 (above) feeds into a related question. Research in the Himalaya, at present, has a *marketing problem*. Problems ('supercrisis', for instance, and ubiquitous 'vicious circles') are being overplayed whilst capabilities (genuine external concern, for instance, the fact that no interested party is intent on destroying the Himalaya, the potential of the International Park idea ... the presence here and there of 'positive sum pockets') are being correspondingly underplayed. Only martyrs (and Oxford men) are attracted to hopeless causes. Since our main conclusion is that the problems of the Himalaya are *serious but not insoluble*, our research strategy should strive to rectify this negative bias.

8. The Usable Synthesis of 'Hard' and 'Soft' Science

By putting the emphasis on all the uncertainty the Conference rightly emphasized the complementarity of two seemingly contradictory questions: 'What are the facts?' (the banner under which the 'hard' sciences, such as physics, biology, and physical geography, traditionally advance) and 'What would you like the facts to be?' (the question that enables 'soft' and subjective sciences, such as anthropology, cognitive psychology, and the sociology of perception to move ahead). Problems, though they do involve all sorts of real physical processes, are not defined by those processes. They are defined by people – people, moreover, who are embedded in very different social and cultural contexts and who naturally define their problems very differently. The system, in other words, is a system of mountains *and* people and we must do everything we can to avoid treating it as if it were two separate slices.

9. Research is Driven by Perceived Information Needs

The exploration of uncertainty and its institutional origins, and the accompanying conversion of many facts into myths, suggest that we should try to identify two sets: facts that would be useful to know, and facts that we are likely to be able to know. The intersection of these two sets would then

Figure 10.1 A framework for 'hard' science research (prepared by Michael Thompson in Ives *et al.*, 1987: 335).

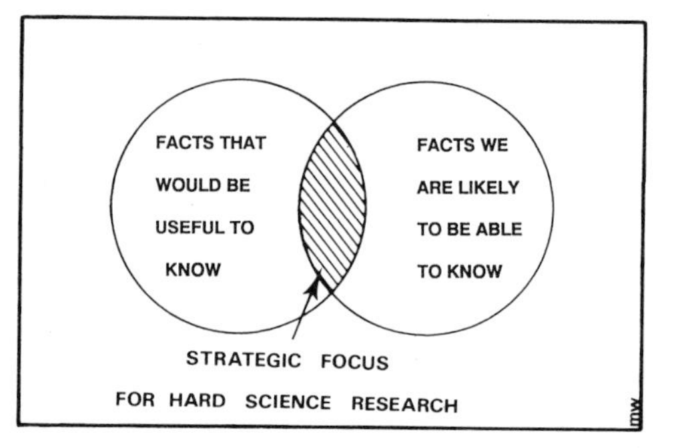

provide both the focus for our 'hard' science research efforts and a framework that will tell us where further research is likely to be unsuccessful, or unnecessary, or both (Figure 10.1).

Another way of saying this is that we should go for problems that are: (1) soluble; and (2) worth solving. For instance, the downstream impact of upstream human activity is probably not a problem we should be tackling, although it may be deemed politically and practically desirable to lay down a process whereby the lowland water-resources manager has a better understanding of mountain–plains linkages than is presently available.

Many of our deliberations could usefully be mapped on to this simple diagram (Figure 10.1). Many of us are concerned with redefining this strategic focus for 'hard' science research. It would be well worth while to go through this mapping exercise, discipline by discipline, to provide detailed *before* and *after* descriptions of this research focus.

10. Making Room for the Home-grown Wisdom

This has been a recurring theme: Jack Ives, for instance, stressed the astonishing level of restorative work that, within several years, completely removed landslide scars but which was simply not visible within the short time-scale of expert scientific studies. This effort, in turn, depended crucially on all kinds of ethno-science and ethno-risk assessment (as the UNU Kakani Project's work on indigenous land-use categories, hazard perception and risk-handling strategies made clear (Johnson *et al.*, 1982; Kienholz *et al.*, 1983)). A nice insight, in other words, into resilience and its micro-social origins, and one that, alas, is too easily missed.

Sunderlal Bahuguna pointed out that, often enough, the local people know exactly what needs to be done (often better than do outside experts) and, moreover, are quite capable of doing it. External aid directed at resilience enhancement, therefore, should be concerned simply with making it easier for them to get on and do it (by paying their taxes while they do the work, for instance, or by providing food for labour, and so on).

Deepak Bajracharya showed us how the widespread Nepalese institution of the 'go-between' internalized a micro-level negotiating wisdom that, even now, is lacking from many research and aid programmes.

David Griffin showed us how effective a candid admission of ignorance on the part of the expert (a 'blank sheet' rather than a fully specified blueprint) was in harnessing the outsider's injection of resources to the local expertise.

Hementa Mishra, in reflecting on his experiences in the Royal Chitwan National Park, teased his way through the viable and unviable ways of making the relationship between the local tigers and the local people less antagonistic and more symbiotic.

Learning, of course, is the crucial ingredient of all these success stories (many of which are implicitly highly critical of much large-scale development work). This suggests that the time is now ripe for a specific research effort aimed at pulling together the valuable experiences of individuals with 'hands-on' local expertise and then tabulating what they have learned into general *Principles for Good Practice*. This 'learning-from-success' exercise would then complement the 'learning-from-failure' exercises that will be proposed presently.

11. How to Promote Flexibility?

Long before the hydroelectric project at Thame in the Khumbu was physically swept away, most of its institutional supports had been extensively eroded. It was extraordinarily difficult to find anyone who could speak up enthusiastically in its favour, either in Nepal or in Austria (the aid donor). Concern in Austria had developed to such a pitch that even the then-Chancellor, Bruno Kreisky, was convinced of its inappropriateness, yet nothing could be done physically to stop it. This required the actual catastrophic outbreak of a glacier lake with its attendant loss of resources (Ives, 1986; Vuichard and Zimmermann, 1987).

This is a particularly vivid example of a widespread and inevitably wasteful phenomenon and there is a real need to develop and apply techniques for distinguishing *at an early stage* between inflexible and flexible development paths. This is the now familiar problem of *entrenchment* through institutional, financial, and technical commitment: the pathology, it could be said, of the anticipatory mode (Collingridge, 1981).

The techniques for minimizing entrenchment, both in technical and institutional terms, do now exist and they should be applied to many of the existing and proposed developments in the Himalayan region.

12. Problems of Scale

Much debate over the appropriateness or inappropriateness of projects (and much of the concern over the inadequacy of the evaluation they at present receive) centred on scale. Indeed, considerations of scale permeated our discussions to such an extent as to suggest that it (like the anticipation: resilience trade-off) should become one of our major organizing themes.

In the course of a casual conversation, on the way to the airport after the Conference, one of the World Bank delegates commented that he had managed to visit the Khumbu in connection with the World Bank forestry project that was his responsibility. He was asked, 'where exactly in the Khumbu is your forest?' Patiently he explained that the visit had been something of a self-indulgence, since the Khumbu was really rather 'out on a limb' so far as the project was concerned. One may have *assumed* that his forest was in the Khumbu when, in fact, the Khumbu was in his forest!

It is assumptions – assumptions with which we are unwittingly supplied by our prior institutional and professional involvements – that are currently driving so much of the decision making on scale. 'Big-is-best' versus 'small-is-beautiful' seldom receives the attention it deserves because the answer has already been given by the very organizational nature of the agencies involved.

In such settings scale decisions are not debated; they come with the job!

Hydroelectricity projects, for example, come in all shapes and sizes: from the micro- and mini-hydel schemes in the Khumbu that have been the subject of so much evaluative debate between their protagonists (and so much natural destruction), through the existing medium-sized dams like Kulekhani and Bhakra, to the projected schemes such as the Karnali High Dam and the Arun Cascade that are among the largest, most expensive, and most novel engineering undertakings ever seriously contemplated by man. Yet they receive only a tiny (and biased) fraction of the expert evaluation and institutionalized assessment that is accorded even such relatively routine and modest engineering ventures as the Diablo Canyon nuclear power station in California or the Sellafield Thermal Oxide Reprocessing Plant in Britain.

The diverse expertise necessary for adequate evaluation, and the institutional development required for that evaluation to be constructively focused, simply are not in place in the Himalayan region. However, recent government actions in the Doon Valley, and the fact that the Chipko Movement has been able to take the Tehri Dam proposal to the Indian Supreme Court, show us that they could easily be put in place. Rectifying this omission, surely, must be one of the main aims of future research.

13. Economic Development for Whom?

This very general question was asked in many specific contexts. Jayanta Bandyopadhyay stressed the need to understand the social, physical, and economic interactions between upland and lowland – the wanted stuff *and* the unwanted stuff. For example: the political economy and political culture of soil erosion; the need to relate widespread processes, such as monetarization–leading to cash crops–leading to forest degradation, to fluctuations (in terms of resilience) in the effectiveness of the classical wisdom of the people. Bandyopadhyay urged that the appropriate unit for this sort of analysis was the major river basin as a whole.

14. Ivory Towers Versus Green Fingers

Janos Hrabovszky confessed to schizophrenia between *pure science, social interventionism*, and *doing*, and urged a three-fold research package aimed at: (1) improved land and water management; (2) institutional change; and (3) cash flow to mountain people. Such a package, he conceded, would be full of all sorts of myths and those myths would have to be confronted. Myths, he argued, should be dealt with by the policy actors agreeing, and agreeing to differ. The apparent paradox here could be resolved by new approaches that sought to stimulate, rather than eradicate, perceptual pluralism. (A plea, in other words, for the adoption of the 'multiple problem/multiple solution' approach.) A small example might help explain how this explicit myth-confronting can advance both the formulation and implementation of policy (and alter the kind of research that they call for).

The cash flows to the mountain people – through military service in the Indian and British armies, through tourism, through certain cash crops, and so on – is not disputed. What is disputed, however, is the consequence of all this economic activity. It is at this interpretive level that the myths begin to assert their contradictory influences. Those who subscribe to Adam Smith's 'hidden hand' see it all in an optimistic light: markets are springing up everywhere and, if we can just remove the dead hand of bureaucracy, self-help will do the rest. Others see this influx of cash as the instrument by which the developed world secures its hold on the undeveloped fringe: marginalization, not self-help, is the name of this all-too-familiar game. Still others are convinced that upward progress *is* possible, provided the cash flow is managed: the free-for-all must be brought under control and the whole process pulled within a planned and integrated framework of development. So Hrabovszky's 'package' (pp. 256–272) will simply fly apart if these myths are not confronted and unless each of them is granted its legitimacy as *one* institutionally valid mode of interpretation.

15. Uniting Problems with Solutions Through Sensitive Intervention

Ijaz Hussain Malik expanded on Hrabovszky's plea and explained his own experiences and successes with the Aga Khan's Rural Support Programme, in terms of the 'garbage-can model of decision making' (cf. Thompson and Warburton, 1985b).

Total pluralism would entail a garbage-can in which all shapes and sizes of problems met up with all shapes and sizes of solutions in a totally random way. In Malik's experience, the pluralism is more structured than this, and the distinctive social and cultural perspective of the mountain farmer ensures that only some of these problems and solutions 'belong' to him. Intervention, again in Malik's experience, is a *learning process* in which these particular conflations of problem and solution are gradually revealed. Aid is then the exploratory process by which resources are transferred in such a way that the farmers become more able to solve *their* problems to *their* satisfaction.

Since this intervention and aid (and, in particular, the process by which the resources are actually transferred) brings into existence local organizational arrangements that, until then, were only latent within the village culture, there is a marked increase in self-confidence (that is, resilience) across the entire mountain region. Malik's idea of helping the villagers to help themselves is, therefore, remarkably similar to Bahuguna's. What differs, however, is the assumption about the direction in which such village-level improvement is to be found. Bahuguna's villagers want to pull back into self-sufficiency; Malik's want to get out into the global marketplace (and, since Hunza apricots can be purchased in local shops in Britain, it would seem they are quite capable of doing this).

16. Optimism and its Uses

Tom Hatley expanded on the Himalaya's marketing problem. Scientists are trained to define and solve problems rather than to identify and expand capabilities. They need to retrain themselves by deliberately developing a sense of possibility; mapping benefits, extolling project diversity, and overcoming their timidity in arguing for small-scale, optimistic, human-based projects. Alongside crisis there is opportunity. Opportunity is fostered through long-term, flexible, people-to-people co-operation – *gift exchange*. For example, the opportunities for the symbiotic development of the relationship between local Himalayan people and their local rare plants and animals have often been missed because of the colonial origins of the utilitarian science inherited by the nations of the region (and still sponsored by the major donors of development aid). Rectifying this bias – moving away from utilitarian science toward a more naturalistic science – has the following beneficial consequences:

(a) we begin to put ecosystems first in the design of rural development projects,
(b) biological conservation becomes the leader – the signpost – for sustainable development in general,
(c) we are able, in many instances, to re-couple biotic values (rarity, diversity, interdependence, site-specificity) with the values of the marketplace (Hatley and Thompson, 1986).

17. Honesty and Scientific Gift Exchange

Michael Warburton rephrased 'gift exchange' as 'working together' and pointed out that, for long-term success, working together called for honesty: for a continuous effort to give frank recognition to the patterns of interests, economic and political, that shape the contexts in which scientists have to work, and that so easily (and so insensibly) mould the facts that they discover.

The perhaps surprising consequence of this (and it is a consequence that we should take seriously) is that recent developments in the seemingly rarified areas of the *philosophy and sociology of science* are of central importance for applied scientific research in the Himalaya. One has only to glance at all the myths that were revealed through the hard questioning of hitherto unquestioned assumptions to realize the imperative of instituting some framework for self-awareness and quality control in the conduct and application of research in this most convoluted (physically and institutionally) of regions.

18. What's Wrong with the Shift from Land- to Manufacturing- or Service-Based Industry?

Sandhya Chatterji found the 'too many people' problem definition difficult to apply to Ladakh (Chatterji, 1987). 'Not enough food', however, helped her make more sense of the large and rapid shifts in population and occupation that have occurred (both recently and historically) in this remote and politically and ecologically sensitive region.

Tourism and defence-related employment, of course, can disappear as quickly as it can appear, but she pointed out that its stabilization is something that *can* be worked at, once it is seen as part of the solution rather than as part of the problem. The creation of international parks in such sensitive areas, for example, is one way in which this sort of stabilization might be fostered. If it is accepted that the land-based, petroleum-free economy of Ladakh is gone for ever (and many do not accept this), then Ladakh's shift straight across from subsistence agriculture to service industry, without any intervening manufacturing stage, can be seen as a leap into post-industrial development that would make most of the developed world green with envy.

Problems, of course, abound, but, unlike many areas in the Himalaya, they exist within a sea of opportunity.

(a) What about new energy technologies – bio-gas and solar, in particular?
(b) What about technology transfer? The ability to command resources (food, energy, transport) is increasing rapidly but the acquisition of the technologies appropriate to the taking of that command is lagging behind.
(c) What about the undesirable consequences of this booming economy – corruption, for instance, and the markedly uneven distribution of opportunity? Again, institutional development, in some appropriate blend of central government regulation and local 'ways of doing things', is called for.

19. Are 'Normal Science' and the Himalaya Inimical?

Dipak Gyawali managed to link a wide range of seemingly diverse topics by posing this simple question.

(a) The uncertainties we have been uncovering have been generated by *normal science* – by accepting the slices we are given (science, for instance, asks 'What are the facts?'; something else that is not science asks 'What would you like the facts to be?'). Slice the apple differently – that is switch to *revolutionary* (in T. S. Kuhn's (1970) sense of the word) *science* – and things begin to look quite different. 'Unimaginable projects,' Gyawali pointed out, 'are not very popular,' yet they are precisely what is needed.
(b) 'Anthropologists map all the variety and propose nothing; economists impose uniformity and propose everything.' Again, this distressing and all-too-familiar dead-end can only be circumvented

by imaginative power. A *middle path* between the debilitating extremes of anthropological and economic orthodoxies, is what we need and, in fact, we already have it in such notions as *structured pluralism* and the *garbage-can* in which the apparent anarchy is deftly organized by the distinctive perspectives of the various actors. Normal science would have to insist that just one of these perspectives was valid; the revolutionary science appropriate to the Himalaya would insist that there is something of value in all of them.

(c) The Third World is everywhere. Marginalization is what it is all about, and margins form as readily in the inner cities of the First Two Worlds as they do along the Himalayan Divide. Prosperity and growth in parts of the Punjab far exceed that in Liverpool or Appalachia. Marginal problems demand marginal solutions, and this leads us to query the credentials of many development projects.

(d) Foreign aid; what does it do? To ask this question is to gain some useful insights. We see, for example, that socially irresponsible money wreaks havoc, and that experts who lack commitment do much the same. Nepal is awash with forestry money (as David Griffin pointed out); but this does not necessarily solve the problem of degradation of the quality of the remaining forests. Dipak Gyawali proposed two standards. The first, for the local expert, measures his willingness to go back to the village. The second, for the foreign expert, measures his commitment to those he is supposed to be helping. The unit of measurement of the first is the *Gandhi*; for the second, the *Theresa*. These are the absolute units; the practical units would be the *milli-gandhi* and the *milli-theresa*!

Points from the Floor

The research panel, by accident or by design, was largely packed with 'Young Turks,' impatient of the 'Old Guard,' their normal science, and their unimagination. One advantage of this was that it has been comparatively easy to link the panel's utterances together into a fairly coherent whole. The themes are broad and strong, they thread their way through the whole debate with a fine disregard for national and disciplinary differences, and they are fun – a rumbustious challenge to the familiar theses of the Development Community and the Environmental Establishment – theses that, over the years, have become as all-engulfing (and as unnoticeable) as is water to the fish that swims in it. The remarks (and, in some cases, the passionate speeches) from the floor, though more disjointed than those from the panel, help to redress the balance. The Young Turks certainly developed their anti-thesis but we would not wish to give the impression that they carried the day unopposed.

20. The Malthusian Hypothesis

K. G. 'Tej' Tejwani sprang to the defence of planning and of the research needs that it generates. He stressed that the problem *was* one of 'too many people' – population out-stripping resources – and that it had to be tackled by effective long-term policy commitments. It was from this perspective that the crucial gaps in current knowledge should be identified. (However, it could be argued that the panel's presentations were implicitly hostile to a particular kind of planning rather than to planning *per se*. They can be seen as an argument against an old style, rigid/total control/complete knowledge, *deterministic* planning and for its replacement by a more flexible, more humble, and more *strategic* style.)

21. The Future Place of Historical Research

John Richards contrasted *advocacy* with *detachment* and argued for some division of labour that would enable historians to contribute in both ways. There was room both for dispassionate analysis (probably in the form of comparative regional research) and for collaborative team research (as, for example, in the Doon Valley Ecosystem Project) (Shiva and Bandyopadhyay, 1985).

Richard Tucker, taking up this collaborative aspect, stressed the value of chance – the serendipitous coming together of disciplines and interests that happened at conferences such as this. Follow-ups were needed if these growth points were not to wither.

22. Principles for Practice

David Griffin was unconvinced of the efficacy of the traditional two-pronged attack in terms of *pure* and *applied* research. The linkages between them were not right, and this led to serious and persistent problems of implementation. Participatory Action Research is a far more appropriate research methodology for fostering development (Griffin, 1987). It presented no problems of linkage and placed itself firmly in the context of *implementation*, rather than in *analysis* and *policy formulation* – the inevitable contexts of pure and applied research.

Most research, at present, is focused on the first differential – the *trends*. It should try to get beyond this – to higher-order differentials. If things have got about as bad as they can get perhaps they will begin to get better. Can we get at the *trends in the trends*?

23. The Gandhian Perspective

Sunderlal Bahuguna posed four simple questions: a) How does the professional become devotional? b) What is modern, what is primitive? c) Can we come up with a blueprint for survival? and d) Can we build self-consciousness? There was much sagacious nodding of heads as Bahuguna spoke, which suggests that many of the scientists and administrators present, though they might not have put it quite like this, were quite familiar with these crucial nails that Bahuguna, in his inimitable way, was hitting fairly and squarely on the head.

(a) His first question, for instance, is echoed in Gyawali's plea for responsibility and for some means of measuring *the expert's commitment.*
(b) His second question resonates with the questions 'Who is the client?', and 'Development for whom?' In other words, it calls for the frank recognition of *the political economy dimension.*
(c) His third question ties the elusive concept of *sustainable development* firmly to the welfare of the poorest people of the region. If they are not to be raised up (in their own terms) what hope is there for the totality?
(d) His fourth question, translated into the esoteric language of the expert, calls for a more explicit awareness on the part of the scientist for the whole *policy context* in which, like it or not, he is working. And it emphasizes the all-too-easily neglected option of *resilience enhancement.*

24. The Agency Perspective

Attila Stersky, the representative from the Canadian International Development Agency (CIDA), emphasized *success* and the importance of spotting it. In evaluating projects CIDA looked for a 'track record for success.' Its policy was to 'pick the winners.' In following this strategy it had found that, on balance, it favoured small-scale projects. Small, it would seem from this learning process, was more successful than large.

Stersky was less forthcoming about the failures (associated, one presumes, more with bigness). Some members of the panel, while supportive of this commitment to learning from success, wanted to see the same learning principles applied to the failures as well. (A point that is taken up later.)

25. The Place of Glaciological Research

Gordon Young explained that there was no glaciological institute in the Himalayan region, despite both the importance of the pure science aspects of glaciology *and* the practical downstream consequences of the behaviour of the region's glaciers. He stressed the need for such an institute, the need to stimulate the transfer of information, and the need for a major glaciological conference in the early 1990s. The International Centre for Integrated Mountain Development (ICIMOD), in Kathmandu, could possibly be the institutional niche for all these, and it would be best if the initiatives came from the region itself.

26. Science for Public Policy

Hementa Mishra explained how his research on tiger conservation did not stop there but was pushed right through to the decision makers. Research, he stressed, was a continuous process and had to be kept flexible and responsive to the policy process. Research should not be seen as separate from development.

On the question of setting up institutional channels for information transfer and dissemination he was sceptical. 'Information,' he felt, 'has to be gone and got.'

27. The Future Place of Hydrological Research

Bruno Messerli presented the outlines of a research strategy that built upon present work and that aimed to provide badly needed data from a small number of crucial points on four exemplary rivers across the region. Since water underpins everything else in the region, this understanding (which should be sought in a non-bureaucratic way) should be actively fed into the wider understanding of the biology, agriculture, and land-use patterns of the region.

28. The Research Consumer's View of Research

C. K. Sharma saw the Himalaya as a 'playground for researchers.' Researchers were having a wonderful time but few solutions were being offered to the problems supplied.

On the upstream/downstream question he agreed with Bahuguna that the Himalaya were vital to Indian culture and civilization and that this, rather than the more negative, scientifically dubious and potentially divisive argument about worsening flooding, was the international linkage that should be emphasized.

He wondered whether 'development' had done anything for Nepal. Did Nepal need development aid? Education, he agreed, was vital yet educational programmes had not resulted in economic uplift. *Criteria for appropriateness and inappropriateness* are what is needed.

29. The Villager's View of Research

D. N. S. Dhakal, speaking as a Bhutanese villager (rather than in his role as a postgraduate at the University of Colorado), wanted the research to be useful to him. Piles of papers in the university libraries of Oxford or MIT were of no use to him.

30. Failure into Success

The sea of uncertainty, the endless problems of implementation, and the many, many surprises that lie in wait for even the best-planned projects have rightly served to direct our attention toward learning; learning to be modest, learning that we do not know everything, learning from one another – expert-and-expert, expert-and-villager, researcher-and-client, and so on.

Many stressed the benefits of learning from success, though Janos Hrabovszky said that the FAO experience was that, since the success of a pilot project often turned out to lie in its leader and not in its contents, they were often disappointed when the same project format was applied in multiple locations: the singer, not the song! But, if failure is so widespread, why not learn from it too?

Frank Davidson said that engineers traditionally dissect their failures until they have extracted every ounce of learning they can. His own course on *failure* at MIT had been an enormous success. An *Institute for the Study of Failure* might well do more for the Himalaya than anything else!

A CONVENTIONAL APPROACH FOR THE NATURAL SCIENTIST

Much of the debate running through several of the chapters of this book has highlighted, if not actually centred on, the problem of uncertainty. One tenet of this theme has been our very attempt to expose and dispose of the many 'sacred cows' upon which the Theory of Himalayan Environmental Degradation appears to depend. A second tenet, closely interwoven with the first, is the unreliability of those data that *are* available – whether they be on the role of forest cover in protecting the soil surface from erosion, the sediment load of streams, or calculations of the life expectancy at birth of subsistence peoples, or especially, the impacts on the plains of land-use changes in the mountains.

Before introducing a conventional approach to narrowing the degree of uncertainty enshrouding this highland–lowland, or downstream impacts component, let us emphasize two conflicting positions. It could be argued that the difficulties in the face of obtaining time-series data on hydrology and sedimentation are so enormous that it is best to forget the whole idea as Utopian, or academic. A small illustrative example in this respect is Byers's (1987c) analysis of the reliability of the existing precipitation data from Namche Bazar. The recording of rainfall, even snowfall for the most part, is usually conceived as simple, routine, not requiring any special skills. Byers was very puzzled with apparent conflicts between his assessment of the occurrence of relatively high-intensity precipitation events and the longer-term station data. For instance, a 90 mm water-equivalent event in January at Namche Bazar must be, in fact, a measurement of snow depth that should have been divided by anything between 10 and 20 to provide water equivalent. And when two rain gauges within 500 m horizontal distance record 3 mm and 92 mm respectively, we cannot equate this with high site-specific variability.[3] And stories of small village boys urinating in erosion study plot rain gauges are too numerous to be amusing any more. Thus, we may have to face up to the fact that any data collection programme will not provide a meaningful response to reducing uncertainty.

The conflicting position is the difficulty we expect to be facing in convincing the vested interest group with our argument that the activities of mountain people are insignificant in their effects on the patterns of lowland flooding and siltation. This is the rather more important of the two positions simply because we believe it essential to remove the mountain subsistence farmers from their unenviable role as convenient scapegoat.

Thus, with all the caution and scepticism that we should have culled from the preparation of chapters 1 to 7, we nevertheless recommend a conventional approach to trying to determine the downstream impacts of human activities in the mountains. At least it will result in the training of cadres of hydrologists, glaciologists, climatologists, and geomorphologists, and human sciences scholars. It must also be borne in mind that, should the present rate of population growth and poor land management be maintained over the next two to five decades, then human interventions in the mountains may well begin to have a detectable impact on the plains.

Thus we propose the need for a regional and long-term programme to measure and monitor selected watersheds along the great Himalayan arc. This would involve systematic observation of standard climatic parameters, streamflow, and sediment load, themselves related to different types of surface cover and land use. This should be augmented at relevant localities with a series of site-specific studies that will be enumerated below.

What are the Problems?

Since a great deal of research is being undertaken in the Himalayan region, it is reasonable to ask why this is not sufficient or why it cannot be adapted to meet the stated needs and so obviate the labour and expense of superimposing a new research structure. While much co-operation can surely be anticipated from on-going researchers there are some problems.

1. Most of the existing research projects are too narrowly defined; they reflect a specific interest and often a process of one-way thinking. What is needed is a systematic and more nearly inter-disciplinary approach tilted at seeking to understand the magnitude and intensity of key processes.
2. Most of the existing research is too local, without any real correlation to the broad regional-scale problems. Even if the fieldwork of our proposed scheme will involve point measurements in specific localities, they should fit into a regional concept.
3. Most of the existing research projects are limited in duration. Long-term data series are needed so that natural oscillations, for example, of precipitation, streamflow, and sediment load, can be distinguished from growing human impacts on the environment, in terms of streamflow and sediment load.
4. Research to date has suffered under the pressure of the demands of

international and national agencies for rapid actions – better yesterday than today – that urgent decision making needs an immediate response. This attitude destroys the opportunity for augmenting our basic understanding of key processes and long-term thinking required for *rational* decision making rather than decision making based on the perceived need to spend so many million dollars by next Friday. The question is no longer: what is basic research and what is applied research? The transition becomes more and more fluid; application-oriented basic research is required: the participatory action research of Griffin (1987).

5. In the Himalayan region national borders play a very important, and usually negative role. Rivers carry water and sediment load across these national borders, and scientific and practical problems with them; the political difficulties must not be underestimated since they lie at the core of all other problems. However, a regional research scheme may facilitate the progressive breakdown in the negative role of national borders. Current examples of trans-border research provide grounds for optimism; for instance, the China–Nepal proposed study of the Arun watershed; the China–Pakistan–Royal Geographical Society research in the Karakorum; India–Nepal trans-border research. A supra-national and regional research programme may facilitate the rapid expansion of this process that has admittedly been very slow to develop, and may serve to further reduce border tension. (This notion is the basis for the Mohonk Mountain Conference resolution recommending establishment of international parks.)

Outline for a Regional and Long-term Research Strategy

Any strategy will have to accommodate vastly different scales of research, from the village, to the micro- and meso-watershed, even to the macro-watershed. However, with careful planning and inter-linking of various investigations, this problem of scale interrelationship can be overcome. A broad framework is illustrated in Figure 10.2. Small, site-specific studies can be integrated into a whole if the overall requirements are detailed first.

As a first approach, a group of Himalayan countries should select a meso-scale watershed that includes a section of the Tibetan Plateau, and all the mountain belts, and so down to the Indo-Gangetic Plain. There are many sets of criteria, some conflicting, that can be identified for the selection of such a watershed and some possibly overriding considerations will have to be taken into account. These include existing research and data, accessibility, access to electric power, available topographic maps and satellite imagery, political sensitivity of border-crossing points, and the possibility of currently existing development needs. As an example of the final point, the Arun-Kosi system can be introduced. The prospects for development of a two-billion dollar Arun hydroelectric cascade scheme in Nepal, and the advanced stage of negotiations between China and Nepal for an international expedition would indicate that this watershed should be carefully considered for inclusion in the

Figure 10.2 Schematic representation of three different scales of research.

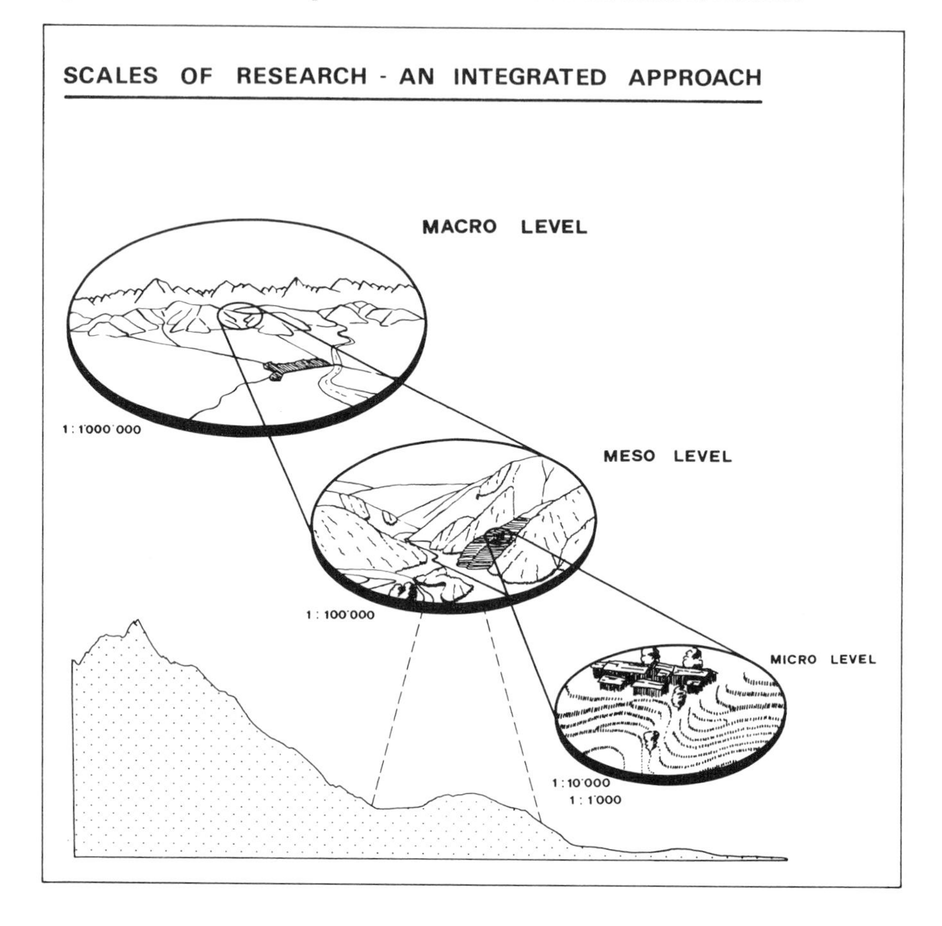

strategy proposed here. We would suspect that any rational planning for the expenditure of huge sums of international aid money on a succession of hydroelectric facilities on the Arun would also stipulate the need for the type of integrated watershed research we are proposing, the more so because an unknown number of a group of fifty or more glacier lakes on the Chinese side of the watershed have the potential for catastrophic break-out for putting at risk any development infrastructure further downstream (Ives, 1986; Vuichard and Zimmermann, 1987).

Regardless of the choice of watershed, one condition of our scheme is that a main research station should be located on the border between the mountains and the plain, a critical point for any long-term investigation of the possible downstream effects of human land-use changes in the mountains. There is no need to describe the myriad of details of such a scheme here. It is obvious that the main station at the transition from mountains to plain will require facilities for long-term, high-quality observations on stream flow and sediment load including solutes and bed load. This would be linked with a minimal network of observation stations in the extensively utilized high mountains and plateaus, and the intensively utilized Middle Mountains, foothills, and plain. A third component would include the organization of experimental watersheds at different altitudes, and with different slope angles and aspect, vegetation, and land-use types. Here routine climatological, hydrological, and geomorphological data should be collected from sites selected to become focal points for a full range of human science research: analysis of *both* the negative and the positive impacts of the subsistence farmer on slope stability; indigenous coping strategies and environmental perceptions, nutrition, and general physical quality of life indicators; migration trends; linkage between perceived needs and appropriate methodologies as well as appropriate technology transfers; land tenure and taxation patterns; and landscape change through time, including the utilization of replicate photography and historical research methods.

Of course, under ideal circumstances, several watersheds should be selected, giving a range of studies from the drier regions of the west to the eastern Himalaya that are most heavily deluged by the summer monsoon. Other vital requisites for success include absolute comparability between field sites in instrumentation, data tabulation, and analysis, and *total* accessibility of all the data and research findings. While different components of this rather large scheme would be initiated by different agencies, national and international, university and private, it should have a scientific co-ordinating body at the international level: it should be closely linked with the International Centre for Integrated Mountain Development in Kathmandu.

We estimate that the scheme would also require the contributions of both regional and extra-regional universities. This should lead to a deliberate and ambitious training and educational component. Other facets would include provision for direct contributions to development agencies and for input by local village-level leadership. Finally, publication of results at various levels would be expected, ranging from the popular news media in various languages, to the scientific literature, to the practical manuals of development agencies. The very establishment of such a scheme should have immediate, world wide significance.

Figure 10.3 provides a generalized graphic representation of the approach we have outlined above. The topography has obviously been simplified. The overall structure, however, lends itself to data acquisition and dissemination and the grafting on of additional specific research tasks identified as the need arises. One such example that we believe to be pressing is the identification and monitoring of glacier- and moraine-dammed lakes, such as the Dig Tsho at the snout of the Langmoche Glacier in Khumbu Himal which broke out and destroyed the Namche small hydel project in August 1985 (p. 70). Further, as indicated by Gordon Young at the Mohonk Mountain Conference, rational utilization of high mountain water resources of the Himalayan region requires a firmly based glaciological programme tied to a

Figure 10.3 Highland–lowland interaction. A schematic outline for a research strategy in the Himalayan region for a better understanding of the downstream effects: precipitation, runoff, sediment load, soil erosion.

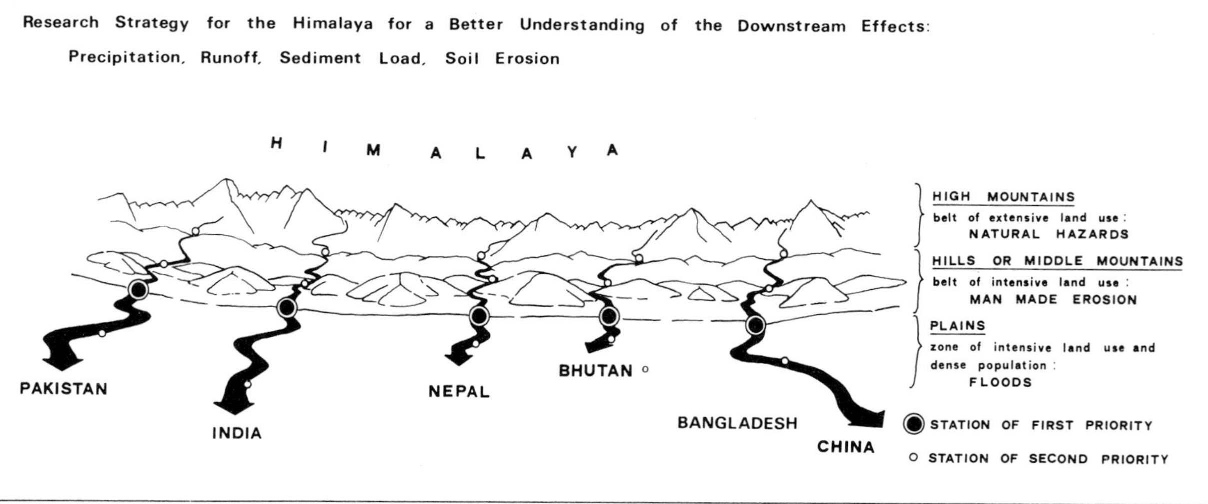

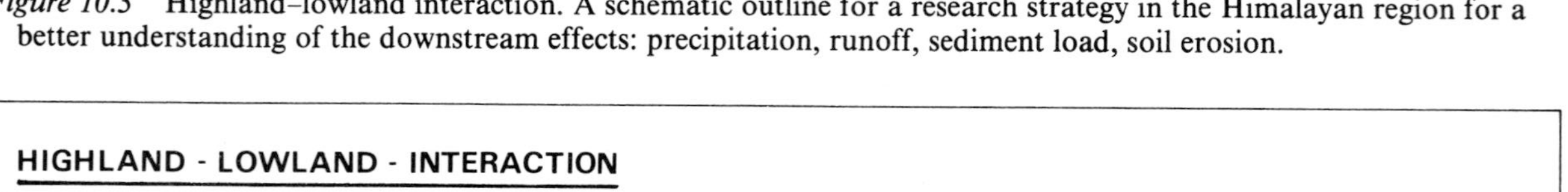

regional institutional research and training structure. Such a development would fit in well with the scheme proposed here.

IMPLEMENTATION FAILURE CAUSED BY INSTITUTIONAL PROBLEMS

One of our great concerns is the apparently large extent of lack of success in the implementation of development projects due to institutional problems. While this, of course, is a further indicator supporting our insistence on multiple problem definition/multiple solution definition, it is appropriate to take a closer look at the specific issue of institutional problems. We are approaching this here through the personal contribution of one of the Mohonk Conference participants, David Griffin (1987).

Institutional problems often lie at the heart of failures to implement satisfactorily a wide range of development activities. Here various components of 'institutions', broadly defined, are discussed along with some problems which are important in a practical sense.

Our aim is not simply to consider as an intellectual exercise the problems of the Himalayan region. It is to identify some of the key features of a developing crisis and then to suggest elements necessary for remedial action. Action programmes require some means of implementation and this usually involves institutions, broadly defined, as key components. However, there is now widespread recognition that institutions are not only a necessary component in problem resolution and in development but are also an important hindrance to both. This 'institutional problem', therefore, is relevant to our strategy.

Let us first consider the issue from the side of those involved in purveying development assistance. These can be grouped in a number of ways but, for our purposes, a simplistic division into: (1) *academics*; (2) *agencies*; and (3) *practitioners* will suffice. By *academics* (1) we do not mean only those employed in universities. Many in research institutions and in a multiplicity of 'think-tanks' also follow the well-established procedures of academic enquiry and analysis. They study a situation and, ideally, reveal much of real value, based on well-validated data. Recommendations for appropriate action are made and too often the task is then thought to be complete. The more action-oriented academic will follow up the recommendations to see if they have sunk without trace or have been implemented by some agency somewhere. In the latter case, implementation will usually have been found to be a dubious success or an outright failure.

Often, the good intentions of recommendations fail to have the desired effect because institutional problems were ignored. In most cases, if a recommendation is to have effect, there must be an institutional system to implement it, or at least to commence implementation. Here naivety is rampant. Consider the following examples. Many major environmental problems are undoubtedly trans-national. In an ideal world, the solution would come from whole-hearted co-operation between nations, but this cannot be assumed. Indeed, it would be better to assume the reverse. A reluctance to tackle environmental problems trans-nationally, in fact and not just on paper, is not restricted to developing countries as shown by problems with whale-hunting and so-called acid rain. National interests in a trans-national problem are varied and altruism is limited. With such a background, it is not surprising that trans-national institutions to implement action are something of a rarity - at least in comparison with such organizations that undertake research, for research involves no commitment to implementation. The writer of resolutions should be wary, therefore, of routes assuming trans-national implementation.

A second example relates to integrated development. There is such intellectual force behind the idea that the development of a region depends upon joint activity in, say, agriculture, forestry, water management, health, and physical infrastructure that it is no wonder that integrated development has been vigorously advocated. Yet how many examples of real success are there – far fewer than of failure to attain anything like the original goal. Is this surprising when in most countries there are no powerful institutional arrangements for such programmes? In developed countries, conflicts within the bureaucracy are almost standard and an inter-departmental committee is as good a way as any to kill an idea when it is to the clear advantage of no single department to foster it. Why should anyone expect developing countries to be better in this regard? Even if a 'mega-department' is created, only very strong guidance from the top will prevent sectoral differences resurfacing.

The lesson to be learned from all this is that, in most cases, programmes should be designed for implementation by existing institutions of appropriate power. This is not to say that such institutions at the outset need be adequate technically or administratively, for they can be improved. (We will return to this point later.) They must, however, have sufficient power, politically and in a bureaucratic sense, to permit effective initiation.

It has been said that it is unjust when a beautiful theory is destroyed by a miserable fact. This surely applies to the 'trickle-down' theory of development. Much economic argument showed that financial aid applied at the top (factory construction, infrastructure, etc) would sooner or later benefit the poor as economic development was stimulated: a beautiful theory destroyed in part by the miserable facts of self-interest and corruption, institutional or individual, which provide an effective filter against such benefits 'trickling-down' very far. This leads to the thought that institutional power must also be linked to a certain minimal amount of institutional will in key places. If corruption, or any other negative factor, exceeds dedication to the task, nothing will be accomplished (a reminder of Gyawali's milli-gandhis and milli-theresas).

The second group of actors on the donor side are the development assistance agencies (2), including specialized banks (included here are official and semi-official agencies; NGOs are excluded from this discussion, although this is not intended to under-rate their importance). These are bureaucracies in their own right and might be expected to see with clarity institutional problems for development. There is little evidence that this is so in practice, again in contrast to theory. Large schemes presupposing inter-departmental co-operation are frequent even though a few minutes' thought would show that a comparable programme would have scant chance of success in the proposer's own developed country. We suspect that the force of fashion is at least partly to blame for such schemes getting off the ground. An idea originates in academe, gains strength and eventually political support, and then has a momentum of its own. Institutional appropriateness seems to disappear.

Another set of institutional problems relating to agencies arise from what can be considered as issues of scale. Here there is a fundamental contradiction. As an example, let us take deforestation and forest degradation. This problem is widespread as we have strongly emphasized yet its causation is various even within one nation: the strategies, and especially the precise tactics, for its solution are still more diverse, especially in the Himalayan region with its great differences in altitude, aspect, climate, population density, history, and ethnicity. Plans of action therefore need to be as 'fine-grained' as the exact nature of the problem they address. Effective programmes will be characterized by proper attention to this matter of scale and may need to be appropriately localized.

If, however, a policy is adopted which is based on the above argument, it can be argued that it will never meet the widespread and critical need within an acceptable time-frame. Only large-scale action will suffice, and this latter route appears to be that favoured by the multilateral and large bilateral agencies. Indeed, it is easy to understand why, by their very nature, they are unsuited to a 'fine-grained' approach. Yet there must be grave doubts whether a coarse-grained, large-scale philosophy can ever satisfactorily solve a fine-grained problem. This issue of appropriateness of scale (also raised in a somewhat different context in Figure 10.2) is highly contentious but, nevertheless, is of fundamental importance to development. A corollary to this matter of scale is that the large-scale route seems in practice to rule out an evolutionary approach. Worthwhile development must have, in our view, a large component of venturing into the unknown. (Please note that throughout we have avoided the term 'technical assistance'.) If the pathway were truly clear, there would probably be no problem and development would have already occurred. In fact, development *is* an evolution and its pathway has unexpected twists and turns and its useful goal may not be exactly the one first set. Yet flexibility is inversely related to scale.

Simplistic development projects – and they are simplistic however complex the computerized pathways – usually ignore institutional problems. Recipient government departments, however powerful, are often in no state at the outset to implement and sustain a given development activity. Suppose that the territorial divisions of the forestry sector in a country have traditionally had two predominant roles, the regulation of timber harvesting from natural forest for commercial purposes and the protection (largely through prosecution) of the forest from peasant incursions. Suppose further that both that country's government and a donor agency have seen a new light. *Community forestry* is to be the new order and a project for its implementation is devised. If evolution over years is permitted and fixed quantitative targets are eschewed, all may be well. If, however, a slow beginning and gradual development are not part of the plan, pity the poor Divisional Forest Officer! He probably has little knowledge of planted forests, even less of the complex socio-economic milieu of community forestry. His expatriate counterparts are probably in no better state in regard to the latter issue. Worse, the majority of his field staff have been little more than forest policemen throughout their employment. Most lack all appropriate technical knowledge for forestation and are now in any case temperamentally unable to make the large jump from policeman to extension agent (from *control* to *assistance*). Administratively things are no better, for the existing rules and regulations, accounting procedures and budget were designed when no one dreamt of an active afforestation programme. Morale is not helped by the fact that funding is sufficient to permit staff to be out of their office and in the forest for only a few days a month. (Extra walking requires more food and occasions more wear and tear on clothes and this necessitates *per diem* supplements for fieldwork.)

However sympathetic the senior departmental staff, such a situation takes years to remedy. Many of the issues cannot really be seen in advance of the commencement of implementation of the project. Then, progressively, they become all too apparent and the institution and the project have to evolve together. A major function of expatriate project staff, supported by their funding agency, should be to offer sympathetic encouragement and advice to help in institutional transformation. Projects with such a component seem rare. This need for institutional development is being more widely appreciated.

> Mere existence or execution of successful programmes or projects is not enough to ensure their continuity. They need to be able to create around them, or within them, an organizational structure and an appropriate institutional base to allow continuing internal and external official support to them.
>
> (FAO, 1985)

Nonetheless, progress is woefully slow and most agencies probably fail to include an appropriate institution-developing component lest they be thought to be endeavouring to interfere in the 'internal affairs' of another country.

Another deficiency in project planning is a failure to get right the appropriate institutional linkages. Suppose, in a certain country, the need for significant improvement in research and development in all aspects of the forestry sector was identified and a project was set in train to establish *de novo* a research institute in forestry. Administratively, let us place the institute within a government forestry department and provide it with equipment relevant to all aspects of the sector, including harvesting, wood science, and timber processing. Further, let us staff it appropriately, only to find that the staff have great difficulty in gaining access to mills, conversion plants, and the economic data of product sales and exports. Why? – because the forestry sector, in fact, is divided in two. The first part, the government department's part, is concerned essentially with forest aspects up to the marking of trees for felling. Harvesting through to product sale is the responsibility of an autonomous corporation: and the department and corporation are separated by a nigh-on impenetrable wall. The *corporation* sees *institute* staff as belonging to the *department*. The staff in the utilization area, and the results of their work, are therefore largely ignored by that component of the sector which ought to benefit. Clearly the initial linkage established was wrong and strong and difficult actions will be needed to remedy the situation.

Practitioners (3) in our terms are those who implement projects. (Wearing another hat, they may also propose and write projects and therefore contribute significantly to the problems discussed in preceding paragraphs.) Most practitioners, whether overtly commercial companies or consortia of tertiary education institutions, are in fact engaging in development activities for primarily commercial considerations. Only a few, outside, non-governmental organizations (NGOs) have other and not necessarily more altruistic motives. Further, most are appointed to implement projects already devised by others and described in formal project documents. Flexibility is often minimal, outside of set mid-project reviews. It is then all too likely that the practitioners will accept the stated project design, warts and all. Beneficial institutional change is unlikely and the project collapses as soon as the expatriates withdraw, if not before.

In all the above, our criticism is aimed at those of us who live in developed countries. We will now speak more briefly on developing countries. Again a tripartite division is possible: (1) *agencies* (usually government departments); (2) *practitioners* (project officers in central administrations and project co-managers in the field); and (3) the *local populace*.

Institutional deficiencies certainly abound in developing countries. In the face of such a situation, it is too easy to throw up one's hands in despair but in most countries this is not warranted. Institutional deficiencies, after all, are at the core of underdevelopment and need to be taken into account and tackled.

In most cases, the social order of a country will be reflected in the very nature of its institutions. This is widely recognized for centrally planned economies but is less so for the rest of the world. All too easily we assume that institutional organization elsewhere largely mirrors that in our own country, when in fact it mirrors its own society. The more we understand the history and society of a country, the more we shall understand its institutions. Institutional change is unlikely to proceed rapidly ahead of societal change.

Most bureaucrats in developing countries are keenly aware of the institutional framework within which they work: their promotion and even survival depends upon this. Knowing the system well, it is no wonder that high senior officials in central offices and potential co-managers in the field look with scant enthusiasum on project proposals which take no account of important practical aspects. Innumerable delays and lost files may then be an almost unconscious response to a proposal that is likely to cause little but trouble because it is essentially not implementable within the existing administrative and financial control systems. An implementation failure may have serious career implications, particularly in a country where it is less reprehensible to do nothing than to make a mistake. Delays and prevarications, therefore, may sometimes be ultimately beneficial if they cause re-thinking or even cancellation of a proposal. If, through donor enthusiasm, an institutionally inappropriate project is implemented, it is likely to draw little more than perfunctory activity from recipient-country staff. Expatriate staff will also soon become frustrated or disillusioned by their inability to make the bureaucratic machinery work in ways for which it was not designed. As before, the answer surely is to let the project and institution evolve together. If this is impossible, then the project should not commence.

Adequacy of developing-country staff, in terms of both quality, quantity, and job classification, is a vexed issue. Quantity and classification are inseparable because numbers required cannot be determined before the necessary balance between professional, technical, and vocational levels has been established. This is by no means a simple issue and there is a marked tendency to establish a balance reflecting past rather than future operations. Improvement in quality must be long term and would seem to be most easily achieved at lower levels. Such an expectation is in fact contradicted by many cases where vocational-level training has proven to be a most intractable problem embedded in larger structural issues. Training at this level is often disastrously divorced from action. It tends to take place in institutions remote from employers and uses ideas and even equipment not current in their place of employment. If, however, training is closely integrated with employment, it can be outstandingly successful.

Some of the ramifications of poor education have great institutional consequences. Thus if the standard of education is generally low in a country, then the base for adequate recruitment becomes extremely narrow: narrow not only in terms of numbers but also in terms of geography. The best educational opportunities will lie in the main cities so that the developing bureaucracy becomes increasingly urban and remote from rural issues. The understanding of such issues by bureaucrats then becomes no better than that of expatriates. Before long, a posting to a remote, or even non-urban, area is looked upon as a punishment – and often is so in fact. Rural development cannot flourish in these circumstances.

Finally, let us reintroduce the concept stipulating that the local populace in a project area are not so much part of a problem as part of a solution. If significant opposition from residents to an environmentally related project is experienced, then there is probably something wrong with project design and often with institutional factors. The fault may be in too rapid a timetable imposed by donor or recipient; in a failure to appreciate local, perhaps informal, institutional realities so that proper linkages are not developed; an invalid assumption on the availability of staff with sympathy for local needs; or a host of other factors (including the possibility that the project is just simply wrong in fundamental design).

We could continue almost indefinitely, but we hope that we have made a case for placing institutional matters *sensu lato* high on the priority list when problems such as those facing the Himalayan region are under consideration.

HIMALAYAN DEVELOPMENT: DEVELOPMENT FOR WHOM?

It is apparent that mountain development is one side of the same coin of which environmental protection is the other: 'development' will never be successful unless it is environmentally (and socially) sound. We feel that, from many perspectives, development policy has largely failed despite very large inputs in time, expertise, financial resources, and goodwill over the past three or four decades. We have concluded that poverty (under-development), however related to a rapid increase in mountain populations, is a central explanation for increased pressure of production on natural resources to the point that serious environmental degradation has been occurring over a period of hundreds, rather than tens, of years, albeit more in some areas than others. Thus government efforts to reduce the pressure on natural resources are set to fail unless a core objective is the improvement in the well-being of the subsistence farmer, which means a reduction in the degree of poverty. Thus we propose that a radical change in access to resources and, consequently, a shift in the direction of the flow of benefits from resource development is a prerequisite to problem solving in the Himalayan region. We believe that this will only be achieved by a radical change in the mind-set of the 'developers', the adoption of the multiple problem definition/multiple solution approach. Sunderlal Bahuguna, in summarizing his impressions from his 5,000 km, 300-day walk through the Himalaya, shows the most sensitive and intimate approach to the problem that we have so far uncovered. We quote a short concluding section from his report to UNICEF:

> The Himalayan crisis is not an isolated event. It has roots in the materialistic civilization, in the spiral of demands, ever-increasing but never satisfied. Even the renewable resources become non-renewable due to over exploitation. The air and water pollution, acid rains and barren stretches, familiar today in many countries, are the gifts of this civilization.
>
> The immediate need is to preserve whatever forests remain but the viable answer to the ecological imbalance is to adopt a new development strategy in which man and nature coexist in harmony. This in turn is possible only if small communities are allowed to meet their own basic needs. The perils of centralized production systems were anticipated at the beginning of this century. As we move towards its end, the challenge is to implement a programme of survival, to which the life-supporting role of forests is integral.

> Alternate sources of fuel and energy and efficient ways of using them are known. There is also no lack of expertise in proper forest management. *But there is little identity of interest as between the people and those who control forests and other resources* [our emphasis]. The only way to redeem forests from the combination of corrupting contractors, corrupt politicians and corruptible officials is to vest their control openly with the community, with government in an overseeing role. It would then be possible to protect and 'exploit' them in a socially acceptable manner; and to summon the blessings of science in the service of the people.
>
> (Bahuguna, 1983:10)

This does not imply that the way ahead is therefore simple. When the subsistence farmer is hungry, environment and culture become remote considerations. How often have we heard a farmer patiently explain: yes, I know this or that practice is bad, but I have to feed my children tomorrow? Nevertheless, Bahuguna, in broadening the issue to the world at large, re-opens the opportunity for gift exchange – we have so much to learn from the surviving traditional and environmental knowledge of the subsistence mountain farmer. And we are concerned that, even given singular success in the generation of political will and supra-national co-operation and teamwork, many still unquestioned assumptions lay in ambush to snare possible advances. As a corollary to a clear identification that mountain development must be aimed primarily for, and in conjunction with, mountain people, therefore, we state the need to continue to question hitherto unquestioned assumptions.

In conclusion we list a number of specific topics for possible inclusion into a supra-national, region-wide research strategy formulation:

1. The identification and testing of alternative quantitative models for a better understanding of population-resources-environment-development interrelationships, especially their accumulative complex, dynamic and qualitative aspects in mountain regions.
2. Examination of the nature of population pressure in different situations throughout the Himalayan region. How can high population pressures be accommodated? What are the effects and extent of migration and depopulation? What 'carrying capacity' models can be applied?
3. Analysis of current developments in the agricultural system and investigation of possible palliative responses. What are the most promising forms of innovative agriculture? How can subsistence agriculture be strengthened?
4. What are the causes and effects of changes in livestock numbers and quality? What potential is there for traditional pastoralism? What are the various conditions of grazing in different parts of the region?
5. Examination of the current status of shifting agriculture (swiddening) in different parts of the region. What alternatives are there?
6. What are the causes of the deepening poverty situation? What

mechanisms exist for the provision of basic needs? How far is ill health a cause and how far a consequence of this situation?

7. How far are least-developed situations emerging in the Himalaya? What is the relationship of least-developed status to foreign trade? What are the demographic and environmental implications of exports and imports, and what modifications could usefully be made in trade strategies?
8. Identification of the most deprived groups in the region. What is the position of women and young people? How can these groups be incorporated into the development process?
9. What are the costs and benefits of tourism? How can adjustments be effected to reduce the costs and augment the benefits?
10. Examination of institutional constraints. What kind of reforms are needed and what kind can be attempted?
11. How far are early warning and prevention services developed for disasters (man-made and natural)? How can they be improved?
12. What kinds of social forestry schemes assist best in reforestation and in improving the lot of the local people? What problems have these schemes faced?
13. Are multi-national industry activities increasing? What are the implications for the welfare of the poor and for the environment?
14. How can the IUCN national conservation strategy concept be extended and intensified?
15. Establish a catalogue of different cultural and sub-cultural frameworks in relation to the problems of population–resources–environment–development. Identify valuable traditional components.
16. How can grass roots, self-reliant movements and community participation be encouraged and/or strengthened? What roles can women and young people play in them?
17. Establish a systematic process, including centralized archiving, of collection and replication of old photographs and sketches to provide some basic documentation of environmental change.
18. In what ways can the telecommunications revolution be utilized in problem solving in the Himalaya? What forms of communication should receive priority? What applications can be made in the context of education and local culture?
19. Analysis of pragmatic criteria for successful intersectoral and integrated action programmes. What new forms of action research are needed?
20. To what extent can holistic analysis of large-scale resource development projects be initiated?
21. What kinds of innovative planning models and approaches are most appropriate?

We re-emphasize, therefore, that there is no further value to be gained from debating whether we are facing an imminent supercrisis or a large array of regional and sub-regional crises. The socio-economic, political, and environmental problems will tend to come together. There is a present danger that, once a threshold is passed, then malnutrition, warfare, and environmental collapse, will feed upon each other. *This is supercrisis*! Yet we believe the issues can be defined, uncertainty can be both reduced and realigned, and supercrisis can be averted. Above all, as the quotation from Sunderlal Bahuguna would indicate, the Himalayan Problem is not a problem, or set of problems, unique to the Himalaya, it is an aspect of the non-functioning of late-twentieth-century civilization as a whole. Despite the enormous challenge facing the formulation of an effective regional approach, there is opportunity for a new initiative that can have repercussions far beyond this complex region. The essence of the Resolutions of the Mohonk Mountain Conference must carry the final word:

Resolution 1

The Mohonk Mountain Conference recognizes the special role of the Himalaya as a unique part of the world cultural heritage and wishes to draw international attention to the critical importance of its spiritual contribution to the well-being of the world community.

Resolution 2

The Mohonk Mountain Conference reaffirms that a serious situation has been developing in the Himalayan region for several decades. This relates to the progressive environmental deterioration and a pronounced decline in the standard of living for many of the peoples affected, particularly the mountain peoples. One aspect is the rapid increase in total population in relation to available agricultural and forest land. Taking this into consideration, it is resolved that an international conference be convened as soon as possible to further examine the issues and to recommend an urgent course of action.

It is also resolved that a small working group be formed to develop an action research design and to lay the groundwork for the proposed conference. The working group should be formed by the International Mountain Society in co-operation with the United Nations University, the East-West Center, the International Union for the Conservation of Nature and Natural Resources, the International Centre for Integrated Mountain Development (ICIMOD), and other organizations.

Resolution 3

Realizing that nature recognizes no international boundaries and that many of the issues and challenges facing development and conservation cannot be dealt with adequately without co-operation between countries of the Himalayan region, the Mohonk Mountain Conference strongly urges the governments of the Himalayan region to take steps to establish international parks in border areas (Parks for Peace) to promote peace, friendship, and co-operation in research and management, for the optimal sustainable use of the natural and human resources, and to improve the quality of life of all the peoples of the region.

Resolution 4

The Mohonk Mountain Conference endorses efforts by the International Centre for Integrated Mountain Development (ICIMOD) to develop a documentation centre and to improve the dissemination of vital information on the region, particularly relating to hydrology and sediment transfer. It is recommended that these efforts be accelerated and that links be established with other appropriate institutions.

Resolution 5

The Mohonk Mountain Conference welcomes the recent initiative of the World Resources Institute, the World Bank, and the United Nations Development Programme in establishing a tropical forest action plan which should facilitate efforts to deal with the comprehensive land-use aspects of the problem facing the Himalayan region, and calls upon donors to provide the appropriate support.

NOTES

[1]Appropriate, in David Griffin's case, to professional silviculture, to external concern for the Himalaya and to explicit global science and, in Sunderlal Bahuguna's case, to indigenous ethno-silviculture, to internal concern for the Himalaya ('the toad beneath the harrow knows exactly where each tooth point goes' – Kipling), and to implicit knowledge-and-practice. Despite the remarkable divergence between the contexts of these two methodologies, they end up at virtually the same place. The recognition of their convergence, in terms of philosophy, practice, and implementation, was one of the great achievements of the Conference. We hear much of Appropriate Technology; perhaps it is time we heard more of Appropriate Methodology!

[2]For a discussion of the implications of this 'heterogeneity principle,' see Thompson *et al.*, 1986, Chapter 3.

[3]To deflect any accusation of ethno-centric prejudice in these inferences of probable observer incompetence, one of us, during an early experience as 'officer-in-charge' of a certain subarctic weather station (JDI), managed to intercept a practice by a well-trained Canadian graduate student on the midnight-to-8.00 a.m. observer shift: this entailed guestimating temperature, windspeed, and precipitation on winter nights that promised to be clear, preparing a teletype message for seven hours in advance with an excuse that it was sent 'late' because of a local power failure, and going to sleep for the shift to release the 'data' at 7.00 a.m. In this case, as the guestimated pre-dawn breeze picked up, the assumed slight mixing of air was matched with a rise in temperature, such that the false record would not have been detected. This man would not have erred on his snow conversion, as doubtlessly occurred at Namche.

POSTSCRIPT

The devastating Bangladesh floods of August-September, 1988 have been widely reported in the western news media as the result of deforestation in the Himalaya. This reporting has included further indications that extensive barrages and high dams are the likely future technological 'fix'. This once again underscores one of the central themes of this book — the dangers of expending vast sums of money on solving perceived problems based upon unproven assumptions.

REFERENCES

Acharya, M. and Bennett, L. (1983) *Women and the Subsistence Sector*, World Bank Staff Working Papers, No. 526, Washington DC. 140 pp.

ADB (1982) *Nepal Agriculture Sector Strategy Study*, 2 vols, Kathmandu, Nepal: Asian Development Bank.

Ahmad, E. (1960) 'The Indus – A study in river geography', *The Geographer* 8, Islamabad.

Allan, N. J. R. (1987) 'Impact of Afghan refugees on the vegetation resources of Pakistan's Hindu Kush - Himalaya', *Mountain Research and Development* 7(3): 200–4.

Andrews, C. (1983) 'Photographs and notes on tourism and deforestation in the Solu Khumbu, Nepal', *Mountain Research and Development* 3(2): 182–5.

APROSC (1978) *Agrarian Reform and Rural Development in Nepal*, Kathmandu: Agricultural Projects Services Centre.

ARTEP (1982) *A Challenge to Nepal: Growth and Employment*, Bangkok: Asian Regional Team for Employment Promotion.

Bahuguna, S. (1983) 'What man does to the mountain and to man', *Future* 1983(1): 6–11.

Bajracharya, D. (1983a) 'Deforestation in the food/fuel context: historical and political perspectives from Nepal', *Mountain Research and Development* 3(3): 227–40.

Bajracharya, D. (1983b) 'Fuel, food or forest? Dilemmas in a Nepali village', *World Development* 11(12): 1057–74.

Bandyopadhyay, J., Jayal, N. D., Schoettli, U. and Singh, C. (1985) *India's Environment, Crises and Responses*, Dehra Dun, U. P., India: Natraj Publishers.

Bandyopadhyay, J. and Shiva, V. (1984) Ecological constraints on mineral resource utilization in the Himalaya: The case of Doon Valley.

Bansal, R. C. and Mathur, H. N. (1976) 'Landslides - the nightmare of hill roads', *Soil Conservancy Digest* 4(1): 36–7.

Barry, R. G. (1981) *Mountain Weather and Climate*, London and New York: Methuen.

Barsch, D. and Caine, N. (1984) 'The nature of mountain geomorphology', *Mountain Research and Development* 4(4): 287–98.

Beckwith, C. I. (1987) *The Tibetan Empire in Central Asia*, Princeton, NJ: Princeton University Press.

Beenhakker, A. (1973) *A Kaleidoscopic Circumspection of Development*, Rotterdam: Rotterdam University Press.

Benedict, J. B. (1970) 'Downslope soil movement in a Colorado alpine region: rates, processes and climatic significance', *Arctic and Alpine Research* 2: 165–226.

Bishop, B. C. (1978) 'The changing geoecology of Karnali Zone, Western Nepal

Himalaya: A case of stress', *Arctic and Alpine Research*, 10(2): 531–43.
Bjønness, I.-M. (1980a) 'Animal husbandry and grazing, a conservation and management problem in Sagarmatha (Mt. Everest) National Park, Nepal', *Norsk Geografisk Tidsskr*. 34: 59–76.
Bjønness, I.-M. (1980b) 'Ecological conflicts and economic dependency on tourist trekking in Sagarmatha (Mt. Everest) National Park, Nepal. An alternative approach to park planning', *Norsk Geografisk Tidsskr*. 34: 119–38.
Bjønness, I.-M. (1982) 'Kulekhani Hydro-Electric Project: Research Working Paper No. 1: A conceptual framework for analysis of the effects from the Kulekhani Hydro-Electric Project, Nepal', Oslo: University of Oslo, Dept. of Geography.
Bjønness, I.-M. (1983a) 'Kulekhani Hydro-Electric Project: Research Working Paper No. 2: Energy and development policy, planning, ideology and reality', Oslo: University of Oslo, Dept. of Geography.
Bjønness, I.-M. (1983b) 'Kulekhani Hydro-Electric Project: Research Working Paper No. 3: Socio-economic analysis of the effects from the Kulekhani Hydro-Electric Project, Nepal', Oslo: University of Oslo, Dept. of Geography.
Bjønness, I.-M. (1984) 'Kulekhani Hydro-Electric Project: Research Working Paper No. 4: Strategies for survival – subsistence agriculture versus work at the Kulekhani Hydro-Electric Project, Nepal', Oslo: University of Oslo, Dept. of Geography.
Bjønness, I.-M. (1986) 'Mountain hazard perception and risk-avoiding strategies among the Sherpas of Khumbu Himal, Nepal', *Mountain Research and Development* 6(4): 277–92.
Bjønness, I.-M. (1987) 'A conceptual framework for analysis of the socio-economic and environmental effects of the Kulekhani Hydro-Electric Project, Nepal', *Mountain Research and Development* 7(3): 205–8.
Blaikie, P. M. (1985) *The political economy of soil erosion in developing countries,* London: Longman.
Blaikie, P. and Brookfield, H. (1987) *Land Degradation and Society*, London and New York: Methuen.
Blaikie, P., Cameron, J. and Seddon, D. (1980) *Nepal in Crisis: Growth and Stagnation at the Periphery*, Oxford: Clarendon Press.
Bosch, J. M. and Hewlett, J. D. (1982) 'A review of catchment experiments to determine the effect of vegetation changes on water yield and evaporation', *Journal of Hydrology* 55(1): 3–23.
Boughton, W. C. (1970) *Effects of land management on quantity and quality of available water*, New South Wales, Australia: Univ. New South Wales, Water Res. Lab. Rept., No. 120.
Bouhdiba, A. (1982) *Exploitation of Child Labour*, New York: United Nations, 44 pp.
Brunsden, D. and Thornes, J. B. (1979) 'Landscape sensitivity and change', *Trans. Inst. Brit. Geographers* 4: 463–84.
Brunsden, D., Jones, D. K., Martin, R. P. and Doornkamp, J. C. (1981) 'The geomorphological character of part of the Low Himalaya of eastern Nepal', *Zeits. Geomorph. N. F.*, Suppl.-Bd. 37: 25–72.
Byers, A. (1986) 'A geomorphic study of man-induced soil erosion in the Sagarmatha (Mount Everest) National Park, Khumbu, Nepal', *Mountain Research and Development* 6 (1): 83–7.
Byers, A. (1987a) 'Landscape change and man-accelerated soil loss: The case of the Sagarmatha (Mount Everest) National Park, Khumbu, Nepal', *Mountain Research and Development* 7(3): 209–16.
Byers, A. (1987b) 'An assessment of landscape change in the Khumbu region of Nepal using repeat photography', *Mountain Research and Development* 7(1): 77–81.
Byers, A. (1987c) A geomorphic study of man-induced soil-erosion in the Sagarmatha

(Mount Everest) National Park, Khumbu, Nepal. Unpublished PhD thesis, Department of Georgraphy, University of Colorado, Boulder.
Byers, A., Thorn, C. E. and Ives, J. D. (1985) 'Man-induced accelerated surficial erosion, upper Khumbu Valley, Sagarmatha National Park, Nepal', in T. Spencer (ed.), *Abstracts of Papers for the First International Conference on Geomorphology*.
Caine, N. (1974) 'The geomorphic processes of the alpine environment', in J. D. Ives and R. G. Barry (eds), *Arctic and Alpine Environments*, London: Methuen.
Caine, N. and Mool, P. K. (1981) 'Channel geometry and flow estimates for two small mountain streams in the Middle Hills, Nepal', *Mountain Research and Development* 1(3–4): 231–43.
Caine, N. and Mool, P. K. (1982) 'Landslides in the Kolpu Khola drainage, Middle Mountains, Nepal', *Mountain Research and Development* 2(2): 157–73.
Campbell, J. G. (1979) 'Community involvement in conservation: social and organizational aspects', in *Resource Conservation and Utilization Project*, Kathmandu: Agricultural Projects Services Centre.
Caplan, P. (1970) *Land and Social Change in East Nepal*, London: Routledge & Kegan Paul.
Carson, B. (1984) 'Geomorphic research in Nepal and its role in watershed management', Kathmandu, Nepal: unpublished paper.
Carson, B. (1985) Erosion and Sedimentation Processes in the Nepalese Himalaya. International Centre for Integrated Mountain Development (ICIMOD), Occasional Paper No. 1, Kathmandu, Nepal, 39 pp.
CEDA (1979) *The Status of Women in Nepal*, Kathmandu: Centre for Economic Development and Administration.
Central Soil and Water Conservation Research and Training Institute (1976) *Annual Report*, Dehra Dun.
Centre for Science and Environment (1985) *The State of India's Environment, 1984–85: The Second Citizens' Report*, New Delhi, India: Centre for Science and Environment.
Chatra Research Centre (1976). Results of runoff and erosion experiments on small catchments, Annual Report, Kathmandu, Nepal.
Chatterji, S. (1987) 'Development prospects in Ladakh', *Mountain Research and Development* 7(3): 217–18.
Christiansen, T. (1982) 'Bodenerosion und ihre einflussfaktoren im gebiet des Indo-German Dhauladhar Projekts', Diplomarbeit Universität Giessen.
Coburn, B. (1983) 'Managing a Himalayan world heritage site', *Nature and Resources* 29(3): 20–5.
Collingridge, D. (1981) *The Social Control of Technology*, Milton Keynes and London: The Open University Press.
CWCC (Child Welfare Coordinating Committee) of UNICEF (1980) *Status of Children in Nepal*, Kathmandu, Nepal: UNICEF, 300 pp.
Dabadghao, P. M. and Shankaranarayan, K. A. (1973) *The Grass Cover of India*, New Delhi, India: ICAR.
Dani, A. and Campbell, J. G. (eds) (1988) *Rural Institutions and Resource Management*, Kathmandu: International Centre for Integrated Mountain Development (ICIMOD).
Dani, A. A., Gibbs, C. J. N. and Bromley, D. W. (1987) 'Institutional Development for Local Management of Rural Resources', Honolulu: East–West Center Environment and Policy Institute, Workshop Rept. No. 2.
DANIDA (1980) 'Ways and means of improving livestock production and productivity in the DANIDA-assisted integrated rural development project', Noakali District, Bangladesh: internal report.

Das, P. K. (1983) 'The climate of the Himalayas', in T. V. Singh and J. Kaur (eds) *Himalayas: Mountains and Man*, 1–9, Lucknow: Print House (India).
Data Systems Incorporated (1980) 'Survey data from Dwarahat Block, Almora District', Lucknow: Government of Uttar Pradesh report.
Davidson, J. (1986) 'Setting aside the idea that Eucalyptus are always bad', FAO/UN Assistance to the Forestry Sector of Bangladesh, Working Paper No. 10, May 1985, 26 pp.
Defence for Children (1982) *Self Reliance*, Geneva: Defence for Children, 251 pp.
Dhakal, D. N. S. (1987) 'Twenty five years of development in Bhutan', *Mountain Research and Development* 7(3): 219–21.
Dhar, O. N. and Mandal, B. N. (1986) 'A pocket of heavy rainfall in the Nepal Himalayas – A brief appraisal', in S. C. Joshi (ed.) *Nepal Himalaya: Geo-Ecological Perspectives*, Nainital, UP, India: Himalayan Research Group.
Dobremez, J-F. (1976) *Le Népal: Écologie et Biogéographie*, Paris: Centre National de la Recherche Scientifique.
Doherty, V. (1975) 'Kinship and Economic Choice', PhD thesis, University of Wisconsin, Madison.
Donovan, D. G. (1981) 'Fuelwood: how much do we need?' Newsletter (DGD 14), Inst. of Current World Affairs, Hanover, N.H., USA. Unpub. mimeo, 23 pp.
Eckholm, E. (1975) 'The deterioration of mountain environments', *Science* 189: 764–70.
Eckholm, E. (1976) *Losing Ground*, New York: Worldwatch Institute, W. W. Norton & Co., Inc.
Ehrich, R. (1980) *Living Conditions and Potential Economic Activities of Women*, Palampur: German Agency for Technical Cooperation (GTZ).
Elder, J. W. (1974) 'Planned resettlement in Nepal's Terai (a social analysis of the Khajura/Bardia Punarvas projects)', Kathmandu: CEDA, Tribhuvan University.
English, R. (1985) 'Himalayan state formation and the impact of British Rule in the nineteenth century', *Mountain Research and Development* 5(1): 61–78.
ESCAP (1984) *Statistical Indicators (June)*, Bangkok: ESCAP.
FAO/UNESCO (1977) Soil Map of the World, vol. V112 1:5,000,000. Paris, UNESCO.
FAO/UNFPA/ILASA (1982) 'Potential population supporting capacities of lands in the developing world', Technical report of Project FPA/INT/513 Land Resources for Populations of the Future 15 Karten, 1: 10,000,000. Rome, FAO.
FAO (1985) 'Action Programme on Institutions', Rome: FAO, FO:FDT/AP/5.
Fisher, J. F. (1978) *Himalayan Anthropology: Indo-Tibetan Interface*, The Hague: Mouton.
Fisher, J. F. (1986) *Trans-Himalayan Traders: Economy, Society, and Culture in Northwest Nepal*, Berkeley: Univ. of California Press.
Fleming, W. M. (1978) 'Classification of catchments in the Western Development Region of Nepal', Kathmandu: UNDP/FAO Working Paper No. 10.
Fort, M. B., and Freytet, P. (1982) 'The quaternary sedimentary evolution of the intra-montane basin of Pokhara in relation to the Himalaya Midlands and their hinterland (West Central Nepal),' in A. K. Sinha (ed.), *Contemporary Geoscientific Researches in Himalaya*, Vol. 2, pp. 91–6, Dehra Dun, India.
Fox, J. M. (1983) Managing Public Lands in a Subsistence Economy: The Perspective from a Nepali Village. Unpub. PhD thesis, Department of Agric. Econ., University of Wisconsin, Madison, 246 pp.
Fricke, T. E. (1986) *Himalayan Households: Tamang Demography and Domestic Processes*, Ann Arbor, Michigan: UMI Research Press.
Fürer-Haimendorf, C. von (1964) *The Sherpas of Nepal: Buddhist Highlanders*, London: John Murray.
Fürer-Haimendorf, C. von. (1975) *Himalayan Traders*, New York: St Martin's Press.

Fürer-Haimendorf, C. von (1984) *The Sherpas Transformed: Social Change in a Buddhist Society of Nepal*, New Delhi: Sterling Publ. Private Ltd.
Fushimi, H. and Ohata, T. (1980) 'Fluctuations of glaciers from 1970–1978 in the Khumbu Himal', *Seppyo* 41: 71–80.
Galay, V. (1986) 'Glacier lake outburst flood (jökulhlaup) on the Bhote/Dudh Kosi – August 4, 1985', internal report, Water and Energy Commission, Kathmandu.
Gansser, A. (1966) 'Geological research in the Bhutan Himalaya', in *The Mountain World 1964/1965*, Zurich: Swiss Foundation for Alpine Research.
Garg, S. P. (1971) 'Soil erosion in the Sub-Himalayan region – a case study', *Deccan Geographer* 9(1): 7–14.
Garrett, W. E. (1987) Editorial, *National Geographic Magazine*, November 1987, Washington, DC.
Gilmour, D. A. (1977) 'Effect of logging and clearing on water quality in a high rainfall zone of north-east Queensland', in *The Hydrology of Australia*, Institution of Engineers, Australia, National Conference Publication No. 77/5.
Gilmour, D. A. (1986) 'Reforestation and afforestation of open land – a Nepal perspective', in A. J. Pearce and L. S. Hamilton (eds), *Land use, watersheds, and planning in the Asia-Pacific region*, Bangkok: FAO, RAPA Report, 157–69.
Gilmour, D. M. (1988, in press) 'Not seeing the trees for the forest: a re-appraisal of the deforestation crisis in two hill districts of Nepal, *Mountain Research and Development* 8(4).
Gilmour, D. A., Bonell, M. and Cassells, D. S. (1987) 'The effects of forestation on soil hydraulic properties in the Middle Hills of Nepal: A preliminary assessment', *Mountain Research and Development* 7(3): 239–49.
Goldstein, M. C. (1977) 'Culture, population, ecology and development: a view from Northwest Nepal', Paris: Proceedings of the C.N.R.S. International Conference on the Ethnology of the Himalayas: 481–9.
Goldstein, M. C. (1981) 'High-altitude Tibetan populations in the remote Himalaya: social transformation and its demographic, economic, and ecological consequences', *Mountain Research and Development* 1(1): 5–18.
Goldstein, M. C., Ross, J. L. and Schuler, S. (1983) 'From a mountain-rural to a plains-urban society: implications of the 1981 Nepal Census', *Mountain Research and Development* 3(1): 61–4.
Gosh, S. K. (1983) 'Estimation of design period of the estimated design flood in a region', Central Board of Irrigation and Power: *Procedures*, Vol II, New Delhi.
Goswami, D. C. (1983) 'Bramaputra River Assam (India): Suspended sediment transport, valley aggradation and basin denudation', PhD thesis, Johns Hopkins University, University Microfilms International: Ann Arbor (MI).
Goswami, D. C. (1985) 'Brahmaputra River, Assam, India: Physiography, basin denudation, and channel aggradation', *Water Resources Research* 21(7): 959–78.
G.B. Pant University (1976) 'Rural area development, research, planning and action in Naurer Watershed in Bikiasen Block of Almora District, U.P.' Pantnagar: G.B. Pant University report.
G.B. Pant University (1980) 'Integrated natural and human resource planning and management in the hills of U.P.' Pantnagar: Progress report of subproject: Study of Grassland and Livestock Resources Management in the Kumaun Hills.
Greenwold, S. (1974) 'Monkhood versus priesthood in Newar Buddhism', in C. von Fürer-Haimendorf (ed.), *Contributions to the Anthropology of Nepal*, Warminster, England: Aris and Phillips, pp. 129–49.
Griffin, D. M. (1987) 'Implementation failure caused by institutional problems', *Mountain Research and Development* 7(3): 250–3.
Griffin, D. M., Shepherd, K. R. and Mahat, T. B. S. (1988) 'Human impacts on some forests of the Middle Hills of Nepal. Part 5: Comparisons, concepts, and some

policy implications', *Mountain Research and Development* 8(1): 43–52.
Gupta, P. N. (1979) 'Afforestation, integrated watershed management, torrent control and land use development project for the U.P. Himalayas and Siwaliks', Lucknow: U.P. Forest Department, internal report.
Gurung, H. B. (1981a) 'Ecological change in Nepal: A native interpretation', *New Era*, Occnl. Paper 1, Kathmandu.
Gurung, H. B. (1981b) 'Study on inter-regional migration in Nepal', Kathmandu: National Commission on Population, National Planning Commission.
Gurung, H. B. and Khanal, N. R. (1987) 'Landscape processes in the Chure Range, Central Nepal', unpub. rept. submitted to National Committee for Man and the Biosphere, Kathmandu, Nepal.
Gurung, S. B. (1973) 'Causes and impacts of migration (preliminary survey in Surkhet Region)', Kathmandu: CEDA, Tribhuvan University.
Gurung, Sumitra M. (1988) Beyond the myth of eco-crisis in Nepal: local response to pressure on land in the Middle Hills. Unpub. PhD thesis presented to University of Hawaii, Honolulu.
Hagen, T. (1960) *Nepal, Königreich am Himalaya*, Bern: Kümmerly und Frey.
Hagen, T. (1963) 'The evaluation of the highest mountain in the world', in *Mount Everest*, London: Oxford University Press.
Hagen, T. (1969) 'Report on the geological survey of Nepal', *Denkscheiften der Schweizerischen Naturforschenden Gesellschaft*, 86(1): 1–185.
Haigh, M. J. (1982a) 'Road development and rural stresses in the Indian Himalaya', *Nordia* 16: 135–40.
Haigh, M. J. (1982b) 'A comparison of sediment accumulations beneath forested and deforested micro-catchments, Garhwal Himalaya', *Himalayan Research and Development* 1(11): 118–20.
Haigh, M. J. (1984a) 'Deforestation and disaster in northern India', *Land Use Policy*, July 1984: 187–98.
Haigh, M. J. (1984b) 'Landslide prediction and highway maintenance in the Lesser Himalaya, India', *Zeits, für Geomorphologie*, N. F. Suppl-Bd. 51:17–37.
Hamilton, L. S. (ed.) (1983) *Forest and Watershed Development and Conservation in Asia and the Pacific*, Boulder, Colorado: Westview Press.
Hamilton, L. S. (1985) 'Overcoming myths about soil and water impacts of tropical forest land-uses', in S. A. El-Swaify, W. C. Moldenhauer and A. Lo (eds), *Soil Erosion and Conservation*, Ankeny: Soil Conservation Society of America.
Hamilton, L. S. (1987) 'What are the impacts of Himalayan deforestation on the Ganges-Brahmaputra lowlands and delta? Assumptions and facts', *Mountain Research and Development* 7(3): 256–63.
Hammond, June E. (1988) Glacial lakes in the Khumbu region, Nepal: an assessment of the hazards. Unpub. Master's thesis presented to the University of Colorado, Boulder.
Hatley, T. and Thompson, M. (1985) 'Rare animals, poor people, and big agencies: a perspective on biological conservation and rural development in the Himalaya', *Mountain Research and Development* 5(4): 365–77.
Heuberger, H., Masch, L., Preuss, E. and Schröcker, A. (1984) 'Quaternary landslides and rock fusion in Central Nepal and in the Tyrolean Alps', *Mountain Research and Development* 4(4): 345–62.
Hewitt, K. (1964) 'A Karakoram ice dam', *Indus: J. Water and Power Devel. Authority* (Pakistan) 5: 18–30.
Hewitt, K. (1982) 'Natural dams and outburst floods of the Karakoram Himalaya', in J. W. Glen (ed.), *Hydrological Aspects of Alpine and High Mountain Areas.* IAHS Publ. No. 138, 259–69.
Hewitt, K. (1985) *Snow and Ice Hydrology in Remote High Mountain Areas: the*

Himalayan Sources of the River Indus, Snow and Ice Hydrology project, Working Paper No. 1, Waterloo, Ontario: Wilfrid Laurier University.

Hewitt, K. (ed.) (1986) Snow and Ice Hydrology Project, Annual Report, Waterloo, Ontario: Wilfrid Laurier University.

Hewitt, K. (ed.) (1987) Snow and Ice Hydrology Project, Annual Report, Waterloo, Ontario: Wilfrid Laurier University.

Hewlett, J. D. (1982) 'Forests and floods in the light of recent investigations', in Proceed. Hydrological Symposium, 14–15 June 1982. Fredericton, Canada: Natl. Res. Council, 543–60.

Hinrichsen, D., Lucas, P. H. C., Coburn, B. and Upreti, B. N. (1983) 'Saving Sagarmatha', *Ambio* 11(5): 274–81.

Hofer, A. (1979) *The Caste Hierarchy and the State in Nepal*, Innsbruck: Universitatsverlag Wagner.

Horton, D. (1987) *Potatoes: Production, Marketing, and Programs for Developing Countries*. Boulder, Colorado: Westview Press.

Houston, C. S. (1982) 'Return to Everest – a sentimental journey', *Summit* 28: 14–17.

Houston, C. S. (1987) 'Deforestation in Solu Khumbu', *Mountain Research and Development* 7(1): 76.

HMG Nepal (1976) National Forest Plan, Kathmandu, Ministry of Forests.

HMG Nepal (1982) The Decentralization Act of 1982, Kathmandu.

Hrabovszky, J. P. and Miyan, K. (1987) 'Population growth and land use in Nepal', *Mountain Research and Development* 7(3): 264–70.

Hurni, H. (1983) 'Soil erosion and soil formation in agricultural ecosystems: Ethiopia and Northern Thailand', *Mountain Research and Development* 3(2): 131–42.

Hurni, H. and Messerli, B. (1981) 'Mountain research for conservation and development in Simen-Ethiopia', *Mountain Research and Development* 1(1): 49–54.

IDS (1983) *Foreign Aid and Development in Nepal*, Proceedings of a Seminar, 4–5 October 1983: Kathmandu: Integrated Development Systems.

Ikegami, K. and Inoue, J. (1978) 'Mass balance studies on Kongma Glacier, Khumbu Himal', *Seppyo* 40: 12–16.

Imhof, E. (1974) *Die Grossen Kalten Berge von Szetschuan*, Zurich: Orell Fussli Verlag.

Impat, P. (1981) Hydrometeorology and sediment data for Phewa Watershed: 1979 data. Integrated Watershed Management Project, Dept. Soil Conservation and Watershed Management, Ministry of Forests, Kathmandu, Phewa Tal Techn. Rept. No. 14, 87 pp.

Indian Meterological Department (1971) *Monthly and Annual Rainfall and Number of Rainy Days (Period 1901–1950),* Part 1, New Delhi.

Ives, J. D. (1970) 'Himalayan highway', *Canadian Geographical Journal* 80(1): 26–31.

Ives, J. D. (1980) *Geoecology of the Colorado Front Range: A Study of Alpine and Subalpine Environments*, Boulder, Colorado: Westview Press.

Ives, J. D. (1981) 'High mountains and plateaus: an excursion to the Roof of the World', *Mountain Research and Development* 1(1): 79–83.

Ives, J. D. (1985) 'Yulongxue Shan, Northwest Yunnan, People's Republic of China: a geoecological expedition', *Mountain Research and Development* 5(4): 382–5.

Ives, J. D. (1986) *Glacial lake outburst floods and risk engineering in the Himalaya*, Kathmandu: International Centre for Integrated Mountain Development (ICIMOD), Occasional Paper No. 5.

Ives, J. D. (1987) 'Repeat photography of debris flows and agricultural terraces in the Middle Mountains, Nepal', *Mountain Research and Development* 7(1): 82–6.

Ives, J. D. and Ives, P. (1987) *The Himalaya-Ganges Problem*, Proceedings of a Conference, Mohonk Mountain House, New Paltz, NY, 6–11 April 1986, Boulder,

Colorado: International Mountain Society – *Mountain Research and Development* 7(3): 181–344.

Ives, J. D. and Messerli, B. (1981) 'Mountain hazards mapping in Nepal: introduction to an applied mountain research project', *Mountain Research and Development* 1(3–4): 223–30.

Ives, J. D., Messerli, B. and Thompson, M. (1987) 'Research strategy for the Himalayan region: Conference conclusions and overview', in J. D. Ives and P. Ives, Proceedings of the Mohonk Mountain Conference – The Himalaya-Ganges Problem, *Mountain Research and Development* 7(3): 332–44.

Iwata, S., Sharma, T. and Yamanaka, H. (1984) 'A preliminary report of Central Nepal and Himalayan uplift', *Journal Nepalese Geological Society* 4: 141–49.

Jackson, M. J. (1983) 'Livestock in the economy of the Himalayan foothills, Uttar Pradesh, India', Nainital, UP, India: Unpub. paper presented to conference on Himalayan Environmental Degradation, October 1983.

Jeffries, B. E. (1982) 'Sagarmatha National Park: the impact of tourism in the Himalayas', *Ambio* 11(5): 274–81.

Johnson, K., Olson, E. A. and Manandhar, S. (1982) 'Environmental knowledge and response to natural hazards in mountainous Nepal', *Mountain Research and Development* 2(2): 175–88.

Joshi, B. C. (1987) 'Geo-environmental studies in parts of Ramganga catchment, Kumaon Himalayas', Unpub. PhD thesis, University of Roorkee, Roorkee, India.

Joshi, S. C. (ed.) (1986a) *Nepal Himalaya: Geo-ecological Perspectives*, Nainital, UP, India: Himalayan Research Group.

Joshi, S. C. (1986b) 'Nepal Himalaya: A physiographic appraisal', in S. C. Joshi (ed.). *Nepal Himalaya: Geo-Ecological Perspectives*, pp. 51–63, Nainital, UP, India: Himalayan Research Group.

Kanai, H. (1966) 'Phytogeography of Eastern Himalaya, with special reference to the relationship between Himalaya and Japan', in H. Hara (ed.), *The Flora of Eastern Himalaya* (results of the botanical expedition to Eastern Himalaya organized by the University of Tokyo, 1960 and 1963) 13–38.

Karan, P. P. (1984) *Sikkim Himalaya: Development in Mountain Environment*, Monumenta Serindica 13, Tokyo: Institute for the Study of Languages and Cultures of Asia and Africa.

Karan, P. P. (1987a) 'Development Issues in Sikkim and Bhutan', *Mountain Research and Development* 7(3): 275–8.

Karan, P. P. (1987b) 'Population characteristics of the Himalayan region', *Mountain Research and Development* 7(3): 271–4.

Kardell, L., Steen, E. and Fabiao, A. (1986) 'Eucalyptus in Portugal – A threat or a promise?', *Ambio* 15(1): 6–13.

Kattelmann, R. (1987) 'Uncertainty in assessing Himalayan water resources', *Mountain Research and Development* 7(3): 279–86.

Kawakita, J. (1956) 'Vegetation', in H. Kihara (ed.), *Land and Crops of Nepal Himalaya*, Vol. 2, 1–65, Kyoto: Kyoto University.

KHARDEP (1979) *A Study of the Socioeconomy of the Kosi Hills*, Surbiton, England: Kosi Hills Rural Development Project, Land Resources Development Centre.

Kienholz, H., Hafner, H., Schneider, G. and Tamrakar, R. (1983) 'Mountain hazards mapping in Nepal's Middle Mountains with maps of land use and geomorphic damages (Kathmandu-Kakani area)', *Mountain Research and Development* 3(3): 195–220.

Kienholz, H., Schneider, G., Bichsel, M., Grunder, M., and Mool, P. (1984) 'Mapping of mountain hazards and slope stability', *Mountain Research and Development* 4(3): 247–66.

Kollmannsperger, F. (1978/79) 'Long range landscape changes under the influence of man', *Journal Nepalese Research Centre* (Sciences) 2/3: 11–31.

Kuhn, T. S. (1970) *The Structure of Scientific Revolutions*, Oxford: Oxford University Press.

Laban, P. (1978) 'Field measurements on erosion and sedimentation in Nepal', Dept. of Soil Conservation and Watershed Management, Kathmandu (1WM/SP/05).

Laban, P. (1979) 'Landslide Occurrence in Nepal', Phewa Thewa Tal Project Report No. SP/13, Integrated Watershed Management Project, Kathmandu.

Lall, J. S., and Moddie, A. D. (eds) (1981) *The Himalaya: Aspects of Change*, New Delhi: Oxford University Press.

Lauer, W. (ed.) (1984) *Natural Environment and Man in Tropical Mountain Ecosystems*, Routledge & Kegan Paul.

Liu, D. and Sun, H. (eds) (1981) *Proceedings of Symposium on Qinghai-Xizang (Tibet) Plateau*, Beijing: Science Press, and New York: Gordon and Breach (2 vols).

Low, B. L. (ed.) (1968) *Mountains and Rivers of India*, Prep. for 21st Internat. Geogr. Congress, India. New Delhi.

Macfarlane, A. (1976) *Resources and Population: A Study of the Gurungs of Nepal*, Cambridge: Cambridge University Press, 364 pp.

McNeely, J. A. and Pitt, D. (eds) (1985) *Culture and Conservation*, London: Croom Helm.

Mahat, T. B. S. (1985) 'Human Impact on Forests in the Middle Hills of Nepal', Unpub. doctoral dissertation. Submitted to Australian National University, January 1985, Canberra (2 vols).

Mahat, T. B. S., Griffin, D. M. and Shepherd, K. R. (1986a) 'Human impact on some forests of the Middle Hills of Nepal. I: Forestry in the context of the traditional resources of the state', *Mountain Research and Development* 6(3): 223–32.

Mahat, T. B. S., Griffin, D. M. and Shepherd, K. R. (1986b) 'Human impact on some forests of the Middle Hills of Nepal. II: Some major human impacts before 1950 on the forests of Sindhu Palchok and Kabhre Palanchok', *Mountain Research and Development* 6(4): 325–34.

Mahat, T. B. S., Griffin, D. M. and Shepherd, K. R. (1987a) 'Human impact on some forests of the Middle Hills of Nepal. III: Forests in the subsistence economy of Sindhu Palchok and Kabhre Palanchok', *Mountain Research and Development* 7(1): 53–70.

Mahat, T. B. S., Griffin, D. M. and Shepherd, K. R. (1987b) 'Human impact on some forests of the Middle Hills of Nepal. IV: A detailed study in southeast Sindhu Palchok', *Mountain Research and Development* 7(2): 111–34.

Manzardo, A., Navin, K. R. and Dilli, R, D. (1975) 'Hill migration in Nepal: The effects of out-migration on a hill village in Far-Western Nepal', Kathmandu: Centre for Nepal and Asian Studies, Tribhuvan University.

Mathur, H. N. (1976) 'Effect of clearfelling and reforestation on runoff and peak rates in small watersheds', *Indian Forester* 102: 219–26.

Mayewski, P. A., Pregent, G. P., Jeschke, P. A. and Ahmad, N. (1980) 'Himalayan and Trans-Himalayan glacier fluctuations and the South Asian monsoon record', *Arctic and Alpine Research* 12(2): 171–82.

Messerli, B. and Ives, J. (1984) 'Gongga Shan (7556 m) and Yulongxue Shan (5596 m). Geoecological observations in the Hengduan Mountains of Southwestern China', *Erdwissenschaftliche Forschung* 153: 55–77. Franz Steiner Verlag Wiesbaden GmbH, Stuttgart.

Messerschmidt, D. A. (1976a) *The Gurungs of Nepal: Conflict and Change in a Village Society*, Warminster, England: Aris and Phillips.

Messerschmidt, D. A. (1976b) 'Ecological change and adaptation among the Gurungs of the Nepal Himalaya', *Human Ecology* 4(2): 167–85.

Messerschmidt, D. A. (1978) '*Dhikurs:* Rotating credit associations in Nepal', in J. F. Fisher (ed.) *Himalayan Anthropology: Indo-Tibetan Interface*, The Hague: Mouton.

Messerschmidt, D. A. (1981) 'Nogar and other traditional forms of cooperation in Nepal: Significance for development', *Human Organization* 40(1): 40–7.

Messerschmidt, D. A. (1982) 'The Thakali of Nepal: historical continuity and socio-cultural change,' *Ethnohistory* 29(4): 265–80.

Messerschmidt, D. A. (1985) 'Commentary on paper by P. B. Shah and H. Schreier, Agricultural land evaluation for national land-use planning in Nepal', *Mountain Research and Development* 5(2): 147–50.

Messerschmidt, D. A. (1987) 'Conservation and society in Nepal: Traditional forest management and innovative development', in P. D. Little and M. M. Horowitz (eds), *Lands at risk in the Third World*, Boulder, Colorado: Westview Press.

Ministry of Education and Culture (1984) *Recent Educational Development in Nepal*, Kathmandu.

Mishra, H. R. (1973) 'Conservation in Khumbu: the proposed Mt. Everest National Park; a preliminary report', Dept. of National Parks and Wildlife Conservation, Kathmandu.

Moench, M. and Bandyopadhyay, J. (1986) 'People–forest interaction: a neglected parameter in Himalayan forest management', *Mountain Research and Development* 6(1): 3–16.

Mohns, B. (1981) 'Agroforestry practices for improving degraded mountain ecosystems in Nepal', Unpub. thesis, Colorado State University, Fort Collins.

Murthy, Y. K. (1982) Quoted in Asian Development Bank, Nepal Agriculture Sector Strategy Study, Vol. 2, 84.

Myers, N. (1983) 'Tropical moist forests: over-exploited and under-utilized? *Forest Ecology and Management* 6(1): 59–79.

Myers, N. (1986) 'Environmental repercussions of deforestation in the Himalayas', *Journal World Forest Resource Management* 2: 63–72.

Nabarro, D. (1984) 'Social, economic, health and environmental determinants of nutritional status', *Food and Nutrition Bulletin* 6(1): 18–32.

Nakao, S. (1957) 'Ecological Notes', in H. Kihara (ed.). *Fauna and flora of Nepal Himalalya*, Vol. I, 278–90, Kyoto: Kyoto University.

Narayana, D. V. V. (1987) 'Downstream impacts of soil conservation in the Himalayan region', *Mountain Research and Development* 7(3): 287–98.

Narayana, D. V. V. and Rambabu (1983) 'Estimation of soil erosion in India', *Journal of Irrigation and Drainage Engineering* 109(4): 409–34.

Nelson, D. (1978) 'A reconnaissance inventory of the major ecological land units and their watershed condition', Unpub. paper, Department of Soil and Water Conservation, Ministry of Forests, Kathmandu, 12 pp.

Nelson, D. (1980) A Reconnaissance Inventory of the Major Ecological Land Units and their Watershed Condition – Summary Report. Dept. of Soil Conservation and Watershed Management. UNDP, FAO. FODP/NEP/74/020, Rome.

Nield, R. S. (1985) 'Fuelwood and fodder – problems and policy', Working paper for Water and Energy Commission Secretariat, Kathmandu, 44 pp.

Numata, M. (1966) 'Vegetation and conservation in Eastern Nepal', *Journal College Arts and Science, Chiba Univ. Nat. Sci.* 4(4): 559–69.

Numata, M. (1981) 'The altitudinal vegetation and climatic zones of the humid Himalayas', in D. Liu and H. Sun (eds), *Proceedings of Symposium on Qinhai-Xizang (Tibet) Plateau*, Beijing: Science Press, and New York: Gordon and Breach.

O'Flaherty, W. (1975) *Hindu Myths: A sourcebook translated from the Sanskrit*, Harmondsworth: Penguin.

Ohsawa, M., Shakya, P. R. and Numata, M. (1986) 'Distribution and succession of

West Himalaya forest types in the eastern part of the Nepal Himalaya', *Mountain Research and Development*, 6(2): 143–57.
O'Loughlin, C. L., and Will, G. M. (1981) 'The effects of exotic forestry' in *The Effects of Land Use on Water Quality – a review*, Water and Soil Miscellaneous Publication, No. 23, Wellington: National Soil and Water Conservation Organization, 31–33.
Østrem, G. (1974) 'Present alpine ice cover', in J. D. Ives and R. G. Barry (eds), *Arctic and Alpine Environments*, London: Methuen.
Panday, D. R. (1983) 'Foreign Aid in Nepal's Development: an Overview', in IDS, *Foreign Aid and Development in Nepal*, Kathmandu: Integrated Development Systems, 270–326.
Pant, S. C. (1935) *The Social Economy of the Himalayas*, London: George Allen & Unwin.
Pant, Y. P. (1970) *Problems in fiscal and monetary policy: A case study of Nepal*, London: Hurst.
Parthasarathy, B. and Mooley, D. A. (1978) 'Some features of long homogeneous series of Indian summer monsoon rainfall', *Monthly Weather Review* 4(78): 771–8.
Peters, Tj. and Mool, P. K. (1983) 'Geological and petrographic base studies for the Mountain Hazards Mapping Project in the Kathmandu-Kahani area, Nepal', *Mountain Research and Development* 3(3): 221–6.
Pignede, B. (1966) *Les Gurungs*, Paris: Mouton.
Pitt, D. C. (1970) *Tradition and Economic Progress*, Oxford: Clarendon Press.
Pitt, D. C. (1976) *Social Dynamics of Development*, Oxford: Pergamon.
Pitt, D. C. (1983) 'The socioeconomic context of migrants and minorities', *Journal of Biosocial Science*, Supplement 8: 111–28.
Pitt, D. C. (1986) 'Crisis, pseudocrisis, or supercrisis? Poverty, women, and young people in the Himalaya', *Mountain Research and Development* 6(2): 119–31.
Pitt, D. C. and Shah, A. A. (eds) (1982) *Child Labour*, Geneva: Defence for Children.
Poore, M. E. D. and Fries, C. (1985) *The Ecological Effects of Eucalyptus*, FAO Forestry Paper 59, Rome: FAO/UN.
Pradhan, B. and Shrestha, I. (1983) 'Foreign Aid and Women', in I.D.S., *Foreign Aid and Development in Nepal*, Kathmandu: Integrated Development Systems: 99–154.
Qian, N. and Dai, D. (1980) 'The problems of river sedimentation and the present status of its research in China', in *Proceedings of the International Symposium on River Sedimentation, 24–29 March 1980*, Beijing: The Chinese Society of Hydraulic Engineering (Vol. 1), Guanghua Press (UNESCO), 19–39.
Rambabu (1984) 'Soil loss prediction research in India', Dehra Dun: Central Soil and Water Conservation Research and Training Institute (CSWCRTI), internal report.
Rambabu, Tejwani, K. G., Agarwal, M. C. and Bhushan, L. S. (1978) 'Distribution of erosion index and iso-erodent map of India', *Indian Journal Soil Conservation* 6(1): 1–12.
Ramsay, W. J. H. (1985) 'Erosion in the Middle Himalaya, Nepal, with a Case Study of the Phewa Valley'. Unpub. MSc thesis, Dept. Forest Resources Management, Univ. British Columbia, Vancouver.
Ramsay, W. J. H. (1986) 'Erosion problems in the Nepal Himalaya – an overview', in S. C. Joshi (ed.), *Nepal Himalaya: Geoecological Perspectives*, Nainital, UP, India: Himalayan Research Group.
Rapp, A. (1960) 'Recent development of mountain slopes in Karkevagge and surroundings, northern Scandinavia', *Geografiska Annaler* 42: 65–201.
Redclift, M. (1984) *Development and the Environmental Crisis: Red or Green Alternatives*? London and New York: Methuen.
Reid, H. F., Smith, K. R. and Sherchand, B. (1986) 'Exposure to indoor smoke from

traditional and improved cookstoves: comparisons among rural Nepali women', *Mountain Research and Development* 6(4): 293–304.

Regmi, M. C. (1963–68) *Land Tenure and Taxation in Nepal*, Berkeley: University of California (4 vols).

Regmi, M. C. (1964) *The Land Grant System: Birta Tenure: Land Tenure and Taxation in Nepal*, Berkeley: University of California Inst. of International Studies.

Regmi, M. C. (1978) *Thatched Huts and Stucco Palaces: Peasants and Landlords in Nineteenth Century Nepal*, New Delhi: Vikas.

Reiter, E. R. (1981) 'The Tibet Connection', *Natural History* 90(9): 65–70.

Richards, J. F. (1987) 'Environmental changes in Dehra Dun valley, India: 1880–1980', *Mountain Research and Development* 7(3): 299–304.

Rieger, H. C. *et al.* (1976) 'Himalayan ecosystem research mission', *Nepal Report*, Südasieninstitut, Heidelberg.

Rose, L. E. (1971) *Nepal: Strategy for Survival*, Berkeley: University of California Press.

Sacherer, J. (1979) *Practical Problems of Development in Two Panchayats of North-Central Nepal*, Kathmandu: SATA/IHDP (mimeo).

Sacherer, J. (1980) *Health, Education, Nutrition and Family Planning Surveys*, Kathmandu: SATA/IHDP (mimeo).

Schroeder, R. and Schroeder, E. (1979) 'Women in Nepali agriculture', *Journal of Development and Administrative Studies* 1(2): 178–92.

Schweinfurth, U. (1957) 'Die horizontale und vertikale verbreitung der vegetation im Himalaya. Mit mehr farbiger vegetationskarte 1:20 Mio', *Bonner Geogr.* Abh., H. 20.

Schweinfurth, U. (1985) 'F. Kingdon Ward, 1885–1958: A commemorative note', *Mountain Research and Development* 5(4): 379–81.

Seddon, D. (1983) *Nepal: A State of Poverty*, Report to the International Labour Organization, Geneva.

Shah, P. B. and Schreier, H. (1985) 'Agricultural land evaluation for national land-use planning in Nepal: a case study in the Kailali District', *Mountain Research and Development* 5(2): 137–46.

Shanker, K. (1983) 'Hydrology requirements in Nepal', Kathmandu: Proceedings of a Seminar on Applied Remote Sensing Technology.

Sharma, C. K. (1983) *Water and Energy Resources of the Himalayan Block: Pakistan, Nepal, Bhutan, Bangladesh and India*, Kathmandu: Mrs. Sangeeta Sharma.

Sharma, K. K. (1983) 'The sequence of phased uplift of the Himalaya', in R. O. White (ed.), *The Evolution of the East Asian Environment*, Hong Kong: University of Hong Kong Press.

Sharp, D. and Sharp, T. (1982) 'The desertification of Asia', *Asia 2000* 1(4): 40–2.

Shiva, V. and Bandyopadhyay, J. (1985) 'Success for citizens' environmental action in the Doon Valley, Dehra Dun, India', *Mountain Research and Development* 5(3): 294.

Shiva, V. and Bandyopadhyay, J. (1986a) 'Ecology and the Politics of Survival (conflicts over natural resources in India)', Unpub. report for The Peace and Global Transformation Programme, Tokyo: United Nations University, 288 pp.

Shiva, V. and Bandyopadhyay, J. (1986b) 'The evolution, structure, and impact of the Chipko movement', *Mountain Research and Development* 6(2): 133–42.

Shiva, V., Sharatchandra, H. C. and Bandyopadhyay, J. (1981) *Social, economic, and ecological impact of social forestry in Kolar*, Indian Institute of Management, Bangalore, India (mimeo).

Shrestha, B. K. (1981) 'Technical Assistance and the Growth of Administrative Capability in Nepal', in IDS *Foreign Aid and Development in Nepal*, Kathmandu: Integrated Development Systems, 219–67.

Shrestha, B. P. and Mosin, M. (1970) *A Study in the Working of Gaon Panchayats*, Kathmandu: H.M.G. Panchayat Study Series, Volume II.

Shrestha, B. P. (1980) *Geology, Soil Erosion and Litho-Structural Division of Nepal*, Department of Soil and Water Conservation, Ministry of Forests, Kathmandu.

Shrestha, N. R. (1985) 'The political economy of economic underdevelopment and external migration in Nepal', *Political Geography Quarterly* 4(4): 280–306.

Singh, D. R., and Gupta, P. N. (1982) 'Assessment of siltation in Tehri reservoir', *Proc. Internatl. Symp. Hydrological Aspects of Mountainous Watersheds*, Univ. of Roorkee, Roorkee (4–6 November 1982). Mangalik Prakashan, Saharanpur, VIII–60 to VIII–66.

Singh, J. S. (ed.) (1985) *Environmental Regeneration in Himalaya: Concepts and Strategies*, Nainital, U.P., India: Central Himalayan Environment Association.

Singh, T. V. and Kaur, J. (1985) *Integrated Mountain Development*, New Delhi: Himalayan Books.

Spears, J. (1982) 'Rehabilitating watersheds', *Finance and Development* 19(1): 30–3.

Speechly, H. T. (1976) 'Proposals for forest management in Sagarmatha (Mt. Everest) National Park', Dept. of National Parks and Wildlife Conservation, Kathmandu.

Starkel, L. (1972a) 'The role of catastrophic rainfall in the shaping of the relief of the Lower Himalaya (Darjeeling Hills)', *Geogr. Polonica* 21: 103–47.

Starkel, L. (1972b) 'The modelling of monsoon areas of India as related to catastrophic rainfall', *Geogr. Polonica* 23: 151–73.

Sterling, C. (1976) 'Nepal', *Atlantic Monthly* (Oct), 238 (4): 14–25.

Stevens, S. (1986) 'Sherpa forests: forest use, protection, and destruction in the Mount Everest region of Nepal', unpub. MS, Berkeley: Dept. of Geography, University of California.

Stiller, L. F. (1975) *The Rise of the House of Gorkha (1768–1816)*, Kathmandu: Ratna Pustak Bhandar.

Stiller, L. F. and Yadav, R. P. (1979), *Planning for People: A Study of Nepal's Planning Experience*, Kathmandu: Research Centre for Nepal and Asian Studies, Tribhuvan University.

Stone, L., and Campbell, J. G. (1986) 'The use and misuse of surveys in international development: an experiment from Nepal', *Human Organization* 43(1): 1–27.

Sun, H. (1983) 'Land resources and agricultural utilization in Xizang Autonomous Region, China', *Mountain Research and Development* 3(2): 143–8.

Tejwani, K. G. (1982) 'Soils policy in India: Need and direction', *Transactions 12th Internatl. Congress Soil Science*, Symposia Papers III, New Delhi.

Tejwani, K. G. (1984a) 'Biophysical and socio-economic causes of land degradation and strategy to foster watershed reclamation in the Himalayas', in IUFRO Symposium on Effects of Forest Land Use on Erosion and Slope Stability, East-West Center, Honolulu.

Tejwani, K. G. (1984b) 'Reservoir sedimentation in India – its causes, control and future course of action', *Water International* 9(4): 150–4.

Tejwani, K. G. (1986) 'Sedimentation of reservoirs in the Himalayan region and estimates of river load', unpub. report prepared for Mohonk Mountain Conference, April 1986, Boulder, Colorado: International Mountain Society.

Tejwani, K. G. (1987) 'Sedimentation of reservoirs in the Himalayan region', *Mountain Research and Development* 7(3): 323–7.

Thompson, M. (1987) in Ives, J. D., Messerli, B. and Thompson, M., 'Research strategy for the Himalayan region: conference conclusions and overview', *Mountain Research and Development* 7(3): 332–44.

Thompson, M. and Warburton, M. (1985a) 'Uncertainty on a Himalayan scale', *Mountain Research and Development* 5(2): 115–35.

Thompson, M. and Warburton, M. (1985b) 'Knowing where to hit it: a conceptual

framework for the sustainable development of the Himalaya', *Mountain Research and Development* 5(3): 203–20.
Thompson, M., Warburton, M., and Hatley, T. (1986) *Uncertainty on a Himalayan Scale*, London: Ethnographica.
Thornbury, W. D. (1954) *Principles of Geomorphology*, New York: John Wiley.
Tilman, H. W. (1952) *Nepal Himalaya*, Cambridge: Cambridge University Press.
Tinau Watershed Management Project (1980) 'Tinau Watershed Management Plan', Kathmandu: Swiss Association for Technical Assistance, internal report.
Troll, C. (1938) 'Der Nanga Parbat als Ziel deutscher forschung', *Zeitschr. Ges. für Erdkunde*, 1–26, Berlin.
Troll, C. (1939) 'Das pflanzenkleid des Nanga Parbat. Begleitworte zur vegetationkarte du Nanga Parbat Gruppe (NW. Himalaya)', *Wiss. Veröff. Dtsch. Mus. f Landeskunde*, 7: 151–80, Leipzig.
Troll, C. (1959) 'Die tropischen gebirge. Ihre dreidimensionale klimatische und pflanzengeographische zonierung', *Bonner Geogr. Abh.* 25.
Troll, C. (1967) Die Klimatische und Vegetationsgeographische Gliederung des Himalaya Systems. In Khumbu Himal, Ergebnisse eines Forschungsunternehmen im Nepal Himalaya. Bd 1, 353–448. Springer: München.
Tucker, R. P. (1986) 'The evolution of transhumant grazing in the Punjab Himalaya', *Mountain Research and Development* 6(1): 17–28.
Tucker, R. P. (1987) 'Dimensions of deforestation in the Himalaya: The historical setting', *Mountain Research and Development* 7(3): 328–31.
Uhlig, H. (1978) 'Geoecological controls on high-altitude rice cultivation in the Himalayas and mountain regions of Southeast Asia', *Arctic and Alpine Research* 10(2): 519–29.
UNCTAD (1981) *Preparatory Committee for the UN Conference on Least Developed Countries*, A/Conf/104/PC15.
United States Library of Congress (1979) Draft environment report on Nepal prepared by the Science and Technology Division with the US Man and the Biosphere Secretariat, Washington, DC.
Vaidyanathan, A., Narayan, N. K. and Harris, M. (1979) 'Bovine sex and species ratios in India', Trivandrum, Kerala: Centre for Development Studies, internal report.
Validya, K. S. (1985) 'Accelerated erosion and landslide-prone zones in the Himalayan region', in J. S. Singh (ed.), *Environmental Regeneration in Himalaya: Concepts and Strategies*, Nainital, UP, India: Central Himalayan Environment Association.
Valdiya, K. S. (1987) *Environmental Geology: Indian Context*, New Delhi: Tata-McGraw-Hill.
Voeikow Main Geophysical Observatory (1981) *Climatic Atlas of Asia*, Vol. 1, Maps of mean temperature and precipitation. Leningrad: WMO/UNESCO.
Vuichard, D. and Zimmermann, M. (1986) 'The Langmoche flash-flood, Khumbu Himal, Nepal', *Mountain Research and Development* 6(1): 90–4.
Vuichard, D. and Zimmermann, M. (1987) 'The catastrophic drainage of a moraine-dammed lake, Khumbu Himal, Nepal: Cause and consequences', *Mountain Research and Development* 7(2): 91–110.
Wagner, A. (1981) 'Rock structure and slope stability study of Walling area, central west Nepal', *Journ. Nepal Geol. Soc.*, 1(2): 37–43.
Wagner, A. (1983) 'The principal geological factors leading to landslides in the foothills of Nepal: a statistical study of 100 landslides; steps for mapping the risks of landslides', Helvetas/Swiss Technical Cooperation/ITECO, Kathmandu.
Whiteman, P. T. S. (1985) 'The mountain environment: an agronomist's perspective with a case study from Jumla, Nepal', *Mountain Research and Development* 5(2): 151–62.

Wiart, J. (1983) 'Ecosystem villageois traditional en Himalaya Nepalais', Unpub. PhD thesis, Université Scientifique et Medicale de Grenoble, Grenoble.

Wischmeier, W. H. and Smith, D. D. (1958) *Predicting Rainfall Erosion Losses – A Guide to Conservation Planning*, United States Department of Agriculture, Agriculture Handbook N. 537.

Wissman, H. von (1959) 'Die heutige vergletscherung und schneegrenze in Hochasien mit hinweisen auf die vergletscherung der letzten Eiszeit', Mainz, Abh. math.-nat. Kl., Jg. Nr. 14: 1105–407.

World Bank (1979) 'Nepal: Development Performance and Prospects', A World Bank Country Study, South Asia Regional Office, World Bank, Washington, DC. 123 pp.

World Bank (1984a) *World Development Report*, Washington, DC: World Bank.

World Bank (1984b) *Bhutan*, Washington, DC: World Bank.

World Bank Atlas (1981) Washington, DC: World Bank.

WRI (1985) *Tropical Forests: A Call for Action* (3 vols), Washington, DC: World Resources Institute.

Wyss, M. (1988) 'Naturraum Himalaya', Unpub. diploma thesis, Geographical Institute, University of Berne.

Xu, D. (1985) 'Characteristics of debris flows caused by outburst of a glacial lake in Boqu River in Xizang, China, 1981', Lanzhou: Institute of Glaciology and Cryopedology, Academia Sinica. Unpub. manuscript. 24 pp.

Young G. J. (1982) 'Hydrological relationships in a glacierized mountain basin', in J. W. Glen (ed.), *Hydrological aspects of alpine and high-mountain areas*, Int. Assoc. Scientific Hydrology, Publ. No. 138: 51–62.

Young G. J. (ed.) (1985) *Techniques for prediction of runoff from glacierized areas*, I.A.H.S. Pub. No. 149.

Zeitler, P. K., Johnsen, N. M., Naeser, C. W. and Tahirkhedi, R. A. K. (1982) 'Fission-track evidence for Quaternary uplift of the Nanga Parbat region, Pakistan', *Nature* 298: 255–7.

Zimmermann, M., Bichsel, M., and Kienholz, H. (1986) 'Mountain hazards mapping in the Khumbu Himal, Nepal, with prototype map, scale 1: 50,000', *Mountain Research and Development* 6(1): 29–40.

Zollinger, F. (1978) 'Analysis of river problems and strategy for flood control in the Nepalese Terai', UNDP/FAO Working Paper No. 15, Kathmandu.

Zollinger, F. (1979) 'The Sapt Kosi: unsolved problems of flood control in the Nepalese Terai', Dept. of Soil and Water Conservation, Kathmandu.

AUTHOR INDEX

SUBJECT INDEX